职业教育“互联网 +”新形态一体化教材
智能工程机械运用技术专业

装载机维修与服务

（中英双语）

主　编　韦茂志　陈立创　彭智峰
副主编　蔡智泉　邓益民　黄光周
参　编　何海峰　李光辉　李　贝
陈　栋　莫德和　邵鹏刚

机械工业出版社
CHINA MACHINE PRESS

本书采用项目式编写，主要内容包括装载机介绍与交验、装载机保养与维护、装载机液压系统分析与检测、装载机电气系统分析与维护、装载机空调系统维护、装载机传动系统分析、装载机故障诊断与排除，并配有工作页。

本书可作为高等职业院校智能工程机械运用技术专业教材，也可供相关从业人员作为参考用书。

本书配有电子课件，凡使用本书作为教材的教师可登录机械工业出版社教育服务网 www. cmpedu. com 注册后免费下载。咨询电话：010－88379375。

图书在版编目（CIP）数据

装载机维修与服务：汉英对照/韦茂志，陈立创，彭智峰主编．—北京：机械工业出版社，2024.3

ISBN 978-7-111-74799-4

Ⅰ.①装…　Ⅱ.①韦…②陈…③彭…　Ⅲ.①装载机－维修－汉、英　Ⅳ.①TH243.07

中国国家版本馆 CIP 数据核字（2024）第 043446 号

机械工业出版社（北京市百万庄大街 22 号　邮政编码 100037）

策划编辑：刘良超　　责任编辑：刘良超

责任校对：张亚楠　张　征　　封面设计：王　旭

责任印制：张　博

北京建宏印刷有限公司印刷

2024 年 7 月第 1 版第 1 次印刷

184mm×260mm · 20.25 印张 · 493 千字

标准书号：ISBN 978-7-111-74799-4

定价：66.80 元（含工作页）

电话服务　　网络服务

客服电话：010-88361066　　机　工　官　网：www. cmpbook. com

010-88379833　　机　工　官　博：weibo. com/cmp1952

010-68326294　　金　　书　　网：www. golden-book. com

　　机工教育服务网：www. cmpedu. com

前　言

装载机属于铲土运输机械类工程机械，广泛应用于建筑、公路、铁路、水电、港口、矿山、林业、料场及国防等领域。装载机用于铲装土壤、沙石、石灰、煤炭、木材等松散物料，也可对矿石、硬土等进行轻度铲挖作业。通过换装不同的辅助工作装置，装载机还可进行推土、起重、装卸、破碎等作业。

本书的编写以培养技能应用型人才为目标，力求使学生了解装载机装备技术，正确运用、保养和维修装载机。

本书采用项目式编写，主要内容包括装载机介绍与交验、装载机保养与维护、装载机液压系统分析与检测、装载机电气系统分析与维护、装载机空调系统维护、装载机传动系统分析、装载机故障诊断与排除，并配有工作页。

在项目的选择上，本书遵循以应用能力和综合素质培养为主线的指导思想，以任务为引领，以广西柳工机械股份有限公司校企合作项目为平台，精选实际工作的真实任务作为载体，并与整体保持密切关联，使学生能从实践中体会理论对实践的指导作用，实现学做一体化。

本书在内容的选择上始终贯穿实用性原则，理论知识以“必需、够用、实用”为度。不片面追求理论知识的系统和完整性，力求做到理论与实践相统一，特别是与学生的需求和未来发展相统一。

本书采用主教材和工作页分开的编排方式，方便学生开展专项学习。工作页部分有利于加强学生对理论知识的理解，培养学生的动手能力，提高学生对液压系统的分析能力，增强学生的自信心。

本书由韦茂志、陈立创、彭智峰担任主编，蔡智泉、邓益民、黄光周担任副主编，何海峰、李光辉、李贝、陈栋、莫德和、邵鹏刚参与了本书编写。

由于编者水平有限，错漏之处在所难免，恳请广大读者批评指正。

编　者

目　录

项目 1

装载机介绍与交验

任务 1.1　装载机介绍

☞ 学习目标

1）能够为客户介绍装载机的结构和工作性能。

2）能够利用 PPT 向目标客户推介装载机。

☞ 工作任务

了解装载机的相关知识，当客户来购买机械设备时，以专业水准向客户介绍 CLG856H 装载机。

☞ 相关知识

1. 装载机的定义

装载机是一种广泛用于公路、铁路、建筑、水电、港口、矿山等建设工程的土石方施工机械，它主要用于铲装土壤、砂石、石灰、煤炭等散状物料，也可对矿石、硬土等进行轻度铲挖作业。换装不同的辅助工作装置还可进行推土、起重、装卸等作业。

在道路施工，特别是在高等级公路施工中，装载机用于路基工程的填挖、沥青混合料和水泥混凝土料的集料与装料等作业。此外，装载机还可进行推运土壤、刮平地面和牵引其他机械等作业。装载机具有作业速度快、效率高、机动性好、操作轻便等优点，因此是工程建设中土石方施工的主要机种之一。

2. 装载机的结构组成及各部件名称

（1）装载机的结构组成　装载机主要由发动机、变矩器、变速器、前后驱动桥组成。

按系统分，装载机由以下系统构成。

1）动力系统：装载机原动力一般由柴油机提供。柴油机具有工作可靠、功率特性曲线硬、燃油经济等特点，符合装载机工作条件恶劣、负载多变的要求。

2）传动系统：传动系统主要包括行走装置、变速器等。

3）液压系统：液压系统的功能是以液压油为介质，利用液压泵把发动机的机械能转变为液压能，再传送给液压缸、液压马达等执行元件。

4）控制系统：控制系统是对发动机、液压泵、多路换向阀和执行元件进行控制的系统。液压控制驱动机构是在液压控制系统中，将微小功率的电能或机械能转换为强大功率的液压能和机械能的装置。它由液压功率放大元件、液压执行元件和负载组成，是液压系统中的控制核心。

（2）装载机各部件名称　图 1-1 所示为装载机各部件名称。

图 1-1　装载机各部件名称

1—前工作灯　2—驾驶室　3—左前组合灯　4—转斗液压缸　5—摇臂　6—铲斗　7—配重　8—后轮　9—后挡泥板　10—扶梯　11—左前组合灯　12—前车架　13—前挡泥板　14—前轮　15—动臂　16—后工作灯　17—液压油箱　18—发动机罩　19—后右组合灯　20—后车架　21—后左组合灯

3. 装载机的分类

1）按行走方式分为轮胎式装载机和履带式装载机。轮胎式装载机行驶速度快，机动灵活，可在城市道路行驶，使用方便；履带式装载机接地比压低，牵引力大，但行驶速度慢，转移不灵活。

2）按车架结构形式分为整体式车架装载机（图 1-2）和铰接式装载机（图 1-3）。

图 1-2　整体式车架装载机

图 1-3　铰接式装载机

3）按装载方式分为前卸式、后卸式、回转式和侧卸式。

① 前卸式：在其前端铲装和卸料。其结构简单，工作可靠、安全，便于操作，适应性强，应用广泛。

② 后卸式：在其前端装料，后端卸料。该机械运料距离短，作业效率高，但安全性差，故应用较少。

③ 回转式：工作装置安装在可回转 90°～360°的转台上。其侧面卸料时不需要调整机械

位置，作业效率高，但结构复杂，质量大，侧向稳定性差，适用于狭小的场地作业。

④ 侧卸式：在其前端装料，侧面卸料。其装料时，不需要调整机械位置，可直接向停在其侧面的运输车辆上卸料，作业效率高，但卸料时横向稳定性较差。

4）按传动方式分为机械传动、液压-机械传动、液压传动和电力传动。

5）按发动机功率分为小型装载机（<74kW）、中型装载机（74~162kW）、大型装载机（162~515kW）和特大型装载机（>515kW）。

4. 装载机的基本参数与性能

（1）主要性能参数

1）额定载荷：在保证装载机稳定工作的前提下，铲斗的最大载重量。

2）额定功率：发动机不带风扇、空气滤清器、发电机等附件时，所输出的功率。

3）操作质量：装载机配置工作装置及属具、驾驶员（按75kg计）、油水加满时的主机质量。

4）标准斗容：也称为额定斗容，在普通物料下，铲斗平装容量和堆尖部分（四周坡度均为1:2）体积之和，如图1-4所示。

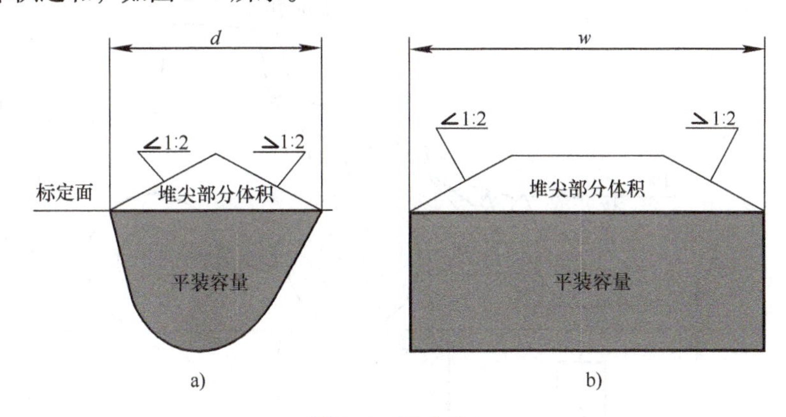

图1-4　标准斗容

5）最大掘起力：铲斗切削刃底面水平高于底面基准20mm时操纵工作装置液压缸在铲斗切削刃后100mm处产生的最大向上铅垂力。掘起力分为转斗掘起力和动臂掘起力两种，如图1-5所示，一般转斗掘起力大于动臂掘起力，故标准掘起力取转斗掘起力。

6）对直倾翻载荷：动臂平伸至离铲斗斗容形心最远，使装载机后轮离地时作用在铲斗斗容形心处允许的最小载荷，如图1-6所示。

7）全转向倾翻载荷：转向角打到最大位置，动臂平伸至铲斗斗容形心最远，使装载机后轮离地时作用在铲斗斗容形心处允许的最小载荷，如图1-7所示。

8）牵引力：装载机除克服由轮胎产生的滚动阻力外，用于牵引或铲土的力。

9）三项和时间：铲斗满载额定载荷时动臂以最大速度从最低处提升到最高处所用的时间、铲斗在最高处卸载物料所用的时间、铲斗空载时动臂从最高处浮动下降到最低处所用时间，这三项时间之和称为三项和时间。

10）最高车速：水平水泥路面、最高档位、最大油门状态下的车速。

（2）装载机主要尺寸参数　如图1-8所示。

1）最小转弯半径（铲斗外侧）：在全转向位置，铲斗外侧轨迹圆的半径，如图1-9

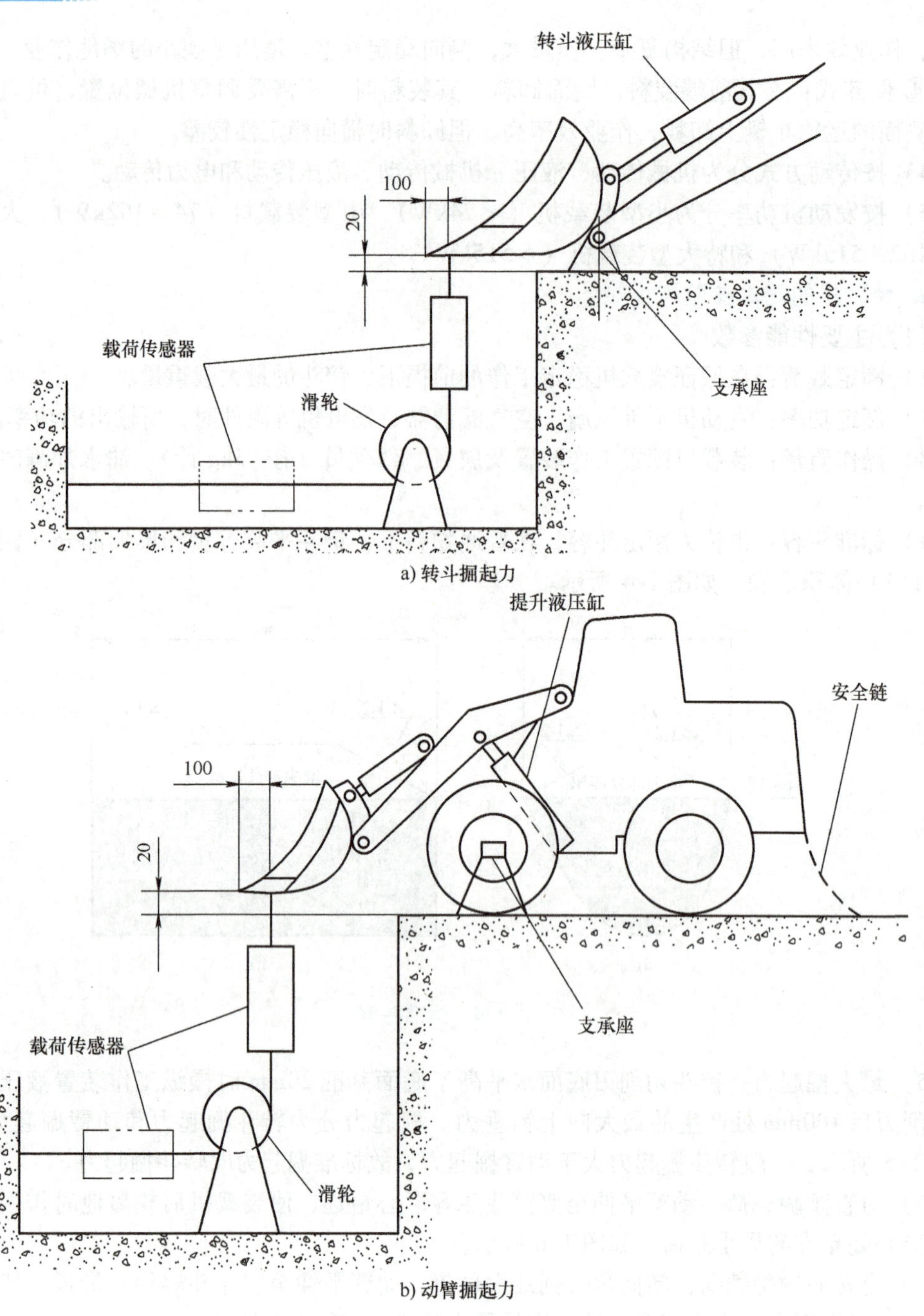

图 1-5　掘起力

所示。

2）铲斗下铰接销最大高度 A：水平地面上，铲斗举到最高处，铲斗下铰接销中心点到地面的距离。

3）卸载高度 B：铲斗举到最高位，且处在最大卸料角度，斗齿尖到地面的距离。

4）卸载距离 C：铲斗举到最高位，且处在最大卸料角度，斗齿尖到前轮外边缘的距离。

5）下挖深度 D：铲斗放平，斗齿下沉到地面以下的深度。

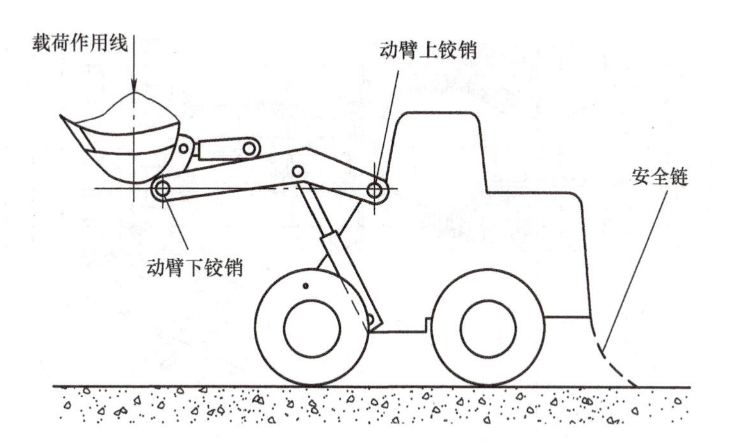

图1-6 对直倾翻载荷

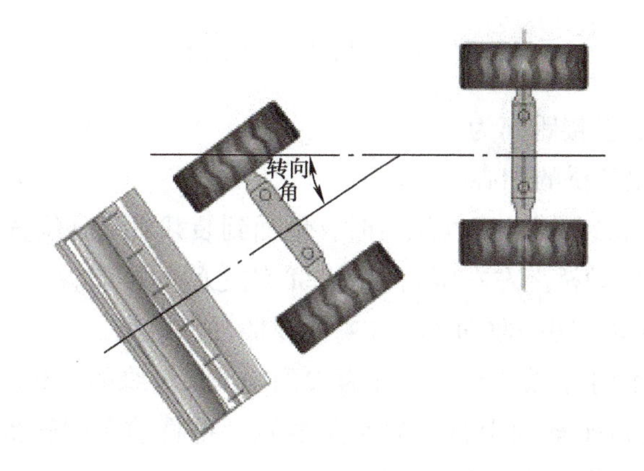

图1-7 全转向倾翻载荷

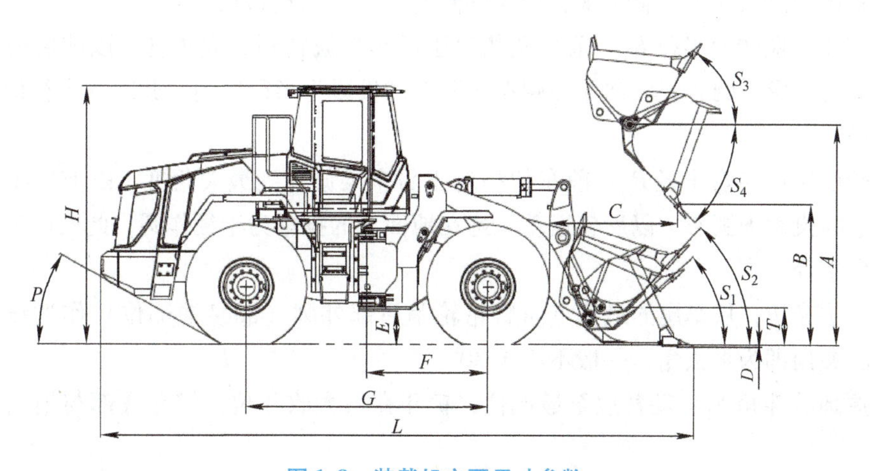

图1-8 装载机主要尺寸参数

6）最小离地间隙 E：水平地面上，底盘最低点（不包括轮胎及工作装置）到地面的垂直距离。

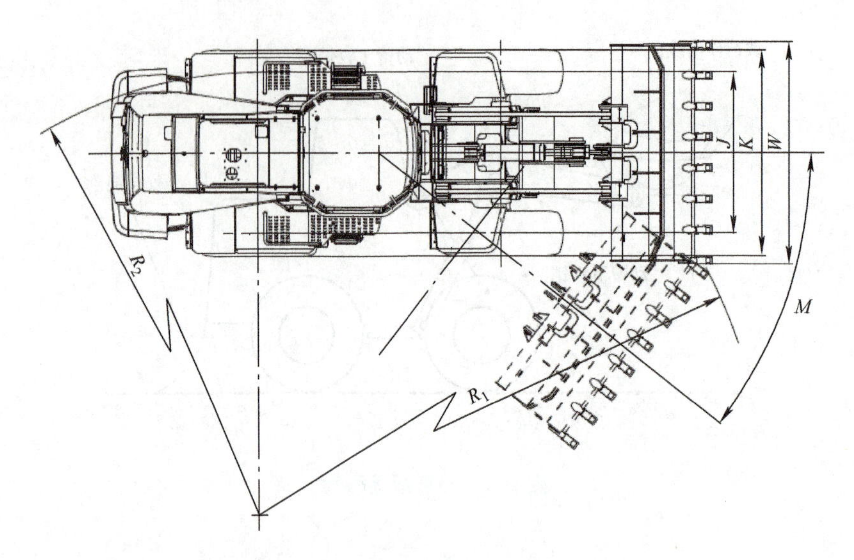

图 1-9　装载机转弯半径

7）前轮中心到中心铰接距离为 F。

8）轴距 G：前后驱动桥的中心距离。

9）整机高度 H：轮胎在标准充气压力时，地面到装载机最高位置处的距离。

10）轮距 J：同一驱动桥上左右两轮胎的中心线之间的距离。

11）轮胎外侧宽度 K：同一驱动桥上左右两轮胎外缘距离。

12）整车长度 L：铲斗平放于地面，斗齿尖到尾部最外缘的距离。

13）转向角 M：在水平地面上将前后车架摆直，再将前车架转到最大角度，此时前车架相对后车架所转动的角度即为转向角。

14）整机宽度（铲斗外侧）W：铲斗外侧的宽度，比轮胎宽一些。

15）铲斗外侧转弯半径 R_1：前车架相对于后车架偏转到最大角度，以前后桥的轴线交点在地面上的投影为圆心，以铲斗外侧在地面上的投影为半径作圆，此圆的半径即为铲斗外侧转弯半径。

16）轮胎中心转弯半径 R_2：前车架相对于后车架偏转到最大角度，以前后桥的轴线交点在地面上的投影为圆心，以后轮运动轨迹在地面上的投影为半径作圆，此圆的半径即为轮胎中心转弯半径。

17）离去角 P：从车尾的最低点向后轮轮胎后部外廓（靠接地部位）作切线，此切线与水平面的夹角即为离去角，一般不小于 30°。

18）原地收斗角 S_1：动臂放至最低位，铲斗在最大收斗位，铲斗底部与地面所形成的角度。

19）运输位置收斗角 S_2：动臂升至高度 T，铲斗在最大收斗位，铲斗底部与地面所形成的角度。

20）最高位置收斗角 S_3：动臂举升至最高位，铲斗收至最大位，铲斗底部与下铰接销中心位所形成的角度。

21）最高位置卸料角 S_4：动臂举升至最高位，铲斗打开至最大卸料位，铲斗底部与下铰接销中心位所形成的角度。

22）运输位置铲斗铰销高度为 T。

5. 装载机安全操作规程

1）上、下装载机。上、下装载机之前要清洁扶手和阶梯上的油迹、污泥；只能在有阶梯和扶手的地方攀上或攀下装载机；上、下装载机时要面对装载机，手拉扶手，脚踩阶梯，保持三点接触（两脚一手或两手一脚）；严禁跳下装载机，严禁在装载机移动时上、下装载机；上、下装载机时绝不能将任何操纵杆当作扶手。携带工具或其他物品时不要攀上或攀下装载机，应用绳子将所需工具吊上操作平台。

2）装载机行驶时，应收回铲斗，铲斗离地面 400 ~ 500mm，在行驶过程中应注意是否有路障或高压线等，除规定驾驶人员以外，不准搭乘其他人员，严禁铲斗载人。

3）装载机在公路上行驶时，应避免突然换向行驶，铲斗带负荷升起行驶时，不准急转弯和紧急制动。

4）装载机在公路上行驶时必须遵守交通规则，谨慎驾驶，下坡时禁止空档滑行。

☞ 任务实施

编写一份介绍 CLG856H 装载机的 PPT，需要包含以下内容：

1）CLG856H 装载机的总体介绍。

2）性能参数。

3）结构。

4）配置。

任务 1.2　装载机交验

☞ 学习目标

1）能够按要求向客户交验装载机。

2）能够编写装载机交验报告。

☞ 工作任务

现有一台新的装载机需要交付给客户，请完成交验工作。

☞ 相关知识

1. 交机检查内容及标准

（1）整机检验和试车中需要重点检验的部位　见表 1-1。

表 1-1 装载机交验点检表

序号	验收项目	结果		序号	验收项目	结果	
		正常	不正常			正常	不正常
1	发动机机油			18	仪表		
2	空调压缩机传动带张紧度			19	空调系统运行		
3	风扇传动带张紧度			20	音响系统		
4	发动机怠速和中速运转			21	充电指示		
5	发动机机油压力			22	蓄电池		
6	发动机水温			23	轮边制动		
7	铲斗动作			24	驻车制动		
8	动臂动作			25	配重连接处		
9	转弯动作			26	门锁		
10	前、后车架铰接装置			27	机罩锁		
11	车轮			28	各油管接头处是否渗漏		
12	行走前进低速、高速			29	散热器组		
13	行走后退低速、高速			30	外观油漆		
14	电气控制系统			31	操控杆是否灵活、到位		
15	灯			32	随车工具、资料及配件		
16	刮水器			33	GPS 报警单元		
17	喇叭			34	其他（上述未列出项目）		

（2）整机检验内容

1）发动机燃油量检查。燃油油位表位于驾驶室仪表总成上，燃油油位表分为绿色和红色两个区域，1 表示燃油量为满油箱，0 表示燃油油位为 0，当燃油油位指示低于 0.2 时，应及时添加燃油。

2）发动机机油量检查。发动机的机油过多或过少都会造成发动机的损坏。

将装载机开到平坦的场地，发动机熄火，等待 10min；让曲轴箱内的发动机机油充分流回发动机油底壳；打开发动机罩，机油油位尺位于发动机右侧；拔出油位尺，用干净的布将油位尺擦干净，重新插入发动机油位口到尽头，再拔出来检查，正常油位应在油位尺的“L”刻度和“H”刻度之间，如图 1-10 所示；如果油位在“L”刻度之下，需补充机油；如果油位在“H”刻度之上，需拧松油底壳底部的放油螺塞，放出部分机油；放回油位尺，关闭发动机罩。

3）液压油检查。检查前，保证液压缸、液压管路、散热器等液压部件都充满液压油；将装载机开到平坦的场地，前后车架对直无夹角；收斗至极限位，发动机全速提升动臂至最高位；整机怠速将动臂操纵杆推至“下降”位，使动臂匀速下降至最低位，把铲斗水平放到地面上，然后熄火并取下钥匙，前后左右推动操作杆卸压；在液位计无气泡的情况下，检

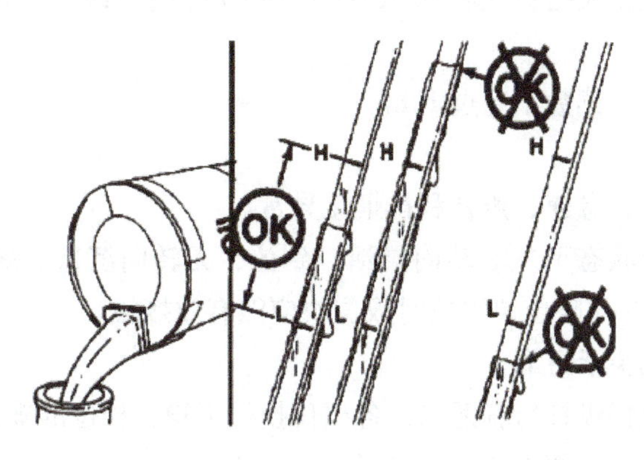

图1-10　油位刻度值

查液压油箱液位计，此时油位应该在液位计的绿色范围内，即MAX线和MIN线之间，如图1-11所示；检查油位时，若发现油液液面高出液位计绿色范围（观察液位计有气泡的情况下），即油位在最高油位线以上，也不能立即放油，需待气泡消除后，按上述步骤要求重新检验；如果油位低于绿色范围，必须及时补充液压油，然后按上述步骤要求重新检验。

MAX
MIN

图1-11　液位计刻度

4）制动液检查。将装载机停驻在平坦的场地上，关闭发动机，拉起驻车制动，以防止装载机移动或转动，旋转打开加力器盖，检查加力器油杯液面高度。如果油杯液面高度低于油杯内滤网面，则需加注合成制动液（DOT4），直至液面高度与油杯内滤网面平齐，油量在油杯约3/4处。

5）变速器传动油检查。冷车起动前，应先进行冷油平面的检查；检查冷油平面，其目的是在起动时确保有足够的油量，这对于长时间停止使用的装载机来说尤为重要。变速器油位的检查步骤如下：

① 将装载机停驻在平坦的场地上，变速操纵手柄置于空档位，实施驻车制动，装上转向架的锁紧装置，以防止装载机移动和转动。

② 起动发动机并怠速运行3～5min，检查变速器液位计。正常变速器油位应处于HOT（热）区域内。

③ 如果油位刻度在HOT（热油）位置之上，应松开变速器底部的放油螺塞放出适量变速器油；如果油位刻度在COLD（冷油）位置之下，则应添加适量变速器油。

6）散热器冷却液检查。打开发动机罩，通过位于副水箱上的液位观察窗检查液位，当冷却液液位低于液位观察窗中心线时，应补充冷却液。

7）仪表检查。

① 发动机起动后机油压力指示灯是否熄灭（灯亮时为低压报警，另外还有制动气压报警）。

② 变速压力。变速器油压表指针应在1.1～1.5MPa之间（绿区范围）。

③ 气压表。制动气压表指针应在 0.5 ~0.784MPa 之间（绿区范围，低于 0.4MPa 时报警灯亮）。

④ 计时表走数正常，读数不超过 10h。

8）操控性检查。

① 发动机运转正常，怠速、声音和排烟无异常。

② 转向性能。怠速状态下左右转向灵活、到位，无转向沉重、滞后、摆动、异响、跑偏、自转、空转现象，转向限位块限位时限位块应能有效接触。

③ 制动性能。制动灵活可靠。

④ 档位操纵。变速操纵杆行程适当，各档位换档平稳，档位准确，无掉档现象。

⑤ 液压操作手柄操纵轻便平稳，无卡滞粘连现象。

9）电器检查。刮水器、空调、倒车报警器等工作正常；前后大灯、转向灯、工作灯、危险警示灯等灯光正常；电器外观无磕碰划伤；蓄电池电量应在绿色部位。

10）外观检查。外观对称，装配匀称、协调，无磕碰划伤；钣金件要平整，无凹凸弯曲、敲击痕迹、磕碰划伤；金属镀层和氧化处理层不得剥落和生锈，表面清洁；各胶管、钢管与液压件连接处无滴油、漏油、渗油现象；无错、漏装，装配合适、端正，外观无明显缺陷。

11）轮胎检查。

① 胎侧：不允许存在深度大于 3mm、长度大于 5mm、宽度大于 1mm 的裂口、划伤或其他缺陷。

② 胎冠：不允许存在深度大于 10mm、长度大于 10mm、宽度大于 1mm 的裂口、划伤或其他缺陷。

③ 胎肩：不允许存在深度大于 5mm、长度大于 10mm、宽度大于 1mm 的裂口、划伤或其他缺陷。

12）随机资料检查。随机资料、备件、随机工具完整齐全。

2. 整机操作、存放注意事项

（1）整机操作注意事项　只有经过专门培训或具有相关操作资质的人员才可以操作装载机；装卸车辆必须有人指挥且操作人员身体状况良好；操作人员操作装载机时须严格按照《操作与保养手册》要求执行，操作油门及手柄时应缓慢，不可用力过猛。

（2）存放注意事项　蓄电池要定期检查，每月起动一次，每次不少于 20min，存放时间超过 3 个月或寒冷地区须将蓄电池拆下单独存放，且定期充电；外露镀层、金属、加工面等要涂抹润滑油脂，防止锈蚀；再次操作时要清除液压缸外露镀层的污物；水箱里的冷却水要放干净，包括发动机机体和变矩器油冷却器的放水阀，或者更换成防冻液；轮胎要保持胎压，前轮胎压为 0.333 ~0.353MPa，后轮胎压为 0.275 ~0.294MPa。

☞ 任务实施

编写交验报告，见表 1-2。

表 1-2　交验报告

<table>
<tr><td colspan="2">验收日期</td><td></td><td>设备型号</td><td></td></tr>
<tr><td>序号</td><td>检查项目</td><td colspan="2">验收内容</td><td>验收结果</td></tr>
<tr><td rowspan="7">一</td><td rowspan="7">外观验收</td><td colspan="2">1. 灯光正常</td><td></td></tr>
<tr><td colspan="2">2. 仪表正常，齐全有效</td><td></td></tr>
<tr><td colspan="2">3. 轮胎螺钉紧固无缺少</td><td></td></tr>
<tr><td colspan="2">4. 传动轴螺钉紧固无缺少</td><td></td></tr>
<tr><td colspan="2">5. 方向机横竖拉杆无松动</td><td></td></tr>
<tr><td colspan="2">6. 无任何部位的漏油、漏气、漏水</td><td></td></tr>
<tr><td colspan="2">7. 全车各部位无变形</td><td></td></tr>
<tr><td rowspan="5">二</td><td rowspan="5">检查各部位油位水位</td><td colspan="2">1. 水箱水位正常</td><td></td></tr>
<tr><td colspan="2">2. 机油油位正常</td><td></td></tr>
<tr><td colspan="2">3. 柴油油位正常</td><td></td></tr>
<tr><td colspan="2">4. 制动液液位正常</td><td></td></tr>
<tr><td colspan="2">5. 液压油油位正常</td><td></td></tr>
<tr><td rowspan="3">三</td><td rowspan="3">发动机部分</td><td colspan="2">1. 水温正常</td><td></td></tr>
<tr><td colspan="2">2. 发动机运转正常无异响</td><td></td></tr>
<tr><td colspan="2">3. 各辅助机构工作正常</td><td></td></tr>
<tr><td rowspan="5">四</td><td rowspan="5">液压传动部分</td><td colspan="2">1. 液压泵压力正常</td><td></td></tr>
<tr><td colspan="2">2. 行走系统正常</td><td></td></tr>
<tr><td colspan="2">3. 动臂液压缸起升正常无下滑</td><td></td></tr>
<tr><td colspan="2">4. 转斗液压缸起升正常</td><td></td></tr>
<tr><td colspan="2">5. 液压油温正常</td><td></td></tr>
<tr><td rowspan="3">五</td><td rowspan="3">操作系统</td><td colspan="2">1. 变速器正常</td><td></td></tr>
<tr><td colspan="2">2. 制动系统正常</td><td></td></tr>
<tr><td colspan="2">3. 行走系统正常</td><td></td></tr>
<tr><td rowspan="3">六</td><td rowspan="3">安全防护</td><td colspan="2">1. 具有产品质量合格证</td><td></td></tr>
<tr><td colspan="2">2. 操作人员持证上岗</td><td></td></tr>
<tr><td colspan="2">3. 驾驶室内挂设安全技术操作规程</td><td></td></tr>
<tr><td>验收结论</td><td colspan="4"></td></tr>
<tr><td rowspan="2">验收人签字</td><td>物资部</td><td>安质部</td><td>工程部</td><td>现场管理人员</td></tr>
<tr><td></td><td></td><td></td><td></td></tr>
</table>

项目 2

装载机保养与维护

任务 2.1　定期保养与维护计划书编制

☞ 学习目标

1）熟悉定期保养与维护的内容与要求。

2）能够编制定期保养与维护计划。

☞ 工作任务

编制装载机1000h的重点保养与维护计划。

☞ 相关知识

装载机的正确保养，特别是预防性的保养，是最容易、最经济的保养。正确的保养可以延长装载机的使用寿命并降低使用成本，因而抵偿了在计划保养中所需的时间和费用。

对装载机进行正确保养与维护，首先必须做好装载机使用过程中的日报工作，根据装载机在使用过程中反映的情况，及时做好必要的调整和修理工作。其次，参照下面介绍的有关内容，并按不同用户的特殊工作情况及使用经验，制订出不同的保养与维护计划。

设备制造商一般要求保养与维护工作应按使用工作计时表或日历（日、周、月等）两种时间中首先到期的时间周期来进行。在极度严酷、多尘或潮湿的工作环境下，需要有比定期维护中规定的更为频繁的润滑保养。在维护保养时，应重复进行原来要求中所列的保养与维护项目。例如，在进行500h或3个月的维护保养项目时，应同时进行250h或1个月、50h或每周、每10h或每天保养与维护中所列的项目。

装载机的保养与维护主要分为每日检查、新车走合和定期保养。

1. 每日检查

每天在装载机工作之前，应当检查以下内容：

1）驾驶室内仪器仪表是否正常，车上的灯光系统、喇叭是否正常工作。

2）发动机的冷却液、柴油和机油的液位是否正常，是否存在渗漏。

3）液压泵、液压缸和液压管路、管接头是否存在渗漏。

4）零部件表面是否覆盖污染物、出现松动、损坏或掉落等。

5）紧固件是否出现松动现象。

6）空调蒸发器的出风口和进风口是否有棉纱、废纸、塑料薄膜等杂物。

2. 新车走合

新车走合对延长装载机的使用寿命、消除故障隐患、避免重大故障的发生具有重要的作用。用户在购买装载机后必须按有关新车走合的规定进行装载机的操作和保养与维护，然后才能正常使用机器。

一般要求装载机的新车走合期为100h。走合期内，装载机的使用应注意以下事项：

1）每次开机后，首先空转5min。装载机应以平稳低速小油门起步，然后逐步提高速度，前进、后退各档位应均匀安排走合。

2）除紧急情况外，应避免突然起动、突然加速、突然转向和突然制动，作业不得过猛过急。尽量装载松散物料，装载重量不得超过额定载重的70%，行驶速度不得超过额定最高车速的70%。

3）注意各部位的润滑，按规定的时间周期更换或添加润滑油和润滑脂。经常注意变速器、变矩器、前后桥、轮毂、驻车制动器、中间支承轴，以及液压油、发动机冷却液、发动机机油的温度，如有过热现象，应找出原因进行故障排除。

4）经常检查各部件螺栓、螺母的紧固情况。

新车走合8h、50h和100h后应做以下项目的检查，见表2-1。

1）检查各部件螺栓、螺母的紧固情况。特别是柴油机气缸盖螺栓、排气管螺栓以及前后桥固定螺栓、轮辋螺母、传动轴连接螺栓、柴油机固定螺栓、变速器固定螺栓、前后车架铰接处螺栓等。

2）检查发动机传动带、风扇传动带、空调压缩机传动带的松紧程度和变速器油位、驱动桥润滑油油位和柴油机的机油位、冷却液液位。

3）检查液压系统、制动系统有无泄漏，各操纵拉杆、油门拉杆的连接是否稳靠。

4）检查电气系统各部件温度及连接情况、发电机供电状态、灯光照明及转向信号灯等的工作情况。

表2-1 新车走合保养与维护项目

项目	8h	50h	100h	备注
螺栓、螺母	◎	◎	◎	防松标识不错位
风扇传动带	◎	◎	◎	指压≤8mm
发电机传动带	◎	◎	◎	指压≤8mm
空调压缩机传动带	◎	◎	◎	指压≤8mm
齿轮油	◎	◎	◎	在刻度范围
液压油	◎	◎	◎	在刻度范围
冷却液	◎	◎	◎	在刻度范围
油水分离器	◎	◎	◎	放水
空调制冷剂	◎	◎	◎	
灯光照明	◎	◎	◎	
工作装置各铰销	•	•	•	加注润滑脂
空气滤清器的主滤芯、安全滤芯		•	◎	
燃油预滤器、滤清器		•	◎	
发动机油滤清器和机油滤清器		•	◎	

注：表中标“◎”者为检查，标“•”者为更换。

3. 定期保养与维护

装载机定期保养与维护分为 10h 或 1 日、50h 或 1 周、100h 或 1 个月、500h 或 1 个季度、1000h 或 6 个月和 2000h 或 1 年等工作时限的保养与维护，一般应按工作计时表或日历（日、周、月等）两种时间中首先到期的时间周期来进行。在极度严酷、多尘或潮湿的工作环境下，需要有比定期保养与维护中规定的更为频繁的润滑保养与维护。定期保养与维护项目见表 2-2。

表 2-2　定期保养与维护项目

项目		10h	50h	100h	250h	500h	1000h	2000h	备注
发动机	燃油油位/滤芯	◎	◎	◎	•	•	•	•	每 50h 排水一次
	机油/滤芯	◎	◎	◎	•	•	•	•	
	空气滤清器/滤芯	◎	◎	◎	◎	◎	•	•	
	冷却液/液位	◎	◎	◎	◎	◎	◎	•	
	油水分离器排水	◎	•	•	•	•	•	•	
	曲轴箱呼吸器管	◎	◎	◎	◎	◎	◎	◎	
	散热器/液位	◎	◎	◎	◎	◎	◎	◎	
	冷却风扇/传动带	◎	◎	◎	◎	◎	◎	◎	指压≤8mm
	发动机传动带	◎	◎	◎	◎	◎	◎	◎	指压≤10mm
	进/排气管	◎	◎	◎	◎	◎	◎	◎	
液压系统	液压油/滤芯	◎	◎	◎	◎	•	◎	•	
	液压泵	◎	◎	◎	◎	◎	◎	◎	
	液压软管/管接头	◎	◎	◎	◎	◎	◎	◎	
	液压缸	◎	◎	◎	◎	◎	◎	◎	
	液压阀/接头	◎	◎	◎	◎	◎	◎	◎	
传动系统	变速器油	◎	◎	•	◎	◎	•	•	
	前、后桥齿轮油	◎	◎	•	◎	◎	•	•	
	轮边减速器油	◎	◎	◎	◎	◎	•	•	
	轮胎胎压	◎	◎	◎	◎	◎	◎	◎	前轮胎压 0.333～0.353MPa 后轮胎压 0.275～0.294MPa
制动系统	气压	◎	◎	◎	◎	◎	◎	◎	
	制动油	◎	◎	◎	◎	•	•	•	
	摩擦片	◎	◎	◎	◎	◎	◎	◎	
	气罐/排水	◎	◎	◎	•	•	•	•	
电气系统	前、后灯/开关	◎	◎	◎	◎	◎	◎	◎	
	熔断器	◎	◎	◎	◎	◎	◎	◎	
	喇叭	◎	◎	◎	◎	◎	◎	◎	
	仪表/指示灯	◎	◎	◎	◎	◎	◎	◎	
	线束/开关/插头	◎	◎	◎	◎	◎	◎	◎	
	电瓶/电解液位		◎	◎	◎	◎	◎	◎	
	刮水器	◎	◎	◎	◎	◎	◎	◎	
	空调/制冷剂/滤芯		◎	◎	◎	◎	•	•	
润滑脂	前、后桥铰接点	◎	•	•	•	•	•	•	
	工作装置各铰接点	◎	•	•	•	•	•	•	
其他	连接螺栓/螺母	◎	◎	◎	◎	◎	◎	◎	
	连杆/拉杆	◎	◎	◎	◎	◎	◎	◎	

注：1. 项目“◎”者为按时检查，按需更换。

2. 项目“•”者为按时更换或操作。

3. 油（液）位在停机 15min 后检查。

4. 时间周期：8h—日检；50h—周检；100h—半月检；250h—月检；500h—季检；1000h—半年检；2000h—年检。

4. 注意事项

添加的冷却液、机油、齿轮油、液压油和柴油等，其牌号应当与原来的一致，并尽量选用装载机制造厂家指定或原制造厂商的产品。随季节或使用环境更换的材料，应按照说明书或使用手册规范进行操作。

☞ 任务实施

针对 CLG856H 装载机编制一份 1000h 定期保养与维护计划书。

在编制保养与维护计划前，应当查询装载机上一次保养与维护的具体时间、保养与维护项目及所用耗材，了解装载机在保养与维护之后的工作状况，以及当前的工作状态。

保养与维护计划书的内容主要包括：

1）检查各部件连接螺栓、螺母是否存在松动。

2）检查发动机传动带、风扇传动带、空调压缩机传动带的松紧程度。

3）检查液压系统、制动系统有无泄漏，各操纵拉杆、油门拉杆的连接是否稳靠。

4）检查前桥及后桥铰接点、工作装置各铰接点并加注润滑油。

5）更换柴油机的机油及滤芯、空气滤清器及滤芯、燃油滤清器滤芯和空调滤清器滤芯。

6）更换变速器、前桥减速器、后桥减速器和轮边减速器等的润滑油。

任务 2.2　工作装置保养

☞ 学习目标

1）能够根据保养要求加注润滑脂。

2）能够根据保养要求更换铲斗斗齿。

☞ 工作任务

更换装载机工作装置的铲斗斗齿。

☞ 相关知识

装载机的铲掘和装卸物料作业是通过工作装置的运动来实现的。工作装置由动臂、动臂液压缸、铲斗、铲斗液压缸、摇臂和拉杆等零部件组成。铲斗用于铲装物料，动臂的后端通过动臂销与前车架连接，前端安装有铲斗，中部与动臂液压缸相连接。当动臂液压缸伸缩时，动臂绕其后端销转动，实现铲斗的提升或下降。摇臂的中部和动臂连接，两端分别与拉杆和转斗液压缸相连。当转斗液压缸伸缩时，摇臂绕其中间支承点转动，通过拉杆使铲斗上转或下翻。

1. 工作装置的等效运动机构

工作装置一般可分为正转六连杆、正转八连杆、反转六连杆，如图 2-1 所示。

装载机的工作装置以反转六连杆为主。反转六连杆又称为“Z”字形结构，主要包括动臂液压缸、转斗液压缸、摇臂、动臂、铲斗、连杆六个部件，又因为转斗液压缸的运动方向

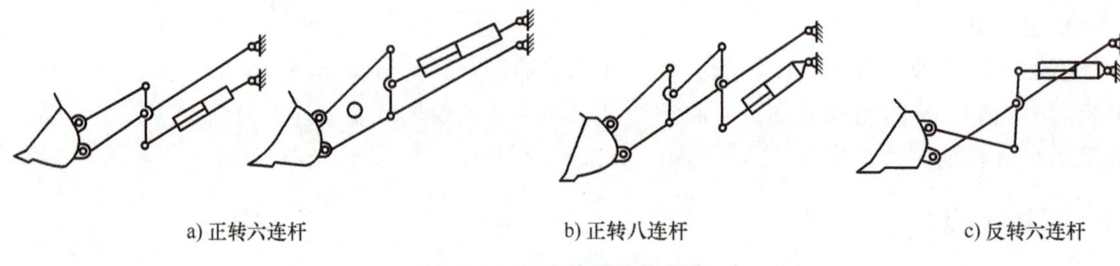

a) 正转六连杆　　b) 正转八连杆　　c) 反转六连杆

图 2-1　工作装置结构示意

和铲斗的运动方向相反，所以称为反转六连杆机构。图 2-2 所示为单摇臂反转六连杆机构的工作装置。

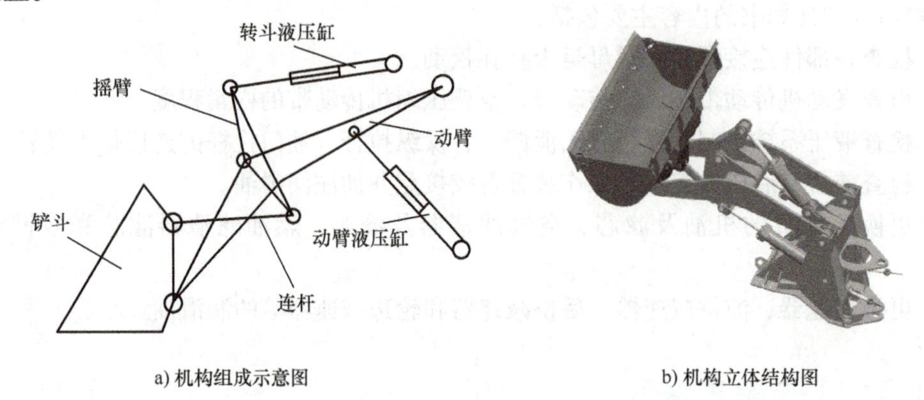

a) 机构组成示意图　　b) 机构立体结构图

图 2-2　单摇臂反转六连杆机构的工作装置

除了常见的单摇臂反转六连杆机构，在某些机型上还采用双摇臂反转六连杆机构，特别是大吨位装载机。双摇臂意味着也有双铲斗液压缸，结构跟单摇臂有所区别，但原理是一样的。

2. 工作装置的销轴配合间隙

装载机的动臂和铲斗之间，以及动臂液压缸、铲斗液压缸与工作装置之间都是采用销连接。销为圆柱形，与销套之间为间隙配合，主要用于两个构件之间的连接，以传递横向力或转矩。在挖掘机作业时，销与套之间总是频繁相对转动。如果润滑不良，就会加剧销与销套之间的磨损，导致配合间隙变大，影响工作装置的工作性能。工作装置各销轴配合间隙见表 2-3。

表 2-3　工作装置各销轴配合状态

序号	销轴	公称尺寸 /mm	装配间隙 /mm	磨损后允许的最大间隙/mm	超过允许值后应采取的措施
1	拉杆与摇臂铰销	$\phi90$	0.220～0.394	0.90	更换销轴或轴套
2	拉杆与铲斗铰销	$\phi90$	0.220～0.394	0.90	更换销轴或轴套
3	动臂与铲斗铰销	$\phi63$	0.190～0.338	0.80	更换销轴或轴套
4	动臂与摇臂铰销	$\phi110$	0.240～0.414	1.00	更换销轴或轴套
5	转斗液压缸与摇臂铰销	$\phi90$	0.220～0.394	0.90	更换销轴或轴套
6	动臂与车架铰销	$\phi90$	0.220～0.394	0.90	更换销轴或轴套
7	动臂液压缸与动臂铰销	$\phi75$	0.220～0.348	0.85	更换销轴或轴套

☞ 任务实施

装载机铲斗为带齿平斗型，主切削刃及侧切削为耐磨板，可延长铲斗的使用寿命。齿体焊在主切削板上，齿套连接销固定，齿套与齿体的连接形式如图2-3中的A—A剖视图所示。当斗齿过度磨损，齿套露出内孔或产生变形时，将严重影响铲斗正常工作，应当予以更换。

更换斗齿的方法与步骤如下：

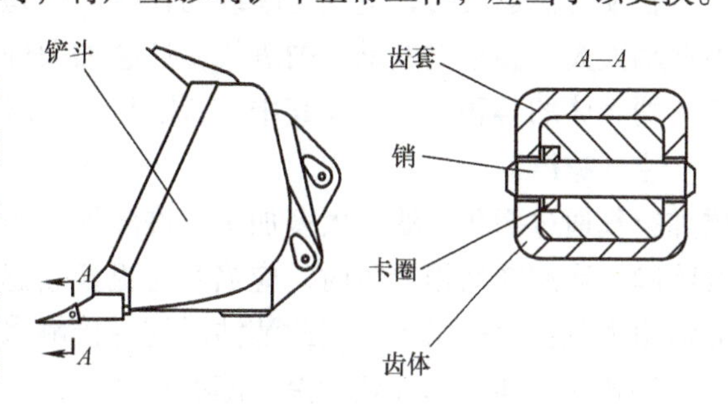

图2-3　齿套安装示意

1）停放铲斗。将装载机安全停放在平坦地面上，在地面放置高度合适的木块或硬质垫块，使铲斗底面平整地落在垫块或木块上，关闭发动机。

2）分离齿套。准备一根外径合适的铜棒冲子，将冲子一端顶靠住斗齿销，使用锤子轻轻敲打冲子的另一端，将斗齿销沿轴线方向推出来之后，拆下齿套和卡圈。

3）清理斗齿。使用抹布或毛刷清理齿体、斗齿销和卡圈表面的污垢、锈垢，必要时可使用柴油清洗，然后晾干。

4）安装新的齿套。先将卡圈安装在齿体侧面的槽内，再将新的齿套安装在齿座上，从卡环的侧面将斗齿销穿入卡圈、齿体和齿套内，使用锤子轻轻敲打斗齿销，使斗齿销与齿套平齐。

任务2.3　发动机保养与维护

子任务2.3.1　空气滤清器检查与更换

☞ 学习目标

1）能根据维修手册正确选用空气滤清器。

2）能按照保养规范要求更换空气滤清器。

☞ 工作任务

按规范要求更换空气滤清器的滤芯。

☞ 相关知识

1. 空气滤清器的作用

发动机在工作过程中需要吸入大量的空气，如果空气不经过滤清，空气中悬浮的尘埃被吸入气缸中，会加速活塞组及气缸的磨损。较大的颗粒进入活塞与气缸之间，会造成严重的“拉缸”现象。

发动机的空气滤清器用于过滤空气中的粉尘和杂质等固体颗粒物，以保证清洁空气进入发动机的燃烧室与燃油发生化学反应。

2. 空气滤清器的分类

空气滤清器一般有纸质和油浴式两种。由于纸质滤清器具有滤清效率高、质量轻、成本低、维护方便等优点，已被广泛采用。纸质滤芯的滤清效率高达99.5%以上，油浴式滤清器在正常情况下滤清效率为95%～96%。

在发动机运转时，进气是断续的，从而引起空气滤清器壳体内的空气振动，如果空气压力波动太大，会影响发动机的进气，还会加大进气噪声。为了抑制进气噪声，可以加大空气滤清器壳体的容积，也可在其中布置隔板，以减小谐振。

空气滤清器的滤芯分为干式滤芯和湿式滤芯两种。干式滤芯材料为滤纸或无纺布。为了增加空气通过面积，滤芯大多加工出许多细小的褶皱。当滤芯轻度污损时，可以使用压缩空气吹净，当滤芯污损严重时，应当及时更换新滤芯。对于干式滤芯，一旦浸入油液或水分，滤清阻力就会急剧增大，因此清洁时切忌接触水分或油液，否则必须更换新滤芯。

湿式滤芯使用海绵状的聚氨酯类材料制造，安装时应滴加一些机油，用手揉匀，以便吸附空气中的异物。滤芯污损之后，可以用清洗油进行清洗，若污损严重则应该更换新滤芯。

如果滤芯阻塞严重，将使进气阻力增加，发动机功率下降。同时由于空气阻力增加，也会增加吸进的燃油量，导致混合比过浓，从而使发动机运转状态变坏，增加燃料消耗，也容易产生积炭。

☞ 任务实施

如果空气滤清器发生严重堵塞，就会造成柴油机进气管前部管路产生负压，此时，报警指示灯将亮起，必须清洁空气滤清器的主滤芯。

按照柴油机保养规范要求，空气滤清器的主滤芯在使用时间达到500h后，应当进行定期检查，如果滤芯表面黏附灰尘或滤芯损坏，应当进行清理或更换。

1. 更换主滤芯的方法与步骤

1）关闭发动机，打开发动机舱侧门，清理空气滤清器外表面的灰尘，然后松开夹子，打开端盖，将主滤芯从滤清器座上轻轻抽出，如图2-4所示。

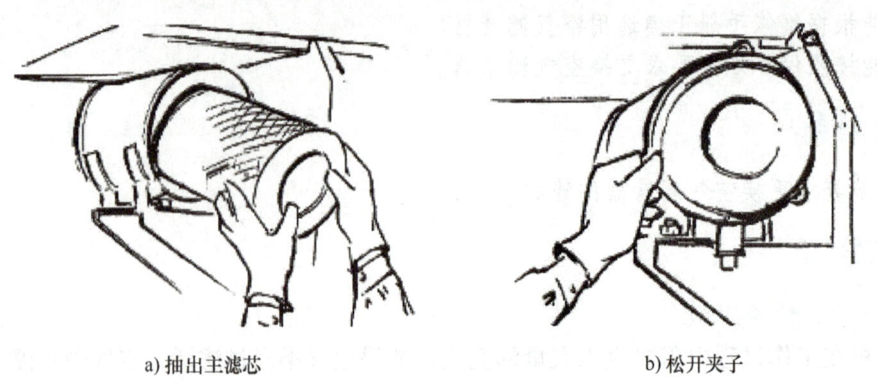

a) 抽出主滤芯　　b) 松开夹子

图2-4　拆卸主滤芯示意

2）使用抹布清洁空气滤清器壳体内壁表面的灰尘，然后使用压缩空气（≤300kPa）沿着壳体圆周吹刷，使灰尘从壳体内部往外散发，如图2-5所示。

3）清洁主滤芯，用压缩空气先从主滤芯内部沿着褶痕往外冲刷，然后再分别从主滤芯外侧和内侧顺着褶痕反复冲刷，将黏附在主滤芯表面的灰尘吹走，如图2-6所示。

图2-5　清洁空气滤清器壳体内壁

图2-6　清洁主滤芯

4）主滤芯清理完毕后，可以利用灯光照明方法来检查滤芯表面是否还残留灰尘。将照明的电灯或手电筒放入滤芯内（图2-7），使灯光经过滤芯的小孔往外散。如果滤芯干净，从滤芯网格散发出来的灯光是均匀的。若发现滤芯表面有小孔或微粒，以及垫圈或密封损坏，应更换新的主滤芯。

图2-7　借光检查

5）将干净的主滤芯安装到空气滤清器壳内，保证主滤芯端部密封均匀接触。不应使用工具，以免损坏主滤芯。

6）清洁并安装空气滤清器的端盖，保证空气滤清器内盖的密封垫与空气滤清器壳体均匀接触。安装端盖时，注意使端盖的粉尘排放口朝下。当端盖完全合上后，卡紧夹子，使端盖稳固在滤清器壳体上。

2. 主滤芯更换的注意事项

主滤芯在清理过6次后应当更换。即使没有清理过6次，每年也应更换一次。在更换主滤芯时，要同时更换主滤芯内部的安全滤芯。

如果清洁了主滤芯以后，空气滤清器阻塞报警灯仍亮，或仍然有黑烟排出，应更换一个新的安全滤芯。

子任务2.3.2　压缩机传动带张紧度调整

☞ 学习目标

1）能够根据技术规范判断传动带张紧程度。

2）能够根据规范要求调整传动带张紧力。

☞ 工作任务

根据规范要求调整压缩机传动带张紧力。

☞ 相关知识

传动带传动是机械传动的一种形式。传动带可以将柴油机曲轴旋转的运动及动力通过传动带传动到空调压缩机的转动轴。由于传动带传动是通过摩擦来传递动力的，所以传动带要调整张紧度以获得合适的摩擦力。传动带张紧度是靠张紧轮进行调整的。如果没有张紧轮装置，就直接通过调整轮距的方式达到调整张紧度的目的。柴油机空调压缩机传动带张紧装置如图 2-8 所示。

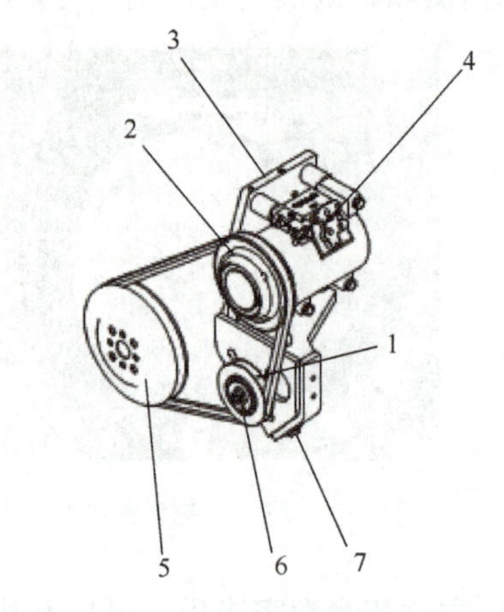

图 2-8　柴油机空调压缩机传动带张紧装置

1—张紧轮　2—带轮　3—压缩机安装支架
4—压缩机　5—发动机带轮　6—锁紧螺母
7—调节螺栓

如果空调压缩机传动带表面与传动带轮槽接触侧面光亮，并且起动空调时有“吱吱”的噪声，说明传动带打滑严重，传动带产生了磨损，应更换传动带或传动带轮。如果传动带过松应给予调整，否则易使空调系统制冷不良。

事实上，空调运行一段时间后，压缩机传动带就会出现开裂、脱线，或者内部芯线达到疲劳寿命时，容易产生断裂现象，所以应当及时更换。

在缺乏专业工具的情况下，传动带张紧度可以用手压的方式来判断。关闭发动机，使用大拇指稍用力往下按压传动带与两传动带轮（空调压缩机传动带轮和发动机上的传动带轮）的切边中点垂直带边，按压力约为 58.5N，此时，传动带产生的挠度 A 应为 5～8mm。否则就需要调整传动带的张紧度。

☞ 任务实施

1. 空调压缩机传动带张紧度的调整步骤

1）使用扳手将锁紧螺母 6 拧松，根据传动带的松紧程度，调整调节螺栓 7 的位置，使传动带的张紧度产生变化。

2）使用大拇指稍用力，作用力约为 58.5N，往下按压传动带与两传动带轮切边中点垂直带边，如图 2-9 所示。

3）如果挠度 $A>8$mm，则拧紧调节螺栓 7，待传动带张紧至合适位置后，将锁紧螺母 6 锁紧。

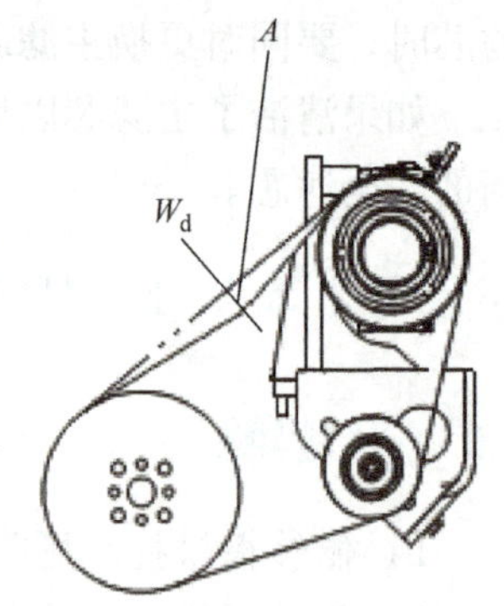

图 2-9　带张紧度调整示意

W_d—调整力　A—挠度

2. 注意事项

当传动带无法张紧时，即在传动带与两传动带轮的切边中点垂直带边加大小为 58.5N 的载荷，如果传动带产生的挠度超过 8mm，则应及时更换传动带。当传动带起毛边或有裂口时也应及时更换传动带。根据整机的使用工况以及传动带的磨损情况，建议每 1000h 更换空调压缩机传动带。

子任务2.3.3 机油更换

☞ 学习目标

1）能正确选用机油。
2）能按规范要求更换机油。

☞ 工作任务

按规范要求更换发动机机油。

☞ 相关知识

机油主要由基础油和添加剂两部分组成。基础油是机油的主要成分，决定着机油的基本性质。添加剂则可弥补和改善基础油性能方面的不足并赋予机油某些新的性能，是机油的重要组成部分。

1. 机油的作用

（1）润滑　发动机工作时，各零部件之间做相对运动，通过机油的润滑，在各运动部件之间形成一层吸附油膜，可以使发动机内部各零部件之间的磨损降低，不但提高了机械效率，也能够延长发动机的工作寿命。

（2）冷却和散热　机油在润滑过程中可以将发动机部件之间摩擦产生的多余热量带走，起着间接的散热作用，避免机件由于温度过高而造成烧坏。

（3）清洁　发动机各部件在反复相对运动过程中会因摩擦产生一些金属碎屑和炭粒，机油可以带走这些杂质，防止出现磨粒磨损，避免加剧机件的老化和变形。

（4）防腐防锈　机油在机件运行过程中均匀分布在其表面，机油中含有防锈添加剂，能够在金属表面形成吸附层，使其与外界的空气、水分、燃气直接隔绝，避免机件产生氧化和锈蚀，减少腐蚀性磨损。

（5）密封　机油具有一定的物理黏性，黏附在零件表面上时，能够把燃烧室内不同的工作环境隔绝开来，起到很好的密封效果。此外，柴油在发动机燃烧室被压燃时会产生高压气体，进而推动活塞做功，机油可以密封活塞和气门之间的间隙，减少压力损失，提高机械效率。

（6）减振　柴油发动机在起动、加速或承受的负荷发生较大变化时，活塞销、曲轴大小瓦、以及连杆的大小端等零部件均要承受振动带来的冲击载荷。适当黏度的机油覆盖在这些机件上所形成的油膜可以吸收部分冲击载荷的能量，减少机件振动对发动机正常工作产生的负面影响，起到缓冲的作用。

2. 机油的选用

柴油机机油质量等级的选择有两个主要依据，一是根据发动机的机械负荷和热负荷的总和，以强化系数来表示；二是根据发动机工况苛刻程度。

柴油机机油一共有11个级别，分别为CA、CB、CC、CD、CE、CF、CF－4、CG－4、CH－4、CI－4、CJ－4。“C”代表柴油机机油。“－4”表示该机油适用于4冲程柴油机。“S”代表汽油机机油。如果“S”和“C”两个字母同时存在，由表示该机油为汽柴通用型。

机油质量级别每递增一个字母，所拥有的性能都会优于前一种。至于柴油机用什么机油好，其实是需要根据具体工作情况来决定的。

3. 机油损耗的原因分析

在发动机工作时，气缸壁、活塞环和活塞之间都充满了机油，当三者配合间隙不正常时，气缸壁和活塞间的机油就会窜入燃烧室，这是引起窜油的主要原因。

发动机工作一定时间后，活塞环径向尺寸减小，弹力减弱，开口间隙变大，密封和刮油作用变差。活塞环与环槽间隙增大，加强了活塞环的泵油作用，活塞上行时机油被刮进燃烧室。

窜油后容易产生积炭，加速气缸磨损，形成恶性循环。活塞环卡死后烧机油，当活塞环卡死在环槽内时，活塞环失去弹性，密封变差，不但在活塞上行时将大量机油带入燃烧室内，而且在活塞做功时有大量的高压气体窜入曲轴箱，使曲轴箱内压力增高，飞溅的机油通过曲轴箱通风装置进入进气道，使机油和空气一起进入燃烧室燃烧。

发动机的大修质量对气缸窜油影响较大，气缸的圆度误差、圆柱度误差及表面粗糙度值过大，活塞环未按技术要求与活塞选配，装配中活塞偏缸以及气缸中心线与曲轴旋转轴线垂直度误差超过规定要求等，都会导致窜油现象。

同时，活塞积炭清除不干净，活塞环装反，环口未合理分布及环口间隙太大，都会不同程度地加速气缸窜油。发动机运行不久就产生窜油，部分原因是由活塞环质量不合格造成的。

☞ 任务实施

机油在使用一段时间后，机油量和特性将会发生变化，所以需要定期检查机油油位和更换机油。更换机油时，应同时更换机油滤清器。

1. 准备工作

将装载机可靠停放在平坦的场地，发动机熄火，等待10min左右，让曲轴箱内的发动机机油充分流回发动机油底壳。

2. 检查机油油位

打开发动机舱罩，找到机油油位尺。发动机机油加注口和机油油位尺位置如图2-10所示。

拔出油位尺，用干净的布将油位尺擦干净，重新插入发动机油位口到尽头，再拔出来检查，正常油位应在油位尺的“L”刻度和“H”刻度之间，如图2-11所示。如果油位在“L”刻度之下，说明机油量过少；如果油位在“H”刻度之上，说明机油量过多。

图2-10　机油油位尺示意

1—机油油位尺　2—机油加注口

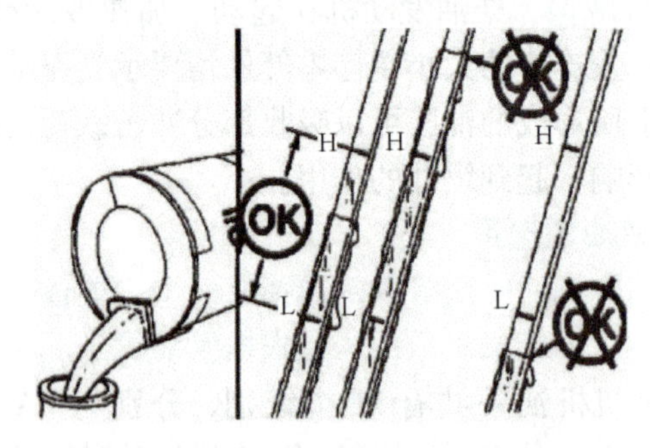

图2-11　机油油位检查示意

H—油位上限　L—油位下限

3. 排放机油

拆下发动机油底壳底部的放油螺塞，将机油排放到事先准备好的容器内，如图2-12所示。机油应当排放干净，必要时可以倒入干净的机油对机油腔进行清洗。容器的容量须大于所盛机油量，避免机油溢出污染场地。

待机油完全排放干净之后，拧上放油螺栓，如图2-13所示。

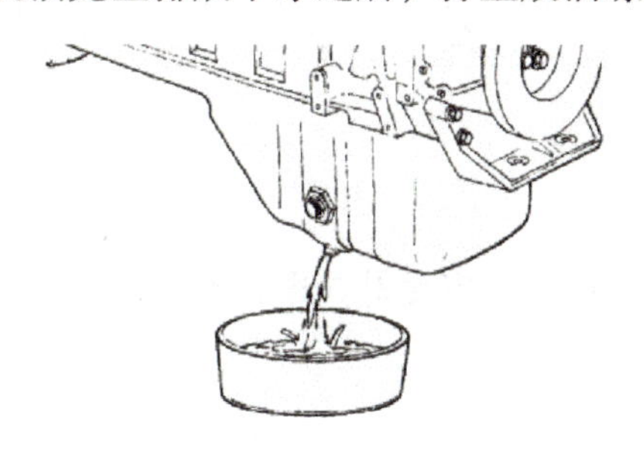
图2-12 排放机油

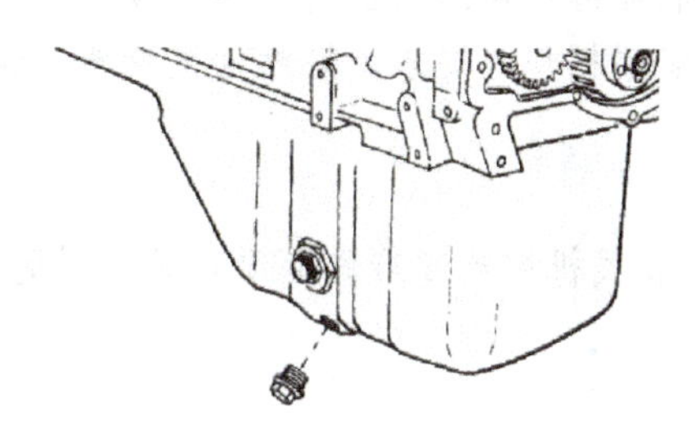
图2-13 拧上放油螺栓

4. 更换机油滤清器

清洁机油滤清器外表面，使用扳手拆下机油滤清器，清理干净安装座上密封垫接触表面，如图2-14所示。如果旧的O形密封圈粘在安装座上，应当使用工具将其去除。

安装新的O形密封圈，在机油滤清器内充满干净的机油，并在密封垫表面涂上一层干净的机油。

将机油滤清器安装到安装座上（图2-15），用手拧紧到机油滤清器密封垫表面接触安装座，再使用扳手拧紧机油滤清器到规定的要求。

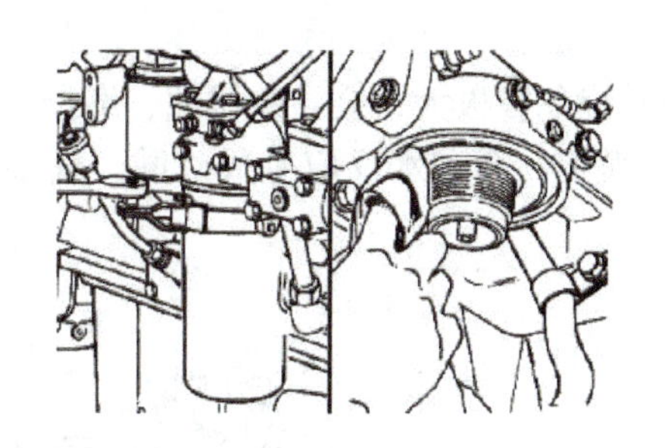
图2-14 清理安装座

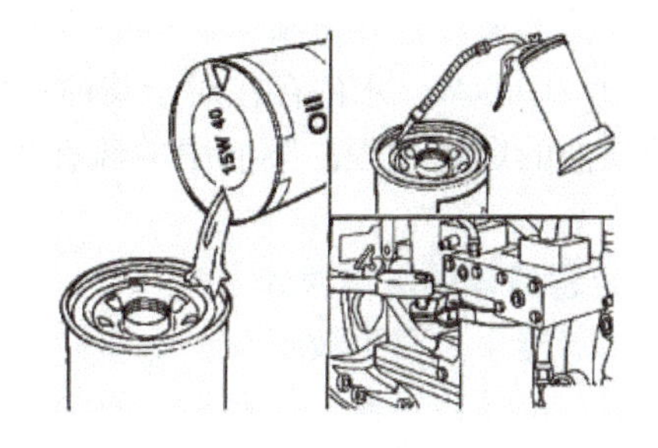
图2-15 安装机油滤清器

需要注意的是，在安装机油滤清器之前，应当在机油滤清器内注满干净的机油，否则容易造成发动机因为缺少机油而损坏。

5. 更换机油

使用抹布清理发动机机油加注口，拧下机油加注口螺堵，从发动机机油加注口加注干净的机油到油位尺刻度“H”处。在怠速下运转发动机，检查机油滤清器和放油螺塞是否有泄漏。发动机熄火，等待10min左右，让机油充分回流到油底壳，再次检查发动机油位，如果机油不足，则补充机油到油位尺刻度“H”处。

6. 整理现场

更换机油滤清器及机油之后，拧上机油加注口螺堵，及时清理机油加注过程中散落的机油，收拾好工具，清理现场。

子任务 2.3.4 燃油滤清器更换

☞ 学习目标

1）能描述燃油滤清器的工作原理。

2）能根据规范要求更换燃油滤清器及预滤器。

☞ 工作任务

一台装载机燃油滤清器及预滤器已到更换保养时间，需要对其进行更换。

☞ 相关知识

1. 燃油滤清器的作用

燃油滤清器的作用是滤除发动机燃油气系统中的有害颗粒和水分，以保护液压泵油嘴、缸套、活塞环等，减少磨损，避免堵塞。

2. 燃油滤清器的种类

根据工作介质的不同，燃油滤清器可分为柴油滤清器、汽油滤清器和天然气滤清器等类型。装载机装用的发动机一般为柴油发动机，所以使用的是柴油滤清器。

柴油滤清器的结构大致与机油滤清器相同，有可换式和旋装式两种。但其承受的工作压力和耐油温要求较机油滤清器低得多，而其过滤效率的要求却比机油滤清器高得多。柴油滤清器的滤芯多采用滤纸，也有采用毛毡或高分子材料的。

柴油滤清器又可以分为预滤器和柴油精滤器。预滤器也称为油水分离器，它的重要功能就是分离柴油中的水。水的存在对于柴油机供油系统危害极大，会导致锈蚀、磨损、卡死甚至会恶化柴油的燃烧过程。柴油中的硫在燃烧过程中，会和水反应生成硫酸，腐蚀发动机部件。

3. 燃油滤清器的工作原理

燃油滤清器主要由手液压泵（滤座）、滤清器部件、水位传感器、温度传感器、加热器、密封圈、管接头等零部件组成，其中过滤核心件为滤芯。图 2-16 所示为一种柴油滤清器。

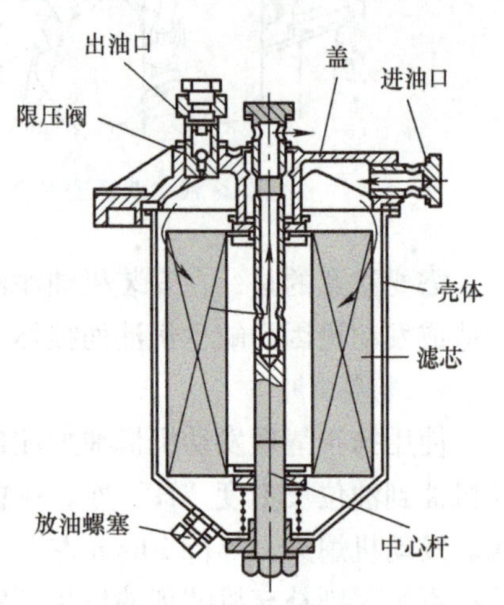

图 2-16 柴油滤清器

滤芯为多孔体结构，当流体通过滤芯时，流体中粒径大的杂质粒子被多孔体拦截在表面；粒径小于多孔体的粒子则随流体通过多孔体。滤材的孔径和杂质粒子的直径不是整圆形的，而是当量孔径和当量粒径。因此，滤芯拦截的粒径比滤材的孔径要小得多。

流体中的杂质微粒做无规则的布朗运动，当其运动到多孔体的孔壁时，被黏附滞留在孔壁上的小区域内，被多孔体滤除。

当带有微小水滴的油流进入滤材时，由于滤材对油的“亲合”性，油流能顺利地从滤材的微孔中通过；由于滤材对水的“疏远”性，微小的水滴不能从滤材的微孔中通过。由于微小水滴对滤材绒毛的浸润作用，使微小水滴滞留在滤材的绒毛上。被拦截的微小水滴逐渐增加，聚结成大水珠。在油压的作用下，水珠被压向滤芯的楔形通路，形成更大的水滴。水滴通过滤材后，便在自身重力的作用下，向下沉降，掉落在集水杯中。

4. 燃油滤清器的更换

燃油滤清器需要定期更换。如果不定期更换燃油滤清器，则燃油滤清器可能会被污染物堵塞并限制燃油流量，从而导致发动机性能明显下降，因为发动机难以汲取足够的燃料以继续正常运行。

柴油中含有水分，而水的密度比柴油的密度大，所以在柴油发动机燃油路上往往安装具有碗状结构的燃油滤清器作为第一道过滤器，让水分沉淀在燃油滤清器底部，然后通过打开燃油滤清器底部的阀门将水分排出，以保证进入燃烧室的柴油尽量少含水分。

有的燃油滤清器安装有油水分离传感器，当燃油滤清器中分离的水分液面达到一定高度时，传感器向发动机控制单元发送信号，提示驾驶员手动放水。

☞ 任务实施

1. 更换燃油预滤器

柴油发动机的燃油预滤器安装在燃油泵一侧的机体上，如图2-17箭头所指。在拧下预滤器之前，要先将预滤器外围及安装座表面清理干净。

使用扳手把预滤器轻缓地从安装座上拧松，拆卸下来，如图2-18所示。

取下安装座螺纹接头上的密封垫，用无纤维布清理干净安装座的密封面，如图2-19所示。检查密封垫，若密封垫损坏则更换。将新的预滤器安装到预滤器座上。

图2-17 燃油预滤器

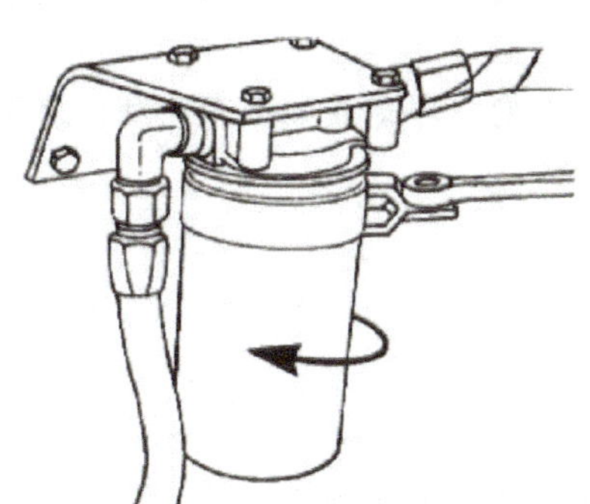

图2-18 拆卸预滤器

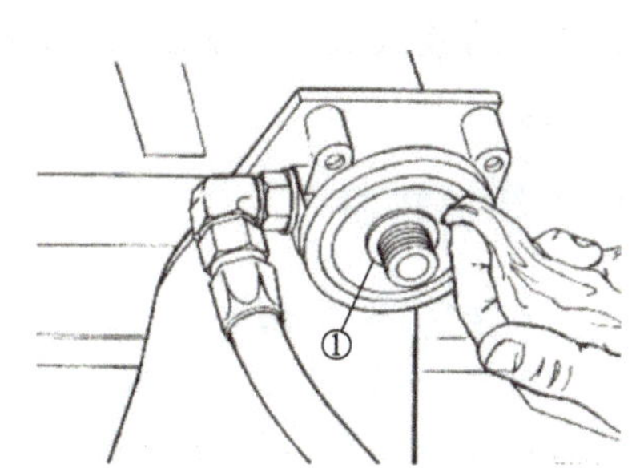

图2-19 清理密封面

2. 更换燃油滤清器

燃油滤清器与预滤器安装在发动机的同一侧，如图2-20所示。拆卸燃油滤清器之前，先将燃油滤清器周围区域及安装座清理干净，然后使用扳手把燃油滤清器从安装座上卸下来。

清理燃油滤清器安装座，清除所有旧密封垫。

安装新的密封垫到燃油滤清器安装座的螺纹接头上，

图2-20 燃油滤清器

在燃油滤清器的密封面涂上一层机油，将燃油滤清器充满干净的燃油（图2-21），再将其安装到燃油滤清器座上，用手把燃油滤清器拧到安装座上，在燃油滤清器的密封垫接触到安装座后，再拧紧1/2～3/4圈即可。不可用机械方法过分拧紧，以免损坏燃油滤清器，如图2-22所示。

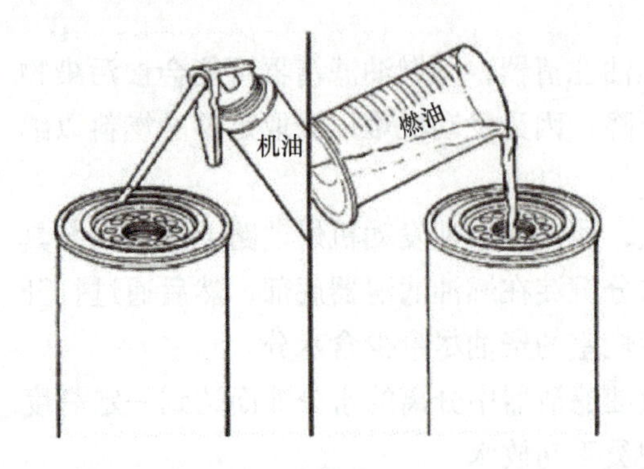

图2-21　涂机油和充燃油

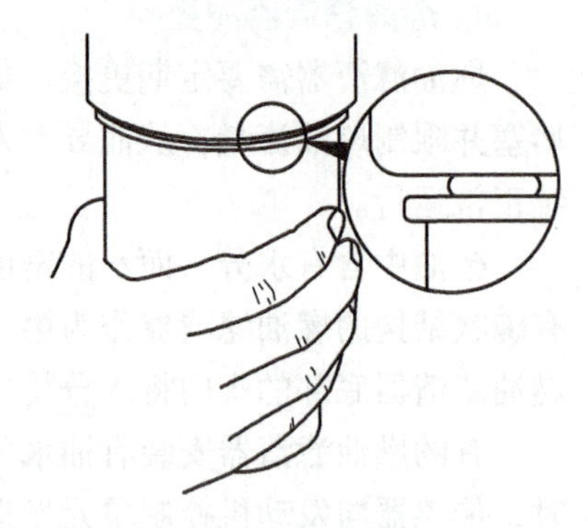

图2-22　安装燃油滤清器

子任务2.3.5　冷却液更换

☞ 学习目标

1）能合理选用冷却液。

2）能按规范要求更换冷却液。

☞ 工作任务

一台装载机显示水温高，经检查发现冷却液减少且已到保养更换时间，需对冷却液进行更换。

☞ 相关知识

1. 冷却液的作用

冷却液具有保护发动机冷却系统免遭锈蚀和腐蚀，有效抑制水垢形成，防止水箱过热，减少冷却液蒸发，为水泵节温器及其他部件提供润滑的作用。

（1）冬季防冻　选用冷却液时，其冰点一般要求低于使用地区最低温度10～15℃，目的是为了防止汽车在冬季停车后，冷却液结冰而造成水箱、发动机缸体胀裂。

（2）缓蚀　冷却系统中散热器、水泵、缸体及缸盖、分水管等部件是由钢、铸铁、黄铜、纯铜、铝、焊锡等金属组成，由于不同金属的电极电位不同，在电解质的作用下容易发生电化学腐蚀，冷却液中都加入一定量的缓蚀添加剂，从而有效防止冷却系统产生腐蚀。

（3）防水垢　冷却液在循环中会产生一定量的水垢。水垢增多可能导致循环管道堵塞，影响冷却系统的散热功能。所以在选用、添加冷却液时，应当根据具体情况选择合适配比的冷却液。

（4）防开锅　冷却液的沸点通常超过105℃，能耐受比纯净水更高的温度而不开锅，在一定程度上满足了高负荷发动机的散热冷却需要。

2. 冷却液的类型

冷却液由水、防冻剂和添加剂三部分组成。按防冻剂成分不同，常见的冷却液可分为乙二醇型和甘油型两种。

乙二醇型冷却液是用乙二醇作防冻剂，并添加少量抗泡沫、防腐蚀等综合添加剂配制而成。由于乙二醇易溶于水，可以配成各种冰点的冷却液，其最低冰点可达－68℃，这种冷却液具有沸点高、泡沫倾向低、黏温性能好、缓蚀和防垢等特点，是一种较为理想的冷却液，目前国内外发动机所使用的和市场上所出售的冷却液几乎都是乙二醇型冷却液。

甘油型冷却液沸点高、挥发性小、不易着火、无毒、腐蚀性小，但降低冰点效果不佳、成本高、价格昂贵，只有少数北欧国家仍在使用。

冷却液在发动机冷却系统中循环流动，将发动机工作中产生的多余热能带走，使发动机能以正常工作温度运转。当冷却液不足时，将会使发动机水温过高，导致发动机机件损坏。冷却液不足时，应当及时添加。

3. 冷却液的选用

冷却液型号一般按照冰点进行分类，主要有－25℃、－30℃、－35℃、－40℃、－45℃、－50℃等。选用冷却液时，要选用冰点低于车辆运用环境最低温度10～15℃的冷却液。

☞ 任务实施

根据定期保养要求，冷却液使用达到2000h后，应当进行更换。

1. 冷却液的检查

正常情况下，发动机冷却液的液面应当处于储液罐上刻度线与下刻度线之间。如果液面很高，说明冷却系统可能出了问题，比如冷却液中混入大量气体或其他液体。如果冷却液降到下刻度线以下，就需要补充冷却液。

当需要检查冷却液时，必须等到发动机冷却液的温度降到50℃以下，再慢慢拧开水散热器加水口盖，释放冷却系统内部压力，以免有高温蒸汽或高温冷却液从加水口喷洒出来。

检查冷却液液位是否位于加水口下10mm范围内，必要时应当补充冷却液。检查冷却液散热器加水口盖的密封，如果损坏则应更换。

2. 冷却液的更换

1）等到冷却液温度降到低于50℃后，慢慢拧开水散热器加水口盖，释放压力。

2）打开水散热器底部的排水阀（图2-23），将发动机的冷却液排出，并用容器盛接。在发动机冷却液排干净后，关上水散热器底部的放水阀门。

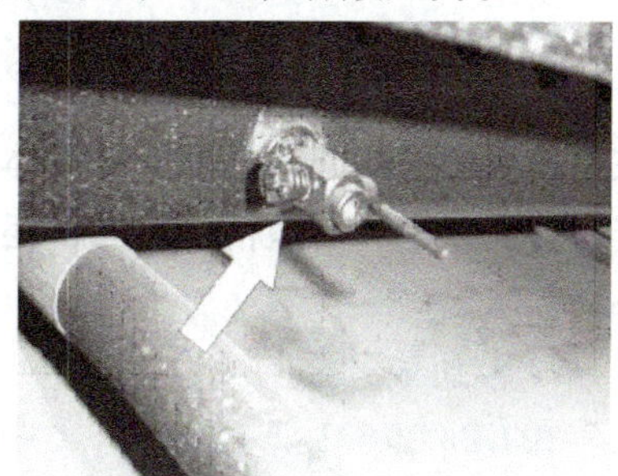

图2-23 冷却液排水阀

3）检查冷却系统的所有水管、管夹是否损坏，如有必要进行更换。检查水散热器有否泄漏、损坏和脏物堆积，根据需要进行清洁和修理。

4）向发动机冷却系统加入用水和碳酸钠混合配成的清洗液，其混合比例是每23L水中加入0.5kg碳酸钠。液位应到达发动机正常使用的液位，并且10min内保持稳定。

5）保持水散热器加水口盖打开，起动发动机，当清洗液温度上升到80℃以上时，再运行发动机5min，然后关闭发动机，泄放清洗液。

6）向发动机冷却系统加入干净水到正常使用液位，并保持10min不变化。保持水散热器加水口盖打开，起动发动机，当水温度上升到80℃以上时，再运行发动机5min，然后关闭发动机，将冷却系统中的水排干净。如果排出的水仍是脏的，必须再次清洗系统直至排出的水变得干净。清洗的水排干净之后，关闭所有的泄放阀。

7）加注冷却液。打开水散热器加水口盖，将冷却液缓慢加入，直至液面到达水散热器加水口下10mm范围内为止，并且10min内保持稳定。

8）保持水散热器加水口盖打开，起动发动机，先在低怠速下运转5min，再在高怠速下运转5min，并且使冷却液温度达到85℃以上。

9）再次检查冷却液液位，如有必要，补充冷却液直至液面到达水散热器加水口下10mm范围内。

10）更换水散热器冷却液的同时，应当更换副水箱的冷却液。

任务2.4　液压系统保养与维护

子任务2.4.1　液压油维护

☞ 学习目标

1）能够制订液压系统的定期保养计划。
2）能够根据要求开展液压油的检查及更换。

☞ 工作任务

按规范要求对装载机的液压油进行维护。

☞ 相关知识

1. 液压油的作用

液压油就是利用液体压力能的液压系统使用的液压介质，液压油在液压系统中起着能量传递、抗磨、系统润滑、缓蚀、防锈、冷却等作用。对于液压油来说，首先应满足液压装置在工作温度下与起动温度下对液体黏度的要求，由于液压油的黏度变化直接与液压系统的动作、传递效率和传递精度有关，还要求油的黏温性能和剪切安定性应满足不同用途所提出的各种需求。

2. 液压油分类

在GB/T 7631.2—2003分类中的HH、HL、HM、HR、HG、HV、HS液压油均属于矿油型液压油，这类油的品种多，使用量占液压油总量的85%以上，汽车与工程机械液压系统常用的液压油多数属于这类。

3. 液压油的选择

根据工作环境和工况条件选择液压油的品种。在选用液压设备所使用的液压油时，应从工作压力、温度、工作环境、液压系统及元件结构和材质、经济性等几个方面综合考虑和判断。环境因素有地上、地下、室内、野外、沿海、寒区、高温、明火。使用工况包括泵的类型、压力、温度、材质、密封材料、运行时间。油品性质主要是指液压油的理化性能特点。经济性主要是指液压油的使用时间、换油期和价格。

4. 液压油的保养

装载机液压系统的液压油应当具备以下性能：

1）适宜的黏度及良好的黏温性能，以确保在工作温度发生变化的条件下能准确、灵敏地传递动力，并能保证液压元件的正常润滑。

2）具有良好的防锈性及抗氧化安定性，在高温高压条件下不易氧化变质，使用寿命长。

3）具有良好的抗泡沫性，使油品在受机械不断搅拌的工作条件下，产生的泡沫易于消失以使动力传递稳定，避免液压油加速氧化。

装载机液压系统的液压油一般选用抗磨型 L－HM 系列液压油，按运动黏度分为 22、32、46、68 四个牌号。抗磨液压油具有良好的防锈、抗氧化性能，在抗磨方面尤为突出。

☞ 任务实施

1. 检查液压油油量

在检查装载机液压油油量之前，正常情况下，应使液压缸、液压管路和散热器等液压部件都充满液压油。

将机器开到平坦的场地，前后车架对直无夹角，收斗至极限位，发动机全速提升动臂至最高位，整机怠速将动臂操纵杆推至“下降”位，使动臂匀速下降至最低位，把铲斗水平放到地面上，然后熄火并取下钥匙，前后左右推动操作杆卸压。

使液压油温度保持在（50±5）℃范围内，在液位计无气泡的情况下，检查液压油箱液位计，此时油位应该在液位计中绿色范围内，即 MAX 线和 MIN 线之间，如图 2-24 所示。

图 2-24 油位计液位刻度

检查油位时，若发现油液液面高出液位计的绿色范围（观察液位计有气泡的情况下），即使油位在最高油位线以上，也不能放油，需待气泡消除后，再检查油位是否在液位计中绿色范围内。如果油位低于绿色范围，必须及时补充液压油，补油后再次检查油位。

2. 补充液压油

装载机的液压油箱一般安装在驾驶室侧面或后面，图 2-25 所示为液压油箱安装在驾驶室后面的结构形式。将装载机安全停放在平坦地面上，打开液压油箱上部通道板的安装螺栓及后罩上空气滤清器盖板的安装螺栓，卸下通道板和盖板，如图 2-26 所示。

清理空气滤清器和加油口法兰表面的污染物。缓慢松开液压油箱空气滤清器盖（如图 2-27箭头所示），以释放压力，拆下加油法兰的安装螺栓，取下法兰（如图 2-28 箭头所示），从加油口添加液压油，并检查液压油液位。

图 2-25　液压油箱安装位置示意

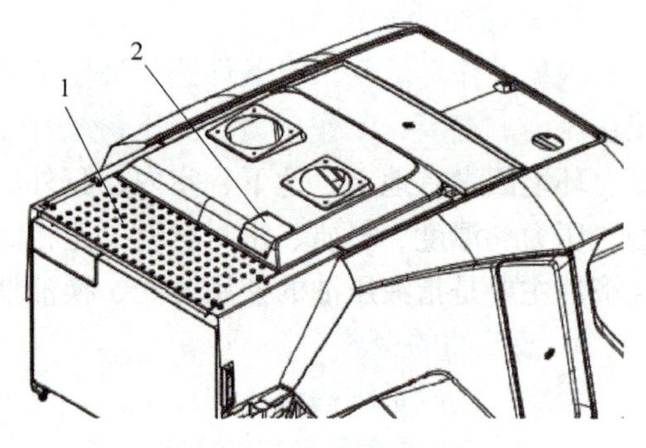

图 2-26　液压油箱覆盖件

1—通道板　2—盖板

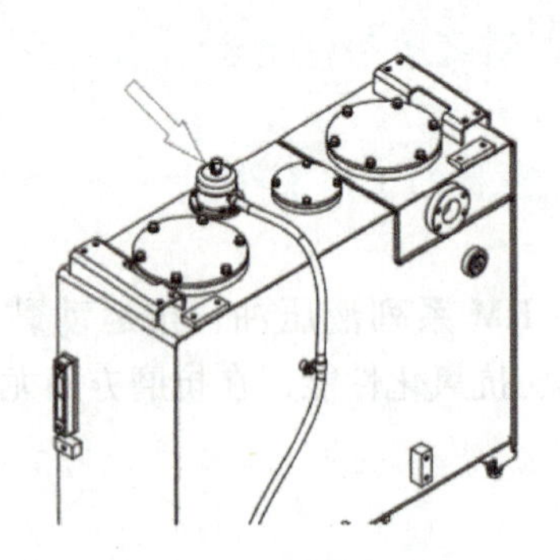

图 2-27　空气滤清器

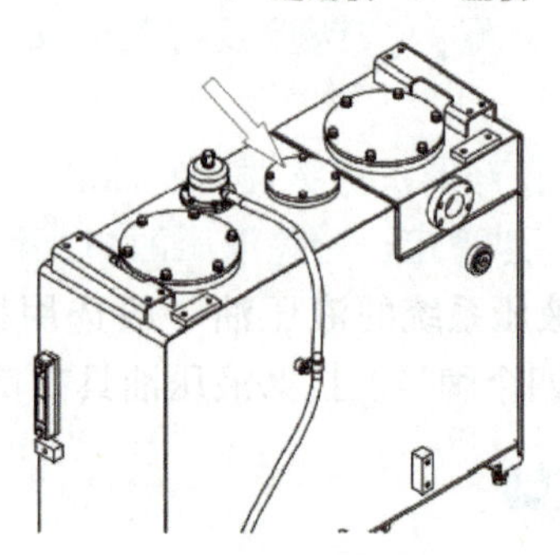

图 2-28　加油口法兰

油位符合要求后，安装液压油箱的加油法兰并拧紧空气滤清器盖，安装液压油箱上部通道板及空气滤清器盖板。

3. 更换液压油

将装载机停放在平坦空旷的场地上，拉起驻车制动，装上转向架锁紧装置，如图 2-29 所示。起动发动机并在怠速下运转 10min，其间反复多次进行提升、下降动臂，或前倾、后倾铲斗等动作。使液压油升温，并保持在 (50 ±5)℃范围内。

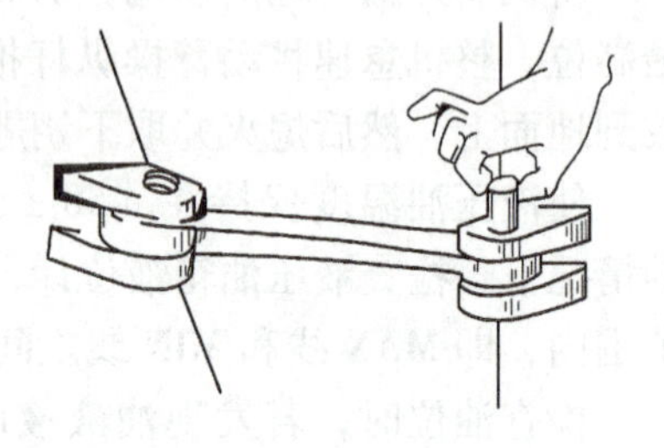

图 2-29　转向架锁紧装置

将动臂举升到最高位置，将铲斗后倾到最大位置，关闭发动机。将工作装置操纵在卸料位，使铲斗在自重作用下往前翻，排出转斗液压缸中的油液；在铲斗转到位后，将工作装置操纵在下降位，动臂在自重作用下往下降，排出动臂液压缸中的油液。

拆开位于后车架左侧腹板内侧的液压油箱放油管固定管夹螺栓，从铰接处引出放油管，拧开放油螺塞，排出液压油，并用容器盛接。同时打开空气滤清器盖，加快排油过程。腹板和固定管夹如图 2-30 所示。

打开液压油散热器的排油阀（图 2-31），排出液压油散热器内残留的液压油。取出液压油箱回油滤芯端盖等部件，取出回油滤芯，更换新滤芯，如图 2-32 所示。

打开加油口法兰，取出内部的加油滤芯。使用不可燃溶剂清洗加油法兰、加油滤芯，甩干或用压缩空气将加油法兰、加油滤芯彻底吹干。

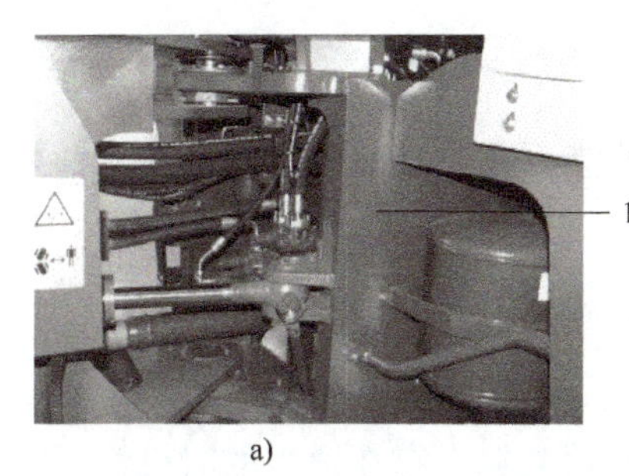

a)

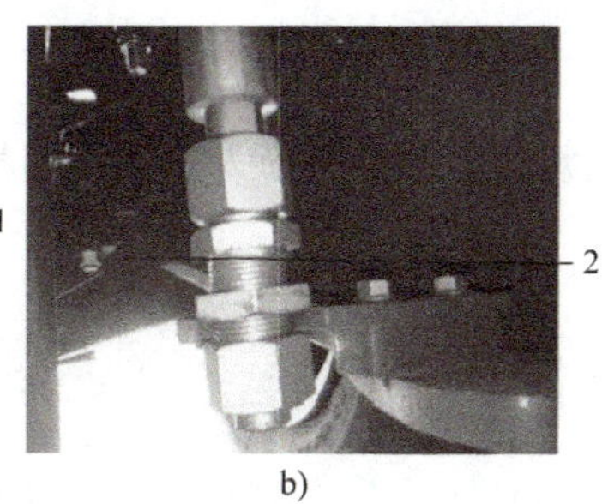

b)

图 2-30　腹板和固定管夹

1—腹板　2—固定管夹

图 2-31　排油阀

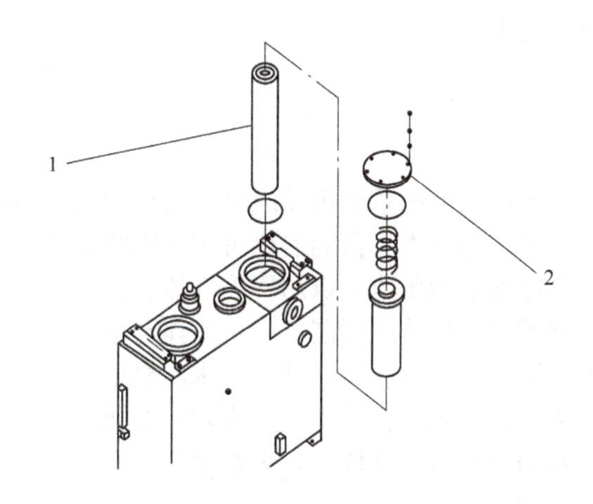

图 2-32　回油滤芯附件

1—回油滤芯　2—端盖

将液压油箱的放油螺塞、回油滤芯、加油滤芯、液压油散热器的排水阀等安装好。从液压油箱的加油口加入干净的液压油，使油位达到液压油箱油位计的上刻度，拧好加油法兰盖。

拆除车架锁紧装置，起动发动机。操作工作装置操纵杆，进行2～3次升降动臂和前倾、后倾铲斗以及左右转向到最大角度，使液压油充满液压缸、油管。然后在怠速下运行发动机5min，以便排出系统中的空气。

发动机熄火，打开液压油箱加油盖，添加干净液压油，直至液面处于液位计的绿色范围内，即MAX线和MIN线之间。

在拆卸任何液压油路或分解任何含有液体的部件时，必须用适当的容器来收集液体，并根据地方法规处置所有液体。在进行换油操作的过程中，应注意液压油清洁度，不能让脏物进入液压系统内。

子任务2.4.2　蓄能器维护

☞ 学习目标

1）能够根据要求更换蓄能器。

2）能够根据要求对蓄能器进行充气。

☞ 工作任务

一台装载机熄火后不能使动臂下降，经检查，先导供油蓄能器无压力，需要对蓄能器进行充气。

☞ 相关知识

1. 蓄能器的作用

蓄能器是液压与气动系统中的一种能量储蓄装置。它在适当的时机将系统中的能量转变为压缩能或位能储存起来，当系统需要时，又将压缩能或位能转变为液压能或气压能而释放出来，重新补供给系统。当系统瞬间压力增大时，蓄能器可以吸收这部分能量，以保证整个系统压力正常。

2. 蓄能器的功能

当低速运动时，载荷需要的流量小于液压泵流量，液压泵多余的流量存入蓄能器，当载荷要求流量大于液压泵流量时，液体从蓄能器放出来，以补液压泵流量的不足。

当停机但液压系统仍需维持一定压力时，可以停止液压泵而由蓄能器补偿系统的泄漏，以保持系统的压力。蓄能器也可用来吸收液压泵的压力脉动或吸收系统中产生的液压冲击压力。蓄能器中的压力可以用压缩气体、重锤或弹簧来产生，因此蓄能器可分为气体式、重锤式和弹簧式。气体式蓄能器中，气体与液体直接接触的称为接触式，其结构简单，容量大，但液体中容易混入气体，常用于水压机上；气体与液体不接触的称为隔离式，常用囊或隔膜来隔离，囊体积变化量大，隔膜体积变化量小，常用于吸收压力脉动。重锤式蓄能器容量较大，常用于轧机等系统中，供蓄能用。

☞ 任务实施

装载机工作时间达到 50h、100h、250h、500h、1000h 时都应该检查蓄能器中氮气的预充压力，以后每 2000h 检查一次。先导供油阀所用蓄能器一般位于装载机左侧、驾驶室的下方，左腹板内侧，如图 2-33 箭头所示。

对蓄能器补充压力前，先将装载机停放在平坦空旷的场地上，将铲斗平放在地面上，变速操纵手柄置于空档位置，关闭发动机。

按下驻车制动手柄，释放驻车制动器，解除制动。任意方向连续扳动工作装置操纵手柄，将蓄能器中的高压油排出。从蓄能器上端卸下充气阀保护帽。将蓄能器充气工具的开关 1 拧紧（右旋），排气阀 2 关闭，止回阀 3 的阀帽旋紧，用带槽螺母 4 将充气工具固定在蓄能器上端的充气阀上，如图 2-34 所示。

图 2-33 蓄能器安装位置示意

图 2-34 充气工具

1—开关 2—排气阀 3—止回阀 4—带槽螺母

缓慢打开充气工具的开关1（左旋），待压力表读数稳定后，该读数即是蓄能器的氮气预充压力，其值应为（1±0.05）MPa。如果压力偏低，应补充氮气；如果压力偏高，则可通过排气阀2卸压到所需的压力值。

关闭充气工具的开关1（右旋），然后从蓄能器上拆下蓄能器充气工具，装上充气阀保护帽。

任务2.5 电气系统保养与维护

子任务2.5.1 蓄电池检查与更换

☞ 学习目标

1）能够正确更换蓄电池。

2）能够选择合适的蓄电器对蓄电池进行充电。

☞ 工作任务

按规范要求对蓄电池进行检查和更换。

☞ 相关知识

装载机一般采用2个标称电压为12V的蓄电池串联。其中一个蓄电池的负极经电源总开关搭铁，正极接到另一个蓄电池的负极；另一个蓄电池的正极接至起动电动机。

1. 装载机蓄电池的作用

1）提供大电流给起动电动机用来起动发动机。因此，装载机的蓄电池一般采用起动用铅酸蓄电池，以便在短时间内能提供很大的电流，用于起动发动机。

2）在发电机不发电时，为车上所有电气负载供电。

3）吸收系统中的瞬变电压，保护电器元件。发电机的转速和负载突然变化以及切换感性负载（如扬声器、电磁线圈等）时，都会在系统中引起瞬变电压（峰值高达100V以上，持续时间为毫秒级），尽管在电源系统中设置有电压调节器，但由于电压调节器对瞬变电压调节的滞后性，致使电源系统中的瞬变电压无法被抑制，从而对电路中的电子元器件形成较强的冲击，甚至使电子元器件损坏。而蓄电池的低阻抗、大电容特性，使其对瞬变电压有较强的吸收作用，从而保护电子元器件。因此，必须保证发电机至蓄电池的充电电路连接可靠，在发动机正常运转时，不得以任何方式断开发电机与蓄电池的连接（如发动机运转期间关闭电锁）。

2. 装载机蓄电池的组成

能将化学能与电能重复转换的装置称为蓄电池。装载机一般采用铅酸蓄电池。铅酸蓄电池一般由外壳、正极板、负极板、隔板、电池槽、电解液和接线端子等部分组成。

任务实施

1. 蓄电池的检查

蓄电池位于整机尾部左侧电瓶箱内，拧开 4 颗盖板螺栓即可看到蓄电池，如图 2-35 中箭头所示。

检查蓄电池压板螺母、蓄电池端子以及电缆接头是否松动。如果线头松动，须将蓄电池压板螺母、蓄电池端子以及电缆接头拧紧，如图 2-36 所示。

图 2-35　蓄电池安装位置示意

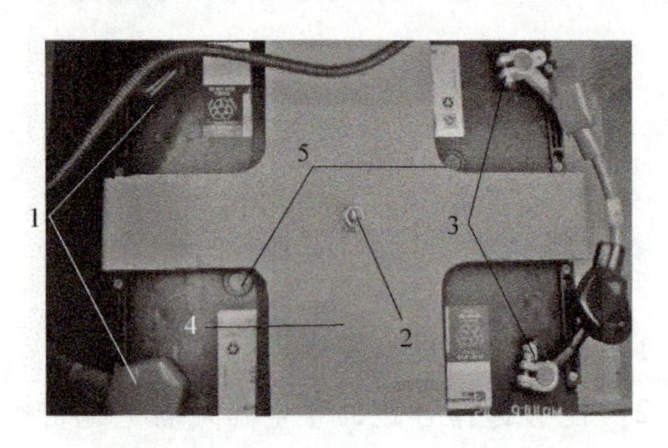

图 2-36　蓄电池电缆连接示意

1、3—蓄电池端子　2—压板螺母　4—压板　5—指示器

检查蓄电池状态指示器（电眼）。如果指示器显示为绿色，表明蓄电池电量充足，可以正常起动车辆；如果指示器显示为黑色，表明蓄电池电量不足，需充电；如果指示器显示为白色，则蓄电池报废，需更换。

检查蓄电池完毕，关闭蓄电池箱盖。

2. 蓄电池的更换

关闭发动机，拔下钥匙，将蓄电池负极开关处于关断状态。

使用干净的抹布清洁蓄电池端子及蓄电池表面。注意不要同时触碰到蓄电池的正极和负极，另外绝对不能用汽油或任何其他有机溶剂或清洁剂。

拆卸蓄电池连接电缆。首先拆卸蓄电池负极连接负极开关一端的电缆接头，接着拆卸蓄电池负极电缆连接负极一端的接头，将负极电缆取下；然后拆卸两个蓄电池之间的跨接线，最后拆卸蓄电池正极电缆。检查蓄电池端子部位是否有腐蚀痕迹，蓄电池电缆是否存在磨损或损坏，如有必要，更换蓄电池电缆。

取出旧的蓄电池，将新的蓄电池放置在安装座上，先连接两个蓄电池之间的跨接线，接着连接蓄电池的正极电缆，然后连接负极电缆，最后将蓄电池负极电缆连接到蓄电池负极开关上，如图 2-37 所示。注意各连接螺栓、螺母须拧紧。

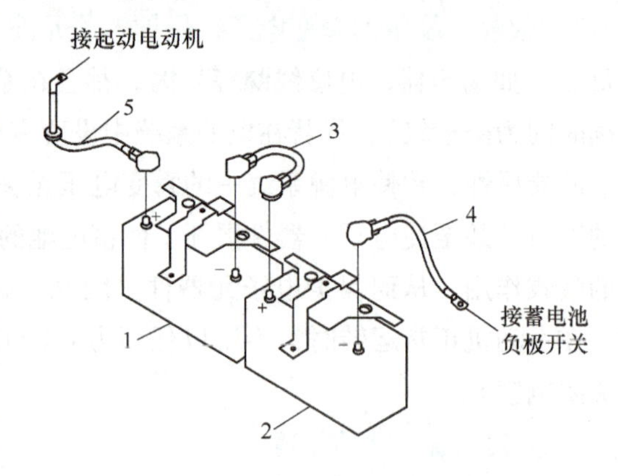

图 2-37　蓄电池电缆连接示意

1、2—蓄电池　3—跨接电缆　4—负极电缆　5—正极电缆

插入钥匙并将蓄电池负极开关置于接通状态，将钥匙开关置于“ON”档，观察仪表状态，然后起动发动机试运转。

子任务 2.5.2 发电机检查与维护

☞ 学习目标

1）能判别发电机的通电状态。

2）能按规范要求对发电机进行维护。

☞ 工作任务

按规范要求检查发电机的输出电压。

☞ 相关知识

工程机械上的发动机均配装有发电机。发电机在发动机的带动下将机械能转化成电能。发电机与蓄电池并联组成整机供电系统，在柴油机未发动之前，由蓄电池向整机用电器进行供电，在柴油机发动之后，主要由发电机对用电器进行供电，同时发电机对蓄电池进行充电，在发电机提供的能量不能满足整机用电器的需要时，发电机与蓄电池共同供电。发电机外形示意如图 2-38 所示。

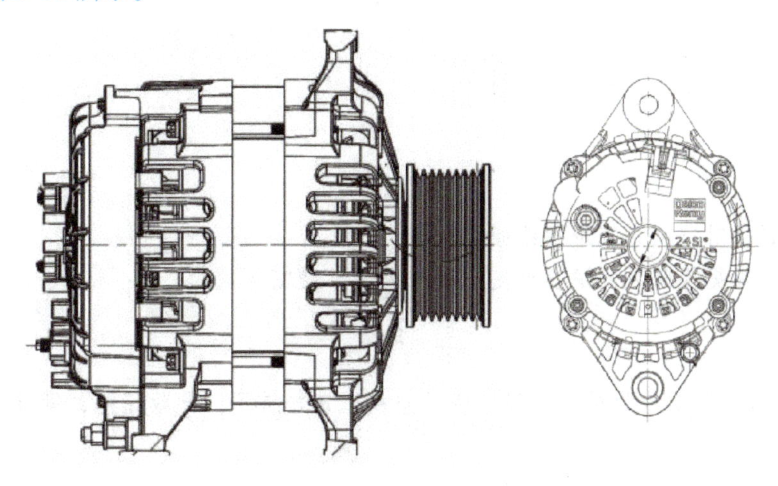

图 2-38 发电机外形示意

发电机有三个接线端子，分别是“B +”端子、Ground 端子和“D +”端子。其中，“B +”端子是发电机正极，为蓄电池充电和车上的用电器供电；Ground 端子是发电机负极，与整车蓄电池负极端子相连；“D +”端子为励磁端，接充电指示灯、起动机保护继电器等，输出电流不超过 1A。

发电机的工作电压为 24V，额定电流为 35A，内置电子稳压调节器。柴油机在运行过程中，应经常检查发电机的接线端子接线是否紧固可靠，无论是正极还是负极接线松动，都会导致整车电气系统无法正常工作，甚至引起严重故障。在检查这两个接线端子接线是否紧固可靠前，请务必关闭起动开关。

判断发电机是否正常发电的方法及处理措施：

1）打开起动开关（电锁），仔细观察电压表的读数，起动柴油机，再观察电压表的读数，后一读数应比前一读数高。

2）用万用表的直流电压200V档进行检查，打开电锁，测量发电机端电压（红表笔搭发电机正极，黑表笔搭地）记下万用表读数。

3）起动柴油机，将柴油机转速提高到额定转速，再次测量发电机端电压，记下万用表读数，两者比较，后一读数应比前一读数高。

4）如果发电机不发电，应检查发电机传动带是否过松，发电机接地是否牢靠，用扳手检查发电机的接线端子连线是否正确、紧固、可靠。

☞ 任务实施

一台装载机的发电机不能正常发电，需要对其进行检查。检查步骤如下：

1）将数字万用表调至电压档200V量程，打开装载机负极开关，将钥匙开关置于“ON”档，数字万用表的红表笔搭接在发电机的“B +”端，黑表笔搭接在发电机的“Ground”端，记下万用表的显示值。正常数值应为24～25V。

2）将钥匙开关置于“START”档，起动柴油机，数字万用表的红表笔搭接在发电机的“B +”端，黑表笔搭接在发电机的“Ground”端，记下万用表的显示值。所测数值应高于25V，否则表明发电机可能存在故障。

任务2.6 空调系统保养与维护

☞ 学习目标

1）能按要求对空调系统进行维护。

2）能按要求更换回风滤清器滤芯、新风滤清器滤芯等。

☞ 工作任务

一台装载机空调制冷效果差，经检查是空气滤清器脏了，导致堵塞，需要对空气滤清器进行更换。

☞ 相关知识

空调的空气滤清器是通过多孔过滤材料的作用从气固两相流中捕集粉尘，并使气体得以净化的设备。它把空气净化处理后送入室内，以保证洁净房间的工艺要求和一般空调房间内的空气洁净度。

驾驶室空气滤清器由外部空气滤清器及内部空气滤清器组成。外部空气滤清器的堵塞程度与工作环境和外气循环的开启时间有关。保持驾驶室内洁净，可延长内部空气滤清器的使用寿命和维护周期。

1. 滤清器的性能指标

（1）过滤精度　指允许通过滤清器的杂质颗粒的最大直径。影响过滤精度的关键是滤芯，可根据后面元器件的需要选择不同的滤芯，使其达到相应的过滤精度。

（2）流量特性 指在一定的进口压力下，通过滤清器的空气流量与滤清器两端压差之间的关系曲线，实际使用时，最好在压力损失小于0.03MPa的范围内选用。在空气滤清器中，影响流量特性的主要是本体和滤芯。

（3）分水效率 指分离出来的水分与进气口空气中所含水分之比，一般要求空气滤清器的分水效率不小于80%，影响分水效率的主要是导流板。

（4）滤清器阻力 在额定风量下新滤清器的阻力称为初阻力；在额定风量下，滤清器的容尘量达到足够大而需要清洗或更换滤料时的阻力称为终阻力。

（5）滤清器的容尘量 在额定风量下，滤清器的阻力达到终阻力时，其所容纳的尘粒总质量称为该滤清器的容尘量。

2. 滤清器的选择原则

根据具体情况合理地选择合适的空气滤清器。

1）根据洁净及净化标准，确定最末级的空气滤清器的效率，合理地选择空气滤清器的组合级数和各级的效率。如室内要求一般净化，可以采用初效滤清器；如室内要求中等净化，就应采用初效和中效两级滤清器；如室内要求超净净化，就应采用初效、中效和高效三级净化过滤，并应合理妥善地匹配各级滤清器的效率，若相邻两级滤清器的效率相差太大，则前一级滤清器就起不到对后一级滤清器的保护作用。

2）正确测定室外空气的含尘量和尘粒特征。因为滤清器是将室外空气过滤净化后送入室内，所以室外空气的含尘量是一个很重要的数据。特别是在多级净化过滤处理中，选择预滤清器时要将使用环境、备件费用、运行能耗、维护与供货等因素综合考虑后决定。

3）正确确定滤清器特征。滤清器的特征主要是过滤效率、阻力、穿透率、容尘量、过滤风速及处理风量等。在条件容许的情况下，应尽可能选用高效、低阻、容尘量大、过滤风速适中、处理风量大、制造安装方便、价格低的滤清器。这是在空气滤清器选择时综合考虑一次性投资和二次性投资及能效比的经济性分析需要。

4）分析含尘气体的性质。与选用空气滤清器有关的含尘气体的性质主要是温度、湿度、含酸碱及有机溶剂的数量。因为有的滤清器允许在高温下使用，而有的滤清器只能在常温、常湿度下工作，并且含尘气体的含酸碱及有机溶剂数量对空气滤清器的性能效率都有影响。

☞ 任务实施

1. 清洁内部空气滤清器

拆卸驾驶室内部左后侧的盖板螺栓，然后取下盖板，并从空调蒸发器上拆卸内部空气滤清器，如图2-39箭头所示。

用压缩空气清洁滤清器。如果滤清器上有油或太脏，可用中性介质冲洗。冲洗以后，应使其彻底干燥才能重新使用。

将清洁后的内部空气滤清器重新装回原位，并将检修门盖板固定好。原则上每隔2000h应更换内部空气滤清器。

2. 清洁外部空气滤清器

滤清器的清洁可以根据机器工作环境的灰尘多少来适当延长或缩短保养周期。如果滤清器内部滤芯堵塞，可能导致驾驶室内的新风风量减少。

空调新风装置位于驾驶室左侧外部，拆卸驾驶室左侧门附件的滤清器外罩固定螺栓，然后取下滤清器外罩及外部空气滤清器，如图 2-40 箭头所示。

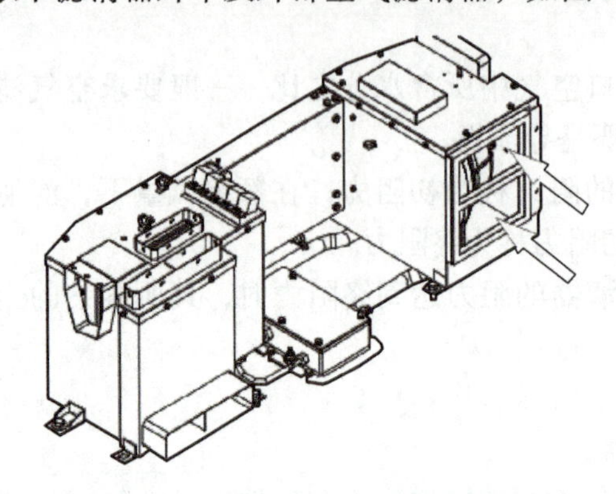

图 2-39　内部空气滤清器位置示意

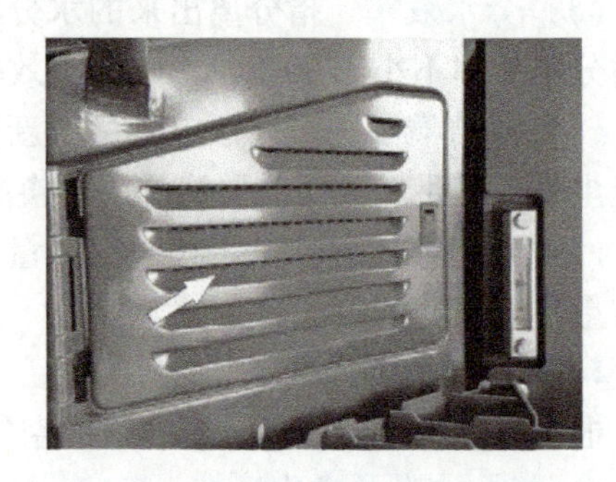

图 2-40　外部空气滤清器位置示意

用压缩空气清洁滤清器。如果滤清器上有油或太脏，用中性介质冲洗。冲洗以后，应使其彻底干燥才能重新使用。

如果污染物对人体有危险，一定要使用护目镜、防尘罩或其他防护用具。

将清洁后的外部空气滤清器放入滤清器外罩的凹槽内，然后将滤清器与外罩一同装回原位，并上紧螺栓。

项目 3

装载机液压系统分析与检测

按照装载机工作装置和各个机构的传动要求，把各种液压元件用管路有机地连接起来的组合体，称为装载机的液压系统。其功能是，以油液为工作介质，利用液压泵将发动机的机械能转变为液压能并进行传递，然后通过液压缸和液压马达等将液压能转换为机械能，实现装载机的各种动作。

一个完整的液压系统由五个部分组成，即动力元件、执行元件、控制元件、辅助元件和液压油。动力元件的作用是将原动机的机械能转换成液压能，如液压系统中的液压泵，它向整个液压系统提供液压油。液压泵的形式很多，一般有齿轮泵、叶片泵、柱塞泵。现代装载机中常用的液压泵为齿轮泵。执行元件的作用是将液压能转换成机械能，如各种液压缸、液压马达。控制元件用于控制液压系统中液压油的流动方向、压力大小以及流量大小，如各种液压阀。辅助元件包括油箱、过滤器、油管及接头、密封圈等。液压油是液压系统中传递能量的工作介质，有各种矿物油、乳化液和合成型液压油等。

装载机液压系统分为工作液压系统和转向液压系统，根据工作泵和转向泵的形式，又可分为定量系统（工作泵和转向泵都是定量泵）、定变量系统（工作泵是定量泵，转向泵是变量泵）和全变量系统（两个泵都是变量泵）。以下主要介绍定量系统。

1. 装载机液压系统工作原理（CLG856H）

如图 3-1 所示，装载机液压系统的传动路径如下：发动机通过变速器取力口，驱动液压泵（工作泵和转向泵）。转向泵输出的油，一路经先导供油阀、转向器为流量放大阀提供先导控制油，一路经流量放大阀为转向液压缸供油。工作泵输出的液压油进入分配阀，操作者通过操纵先导阀来控制分配阀，液压油通过分配阀进入动臂液压缸、转斗液压缸实现工作装置的动作。

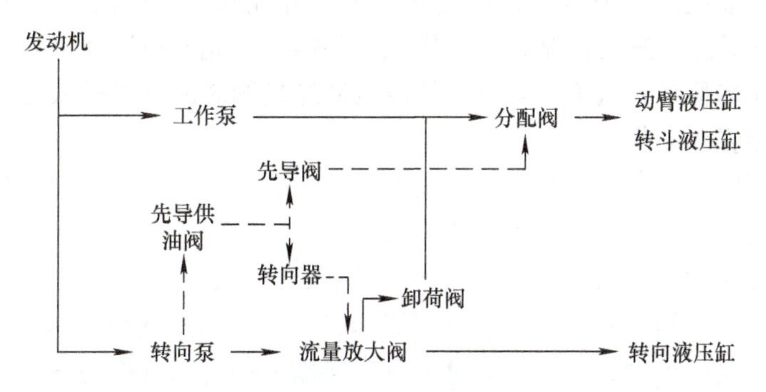

图 3-1 装载机液压系统工作原理

2. 装载机液压系统的组成

1）从组成元件的角度，装载机液压系统分为以下五部分：

① 能源装置（动力元件）：工作泵、转向泵。

② 执行装置（执行元件）：动臂液压缸、转斗液压缸、转向液压缸。

③ 控制调节装置（控制元件）：先导供油阀、先导阀、单向阀、分配阀、流量放大阀、卸荷阀。

④ 辅助装置（辅助元件）：液压油箱、散热器、滤清器、接头、软管、钢管等。

⑤ 传动介质：液压油。

2）从系统的角度（装配图），装载机液压系统包括工作装置液压系统、转向系统、制动系统（湿式桥）、液压管路总成和附属装置等。

任务 3.1 液压系统工作原理分析

子任务 3.1.1 先导液压系统工作原理分析

☞ 学习目标

1）能够描述工作液压系统控制原理。

2）能够分析先导液压系统控制回路。

3）能够通过先导回路原理分析常见故障并排除。

☞ 工作任务

一台装载机出现动臂提升慢的问题，经检查主阀压力及泵都无问题，检查先导压力异常，需要对先导压力进行测试调压。

☞ 相关知识

工作液压系统用于控制装载机工作装置中动臂和转斗以及其他附加工作装置的动作。如控制装载机工作装置完成铲斗收斗、卸料，动臂提升、下降、浮动下降的动作。具体工作原理：工作泵将液压油从油箱中吸出，经过管路进入分配阀，分配阀内有动臂阀芯、转斗阀芯，两阀芯通过先导压力进行控制，当操作先导手柄时，先导液压油进入动臂阀芯或转斗阀芯的一侧，推动阀芯移动，从工作泵过来的液压油经过阀芯的开口进入动臂液压缸或转斗液压缸，实现工作装置的动作。

根据机型的不同，工作液压系统可分为定量液压系统、定变量液压系统和全变量液压系统。以下主要介绍定量液压系统，如图 3-2 所示。

1. 先导液压回路的工作原理

先导操纵阀安装在驾驶室内，驾驶员座椅的右侧。先导操纵阀为单手柄先导阀，由动臂操纵联和转斗操纵联两个阀组组成。通过操纵先导操纵阀的动臂控制杆或转斗控制杆，可以操纵分配阀内动臂滑阀或转斗滑阀的动作，从而实现对车辆工作装置的控制。动臂手柄的操作位置有提升、中位、下降及浮动四个位置，转斗手柄的操纵位置有收斗、中位和卸料三个

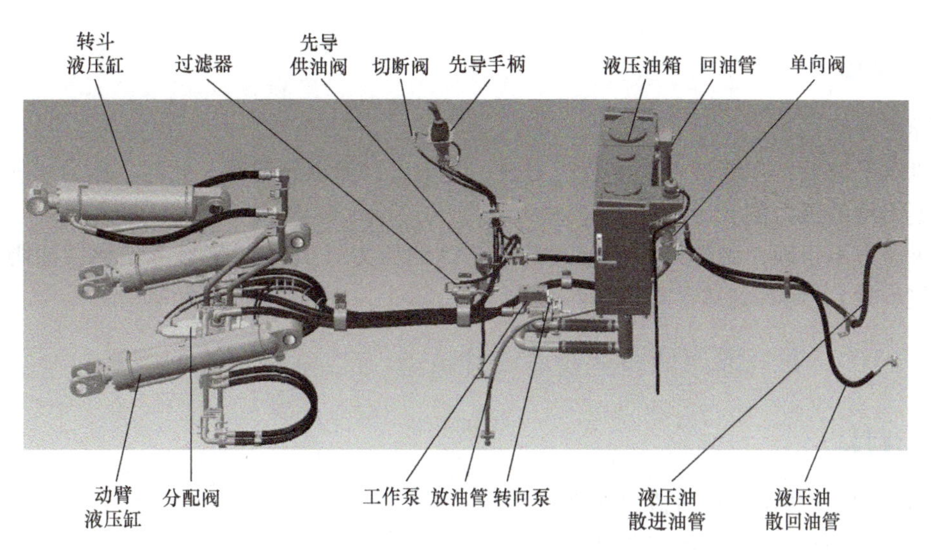

图3-2　定量液压系统

位置。其中在先导操纵阀中，动臂提升、动臂下降、转斗收斗三个位置中设置有电磁铁，通过与前车架和摇臂上的动臂及转斗自动复位装置的连接，可实现动臂高度的自动限位及铲斗的自动放平。

2. 先导压力测试

先导压力测试点如图3-3所示。

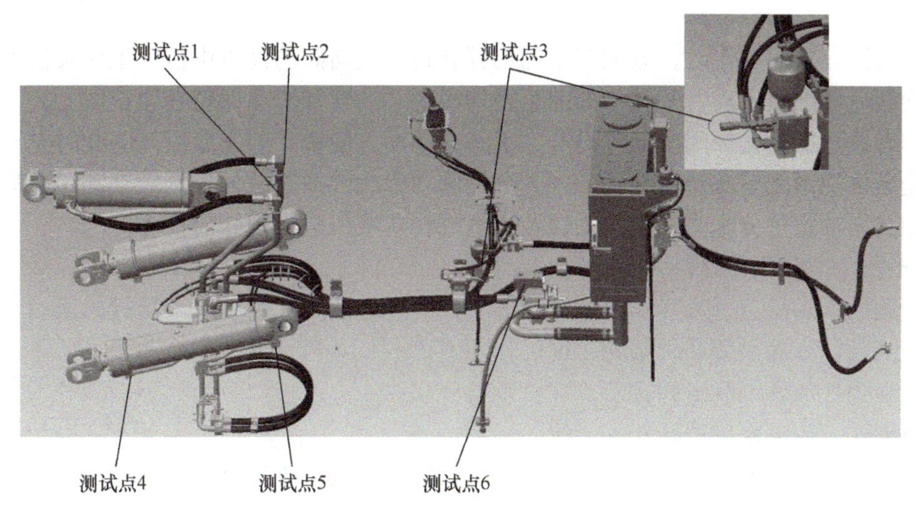

图3-3　先导压力测试点

3. 先导压力测试点说明

先导压力测试点说明见表3-1。

表3-1　先导压力测试点说明

测试点	测试点说明	理论压力值/MPa	接口尺寸
测试点3	先导阀进口（先导供油阀出口）压力	按照机型给出的值	M18×1.5－6g测压接头

☞ 任务实施

先导压力测试：

整机水平放置，在测试点 3 安装压力表；起动后，操纵先导阀，此时，压力表的压力值应为机型给出的压力值。

如果压力不正确，调整先导供油阀的减压阀的压力（具体调整参考先导供油阀的减压阀的调整步骤），使其到达合理值。

子任务 3.1.2 转斗液压回路工作原理分析

☞ 学习目标

1）能够描述转斗液压回路工作原理。

2）能够分析转斗液压控制回路。

3）能够通过工作原理分析转斗液压回路常见故障并排除。

☞ 工作任务

一台装载机出现转斗无力问题，经测试系统压力不满足设计压力值，需要对系统压力测试调压。

☞ 相关知识

装载机没有任何动作时，分配阀中的动臂和转斗换向阀均处于中位，工作泵输出的液压油经分配阀返回油箱，动臂和转斗的前后腔均封闭，动臂和铲斗保持在原位置。

操作者通过操纵先导阀使动臂提升、下降或浮动，使铲斗前倾或后倾。此时先导来油从泵通过先导供油阀之后，由先导阀进入到分配阀动臂或转斗阀芯的控制端推动阀芯移动，从而使主油路来油通过分配阀进入到动臂液压缸或转斗液压缸实现相应动作。

1. 分配阀的作用及结构

分配阀的作用是通过改变油液的流动方向控制转斗液压缸和动臂液压缸的运动方向和动臂、铲斗的动作以满足装载机不同工况的作业要求。

结构形式：整体双联滑阀式。

油路形式：串并联优先转斗。

主要组成：转斗换向阀、动臂换向阀、安全阀，如图 3-4 所示。

2. 分配阀原理

如图 3-5 所示，转斗换向阀是三位置阀，它可控制铲斗前倾、后倾和保持三个动作；动臂换向阀是四位置阀，它可控制动臂上升、保持、下降、浮动四个动作；安全阀是控制系统压力的，当系统压力超过设定值时，油液通过安全阀溢流回油箱，保护系统不受损坏；分配阀上 P 口为进油口，T 口为回油口，A1、B1 分别与转斗液压缸小腔、大腔相通，A2、B2 分别与动臂液压缸小腔、大腔相通。

3. 双作用安全阀工作原理分析

与转斗液压缸大、小腔的油道相通的油路中装有大、小腔双作用安全阀，对转斗液压缸

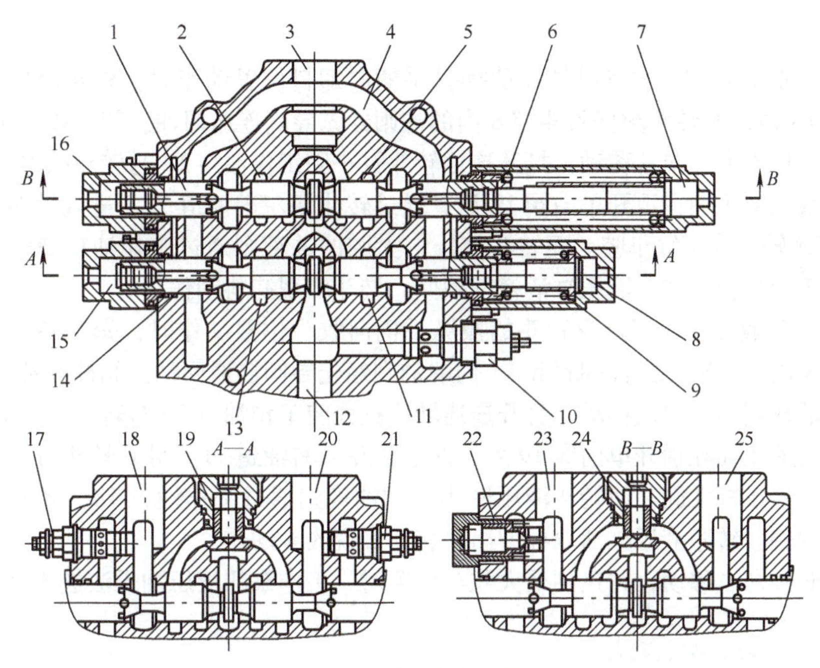

图 3-4 分配阀剖面图

1—动臂阀杆 2—动臂联左进油道 3—回油口 4—回油道 5—动臂联右进油道 6、9—弹簧 7—动臂阀杆提升腔 8—转斗阀杆收斗腔 10—主安全阀 11—转斗联右油道 12—进油口 13—转斗联左进油道 14—转斗阀杆 15—转斗阀杆卸料腔 16—动臂阀杆下降腔 17—转斗缸小腔过载阀 18—转斗缸小腔工作油口 19、24—单向阀 20—转斗缸大腔工作油口 21—转斗缸大腔过载阀 22—补油单向阀 23—动臂缸小腔工作油口 25—动臂缸大腔工作油口

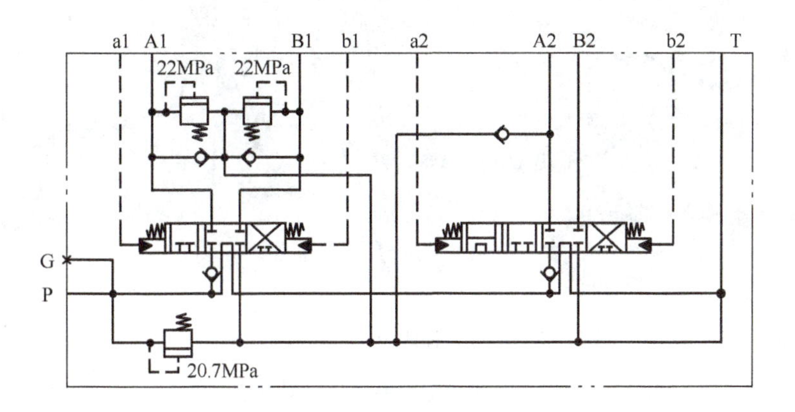

图 3-5 分配阀原理图

的大腔和小腔起过载保护和补油作用，从而解决了工作装置干涉的问题，起到稳定系统工作，保护系统有关元件的作用。

4. 转斗液压缸回路分析

（1）转斗联中位 当分配阀转斗阀杆两端没有先导压力油时，转斗阀杆在弹簧 9 的作用下处于中位。工作泵的来油经进油口 12 进入油道，向动臂联供油。此时转斗液压缸大小腔两端接分配阀的两个工作油口 18 和 20 被转斗阀杆封闭，转斗液压缸保持不动。如果此时动臂阀杆也处于中位，则工作泵的来油经过转斗联和动臂联的两处心形油道，通过回油口 3

流回油箱。

（2）转斗联卸料位　当操纵转斗的操纵手柄向卸料位置动作时，先导液压油进入转斗阀杆卸料腔 15 内，而转斗阀杆收斗腔 8 内的油则经先导阀连通回油。阀杆在油压的作用下，克服弹簧 9 的作用力，向右移动，打开连通转斗缸小腔工作油口 18 与转斗联左进油道 13 的开口。工作泵的液压油在顶开单向阀 19 后，通过转斗联左进油道 13，进入转斗缸小腔工作油口 18，到达转斗液压缸小腔。而转斗液压缸大腔的油液则通过工作油口 20，经回油道 4 通过回油口 3 回油箱。转斗液压缸活塞杆缩回，转斗实现卸料动作。

（3）转斗联收斗位　当操纵转斗的操纵手柄向收斗位置动作时，先导液压油进入转斗阀杆收斗腔 8 内，而转斗阀杆卸料腔 15 内的油则经先导阀连通回油。阀杆在油压的作用下，克服弹簧 9 的作用力，向左移动，打开连通转斗缸大腔工作油口 20 与转斗联右油道 11 的开口。工作泵的液压油在顶开单向阀 19 后，通过转斗联右油道 11，进入转斗缸大腔工作油口 20，到达转斗液压缸大腔。而转斗液压缸小腔的油液则通过工作油口 18，经回油道 4 通过回油口 3 回油箱。转斗液压缸活塞杆伸出，转斗实现收斗动作。

当转斗阀杆向右移动，并达到最大收斗位置时，工作泵的液压油无法进入动臂联，动臂无法工作。

5. 工作液压系统压力测试

工作液压系统压力测试点如图 3-6 所示。

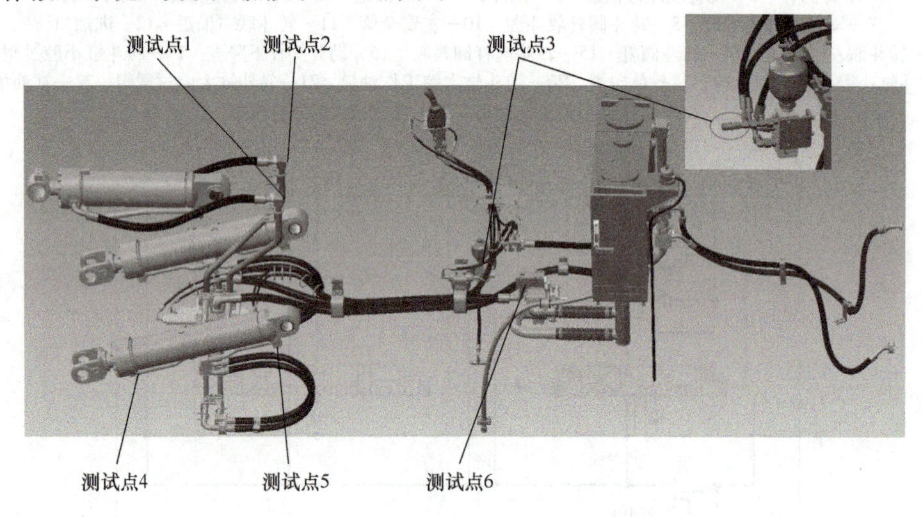

图 3-6　工作液压系统压力测试点

6. 工作液压系统压力测试点说明

工作液压系统压力测试点说明见表 3-2。

表 3-2　工作液压系统压力测试点说明

测试点	测试点说明	理论压力值/MPa	接口尺寸
测试点 1	转斗液压缸小腔压力	按照机型给出的值	M14×1.5-6H
测试点 2	转斗液压缸大腔压力	按照机型给出的值	M14×1.5-6H
测试点 4	动臂液压缸小腔压力	按照机型给出的值	M14×1.5-6H
测试点 5	动臂液压缸大腔压力	按照机型给出的值	M14×1.5-6H
测试点 6	工作泵出口压力	按照机型给出的值	M14×1.5-6H

☞ 任务实施

工作液压系统压力测试：

整机水平放置，在测试点2，按照测试的接口尺寸装上相应的测压接头和压力表；起动后，操纵先导阀进行憋斗，此时，压力表的压力值应为机型的相应压力值。

如果压力不正确，调整分配阀主安全阀的工作压力（具体调整参考分配阀主安全阀的调整步骤），使其到达符合的压力值。

子任务3.1.3 动臂液压回路工作原理分析

☞ 学习目标

1）能够描述动臂液压回路工作原理。

2）能够分析动臂液压控制回路。

3）能够通过工作原理分析动臂液压回路常见故障并排除。

☞ 工作任务

一台装载机出现动臂提升无力问题，经测试系统压力不满足设计压力值，需要对系统压力测试调压。

☞ 相关知识

1. 补油阀工作原理分析

在转斗联接转斗液压缸大、小腔的过载阀21和17中包含有补油单向阀，在动臂联接动臂液压缸小腔也有一个补油单向阀22。其作用是：当转斗液压缸或动臂液压缸活塞杆的速度大于工作泵输出流量所能提供的速度时，液压缸其中一腔中的压力要小于油箱中的压力，此时单向阀打开，从油箱中的来油向液压缸压力较低的一腔补充油液，以确保液压缸中油液的充足，避免在液压缸中产生气穴。

2. 主安全阀工作原理分析

在整体式分配阀的进油道上，集成有控制整个主工作液压系统压力的主安全阀。主安全阀为先导型插装阀，其压力设定值即为整车工作液压系统的最高系统压力。当工作液压系统压力升高并达到主安全阀所调定的压力时，主安全阀开启，工作泵液压油经回油口3溢流回油箱。工作泵的输出油压将被限定在该调定压力或调定值以下。

通过增加或减小先导阀芯上的初始弹簧压力，可以增大或降低主安全阀的调定压力。

3. 动臂液压缸油路分析

（1）动臂联中位　在转斗联不工作的情况下，当分配阀动臂阀杆两端下降腔16和提升腔7没有先导压力油时，动臂阀杆在弹簧6的作用下处于中位。工作泵的来油经进油口12经转斗联，向动臂联供油。此时动臂液压缸大、小腔两端接分配阀的两个工作油口25和23被动臂阀杆封闭，动臂液压缸保持不动。工作泵的来油经过转斗联和动臂联的两处心形油道，通过回油口3流回油箱。

（2）动臂联提升位　在转斗联不工作的情况下，当操纵动臂操纵手柄向提升位置动作时，先导液压油进入动臂阀杆提升腔 7 内。而动臂阀杆下降腔 16 内的油则经先导阀连通回油。动臂阀杆在油压的作用下，克服弹簧 6 的作用力，向左移动，打开连通动臂缸大腔工作油口 25 与动臂联右进油道 5 的开口。工作泵的液压油在顶开单向阀 24 后，通过动臂联右进油道 5，进入动臂缸大腔工作油口 25，到达动臂缸大腔。而动臂液压缸小腔的油液则通过工作油口 23，经回油道 4 通过回油口 3 回油箱。动臂液压缸活塞杆伸出，动臂实现提升动作。

（3）动臂联下降位　在转斗联不工作的情况下，当操纵动臂的操纵手柄向下降位置动作时，先导液压油进入动臂阀杆下降腔 16 内。而动臂阀杆提升腔 7 内的油则经先导阀连通回油。动臂阀杆在油压的作用下，克服弹簧 6 的作用力，向右移动，打开连通动臂缸小腔工作油口 23 与动臂联左进油道 2 的开口。工作泵的液压油在顶开单向阀 24 后，通过动臂联左进油道 2，进入动臂缸小腔工作油口 23，到达动臂液压缸小腔。而动臂液压缸大腔的油液则通过工作油口 25，经回油道 4 通过回油口 3 回油箱。动臂液压缸活塞杆缩回，动臂实现下降动作。

（4）动臂联浮动位　当操纵动臂的操纵手柄从下降位置继续向前动作时，动臂阀杆下降腔 16 内的先导油压继续升高，推动动臂阀杆继续向右移动。此时，工作泵来油通过阀杆中部台肩的节流槽与回油接通，动臂缸大腔工作油口 25 与回油道 4 之间的阀口开度增大；动臂联右进油道 5 与回油接通，因动臂联左、右进油道是相通的，故此时，动臂液压缸大、小腔以及工作油口来油都接通油箱。在工作装置自重作用下，动臂实现浮动下降。

4. 工作液压系统压力测试

工作液压系统压力测试点如图 3-7 所示。

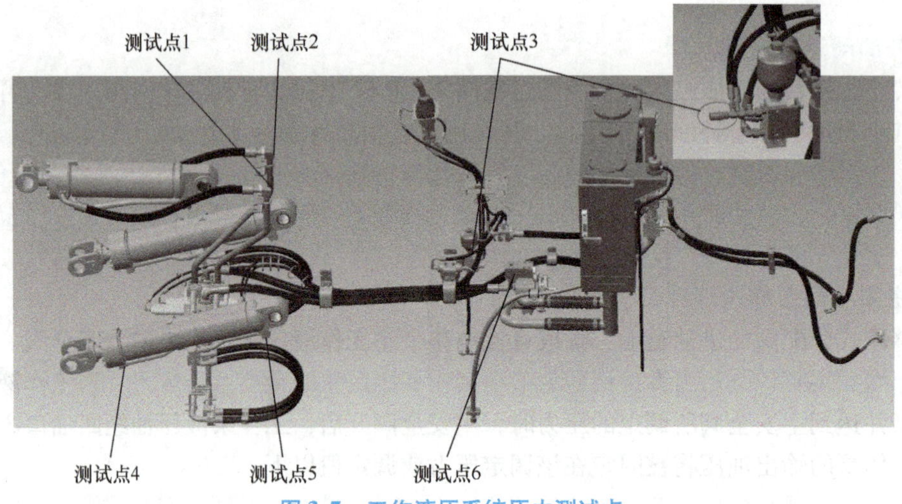

图 3-7　工作液压系统压力测试点

5. 工作液压系统压力测试点说明

工作液压系统压力测试点说明见表 3-3。

表 3-3　工作液压系统压力测试点说明

测试点	测试点说明	理论压力值/MPa	接口尺寸
测试点 1	转斗液压缸小腔压力	按照机型给出的值	M14×1.5－6H
测试点 2	转斗液压缸大腔压力	按照机型给出的值	M14×1.5－6H

（续）

测试点	测试点说明	理论压力值/MPa	接口尺寸
测试点4	动臂液压缸小腔压力	按照机型给出的值	M14×1.5-6H
测试点5	动臂液压缸大腔压力	按照机型给出的值	M14×1.5-6H
测试点6	工作泵出油口压力	按照机型给出的值	M14×1.5-6H

☞ 任务实施

工作液压系统压力测试：

整机水平放置，在测试点5，按照测试的接口尺寸装上相应的测压接头和压力表；起动后，操纵先导阀进行憋斗，此时，压力表的压力值应为机型的相应压力值。

如果压力不正确，调整分配阀主安全阀的工作压力（具体调整参考分配阀主安全阀的调整步骤），使其到达符合的压力值。

子任务3.1.4 转向液压系统工作原理分析

☞ 学习目标

1）能够描述转向液压系统工作原理。

2）能够分析转向液压系统控制回路。

3）能够通过工作原理分析转向液压系统常见故障并排除。

☞ 工作任务

一台装载机转向沉重，经检查转向液压系统压力不满足设计压力值，需要对转向液压系统压力进行测试调整。

☞ 相关知识

1. 装载机转向液压系统

在此以CLG856H装载机为例介绍装载机转向液压系统。转向液压系统采用流量放大系统，主要用于控制整机行车时转向。该系统主要分为转向控制油路和主工作油路两部分。主工作油路的动作由转向控制油路进行控制，以实现小流量、低压力控制大流量、高压力。整个转向液压系统主要由转向泵、转向器、流量放大阀、卸荷阀、转向液压缸以及相关管路等组成，如图3-8所示。

转向液压系统可以实现转向优先及双泵合流功能。转向优先指的是转向泵来油优先供给转向系统，多余的油供给工作液压系统，它是通过内部集成有优先阀的流量放大阀来实现的，优先阀实际上是流量放大阀内的一根浮动阀芯，它通过感应转向系统的负载压力调整对转向系统供油的流量，供入转向系统的油的流量与转向负载及速度相适应，继承了普通流量放大阀的负载感应功能。

系统合流主要有三方面的优点：

1）降低了工作液压系统非作业工况时的功率损失。

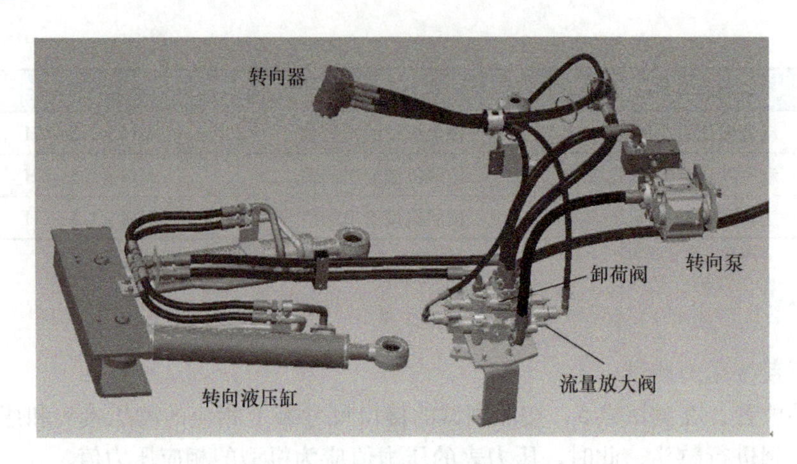

图 3-8 装载机转向液压系统

2）解决了装载机工作液压系统与转向液压系统同时工作时柴油机容易熄火的故障。

3）降低成本和提高可靠性。

2. 转向液压系统的工作原理

转向液压系统采用流量放大系统，系统油路由控制油路与主油路组成，所谓流量放大，是指通过全液压转向器以及流量放大阀，可保证控制油路的流量变化与主油路中进入转向缸的流量变化具有一定的比例，达到低压小流量控制高压大流量的目的。其优点是驾驶员操作平稳轻便，系统功率利用充分，可靠性好。

转向器为闭芯无反应型，方向盘不转动时中位断开。此时，流量放大阀主阀杆在复位弹簧作用下保持在中位，转向泵与转向液压缸的油路被断开，主油路经过流量放大阀中的流量控制阀卸荷回油箱。转动方向盘时，转向器排出的油与方向盘的转速成正比，先导油进入流量放大阀后，作用在流量放大阀的主阀杆端，控制主阀杆的位移，通过改变开口的大小控制进入转向液压缸的流量。由于流量放大阀采用了压力补偿，因而进入转向液压缸的流量与负载基本无关，只与阀杆上开口大小有关。停止转向后，进入流量放大阀主阀杆一端的先导压力油通过节流小孔与另一端接通回油箱，阀杆两端油压趋于平衡，在复位弹簧的作用下，阀杆回复到中位，从而切断主油路，装载机停止转向。通过方向盘的连续转动与反馈作用，可保证装载机的转向角度。系统的反馈作用是通过转向器和流量放大阀共同完成的。流量放大阀回油一部分通过节流孔回油箱，一部分经散热器回油箱。

3. 装载机流量放大阀左右转向回路分析

如图 3-9 所示。

1）流量放大阀主要由转向阀芯、优先阀芯、主安全阀、过载补油阀等组成。

2）转向阀芯实现改变油路方向及流量放大功能；优先阀芯实现转向优先、双泵合流功能；主安全阀限制系统最高压力，保护系统元件；过载补油阀用于整机行走过程中克服行驶障碍等引起的转向缸大、小腔局部压力突然升高，保持行驶稳定。

3）当右转向时，来自 Pa 口（接转向器 R）的先导油推动流量放大阀转向阀芯右移，来自转向泵的液压油经阀芯及阀体 A 口进入左转向缸大腔和右转向缸小腔实现右转向。同理，当左转向时，来自 Pb 口（接转向器 L）的先导油推动流量放大阀阀芯左移，来自转向泵的液压油经阀芯及阀体 B 口进入左转向缸小腔和右转向缸大腔实现左转向。

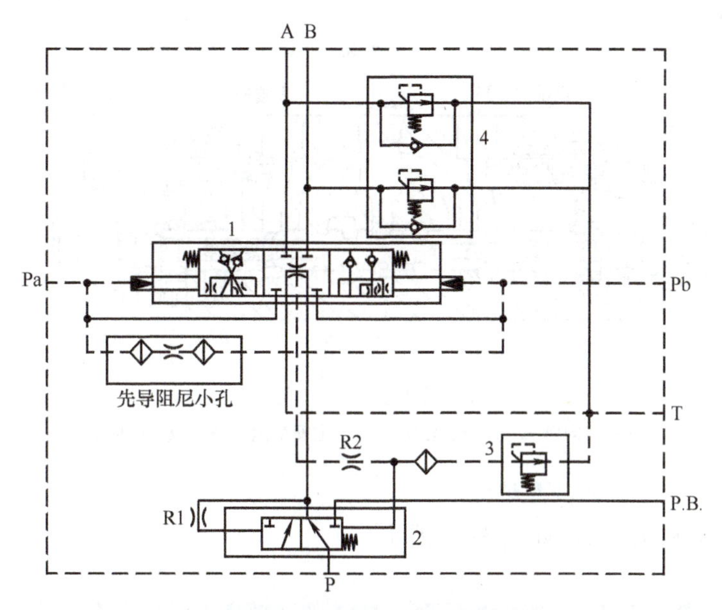

图 3-9 流量放大阀原理图

1—转向阀芯 2—优先阀芯 3—主安全阀 4—过载补油阀

4）单独转向过程，优先阀芯工作在右位，来自转向泵的液压油全部进入流量放大阀；当工作时，优先阀芯工作在左位，大部分油液从 Pb 口进入工作液压系统实现双泵合流功能，如图 3-10 所示。

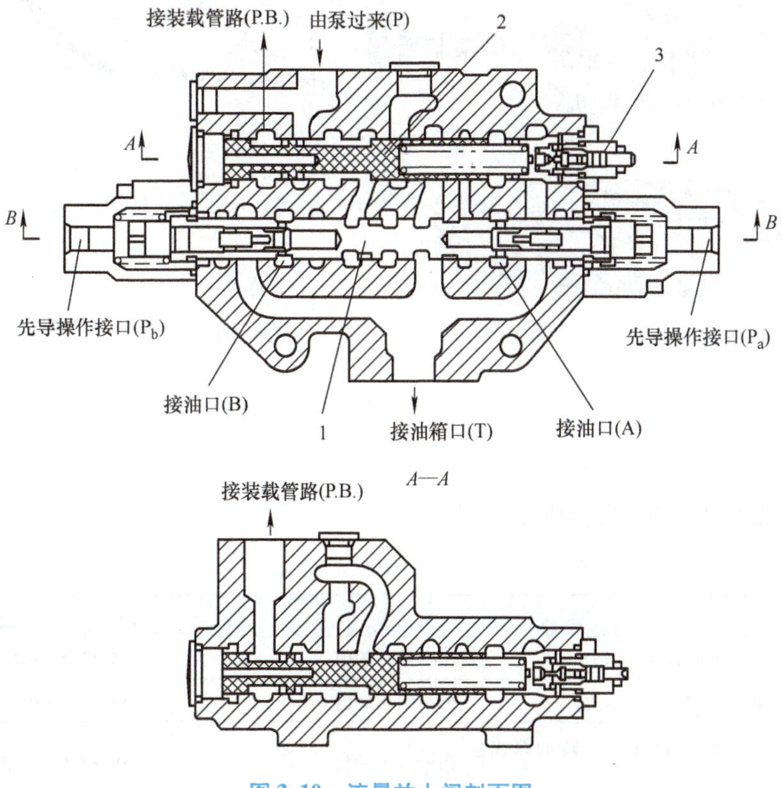

图 3-10 流量放大阀剖面图

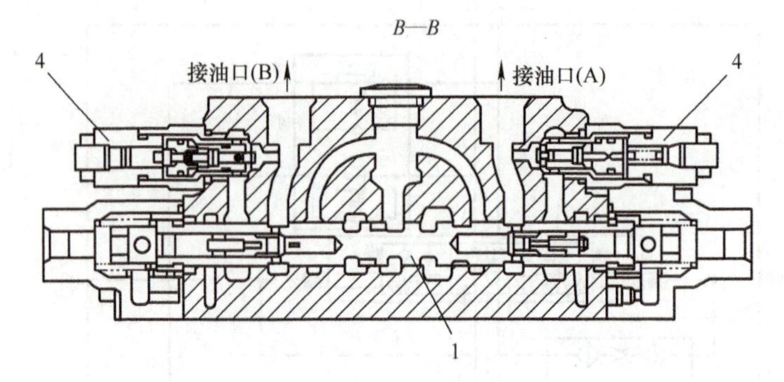

图 3-10　流量放大阀剖面图（续）

1—转向阀芯　2—优先阀芯　3—主安全阀　4—过载补油阀

4. 转向液压系统压力测试点位置

转向液压系统压力测试点如图 3-11 所示。

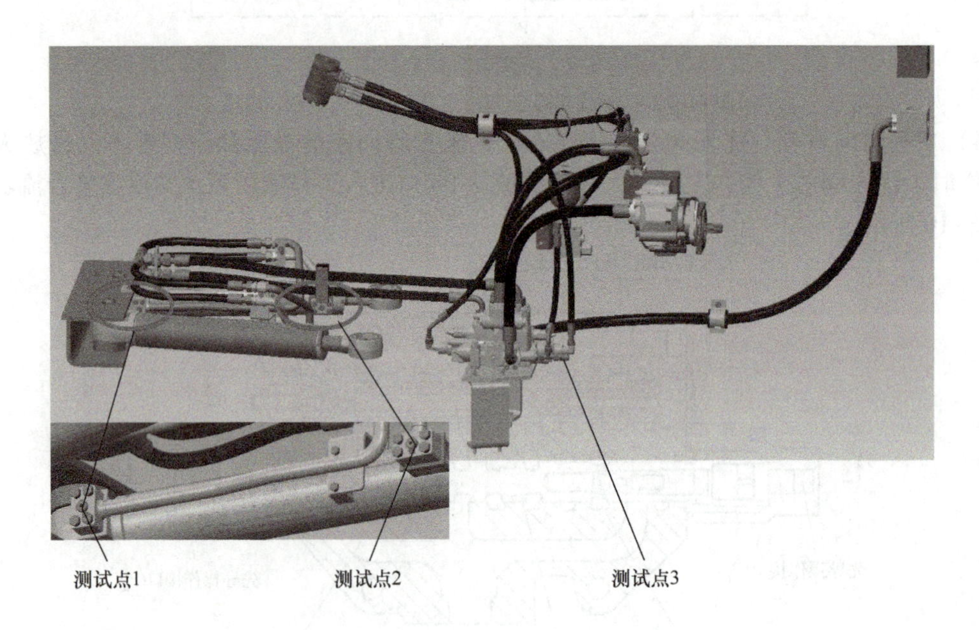

图 3-11　转向液压系统压力测试点

5. 转向液压系统压力测试点说明

转向液压系统压力测试点说明见表 3-4。

表 3-4　转向液压系统压力测试点说明

测试点	测试点说明	理论压力值/MPa	接口尺寸
测试点 1	转向液压缸大腔压力	机型给出的值	M14 × 1.5 – 6H
测试点 2	转向液试缸小腔压力	机型给出的值	M14 × 1.5 – 6H
测试点 3	流量放大阀进口（转向泵出口）压力	机型给出的值	M18 × 1.5 – 6g 测压接头

☞ 任务实施

转向液压系统压力测试点如图3-12所示。

安全阀压力调整步骤如下：

1）将前后车架保险杆安装好，使车架不能相对转动，或将方向盘打至极限位置。

2）在流量放大阀上测试点3处装上压力表（量程为25MPa）。

3）将整机发动，并使其高速空转。快速转动方向盘，直至安全阀打开，此时压力表指示应为机型给出的压力值。

4）如果压力不正确，可用内六角扳手在安全阀压力调整点进行调节（拧紧为增压，拧松为减压）。

5）压力调整合适后，拆去压力表、保险杆。

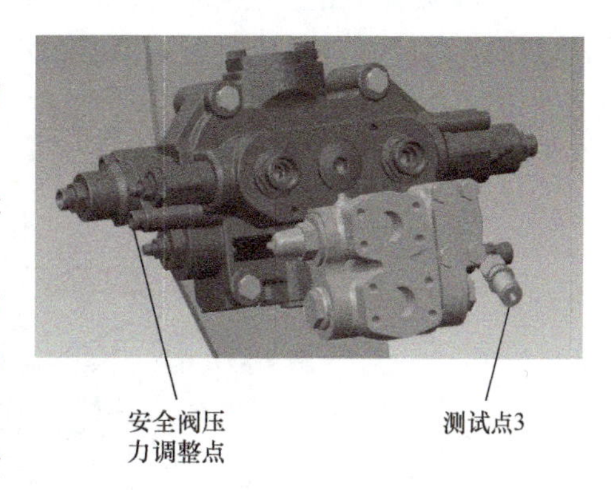

图3-12　转向液压系统压力测试点

子任务3.1.5　制动系统工作原理分析

☞ 学习目标

1）能够描述制动系统工作原理。

2）能够分析制动系统控制回路。

3）能通过工作原理分析制动系统常见故障并排除。

☞ 工作任务

一台干式制动的装载机，用户反馈制动软，需要对制动系统压力进行测试检查。

☞ 相关知识

1. 制动系统的基本组成及工作原理

装载机的制动系统用于行驶时的降速或停止，以及在平地或坡道上较长时间停车。其分为两部分，一部分是行车制动，另一部分是驻车制动。行车制动用于经常性的一般行驶中速度控制及停车。驻车制动用于停车后的制动，或在行车制动失效时作为紧急制动。

装载机制动系统又可分为干式制动系统和湿式制动系统。干式制动也称为气顶油制动，目前大部分装载机采用这种系统；湿式制动也称为全液压制动，用在高端机和大型机上。干式制动具有制动平稳、安全可靠、结构简单、维修方便、沾水复原性好等特点；湿式制动具有制动平稳、响应时间短、操作轻便、安全可靠、制动性能不受作业环境影响等优点。在此主要介绍干式制动系统。

2. CLG856H装载机干式制动系统的构成

CLG856H装载机行车制动采用气顶油四轮钳盘式制动，驻车制动采用蹄式制动器，其制动的位置在变速器的输出轴前端。驻车制动的驱动方式为手拉软轴控制，不同于 CLG856 装载机和 ZL50C 装载机的气动控制，故没有紧急制动功能。

行车制动系统由空气压缩机、组合阀、气罐、制动阀、加力器、钳盘式制动器及附属管路等组成，驻车制动系统由手制动器和软轴构成。制动系统元部件布置如图 3-13 所示。

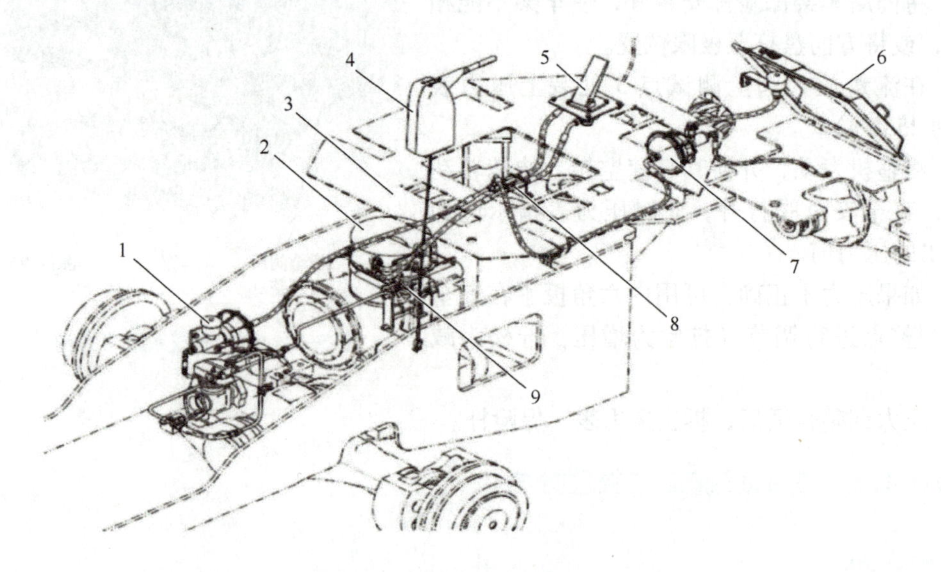

图 3-13　制动系统元部件布置

1—后加力器　2—气罐　3—软轴　4—手制动器　5—制动阀
6—罩板　7—前加力器　8—接头　9—组合阀

3. CLG856H 装载机干式制动系统工作原理

1）行车制动系统的工作原理（图 3-14）。空气压缩机由发动机驱动将空气转化成压缩空气，压缩空气经过组合阀后存储在气罐中。当气罐内的压力达到制动系统最高工作压力时（一般为 0.78MPa），组合阀就打开卸荷口，将空压机输出的压缩空气直接排向大气。当气罐内的压力低于制动系统最低工作压力时（一般为 0.71MPa），组合阀就打开通向气罐的出口，关闭卸荷口，补充气罐内的压缩空气，直到气罐内压力达到最高工作压力。在制动时，操作者踩下制动踏板后，压缩空气通过制动阀进入加力器的气缸，推动加力器活塞，活塞将加力器液压缸的制动液送进轮边制动器的夹钳分泵，挤压摩擦片实现制动。

2）驻车制动系统的工作原理（图 3-15）。操作者拉起驻车制动器的手柄，拉动软轴向上移动，软轴拉动制动器的制动蹄总成制动；同时按钮被压下，通过尼龙管进入变速器变速操纵阀的压缩气体被切断，变速器动力切断；拉起驻车制动器的手柄时，手制动器还会输出一个电信号给制动灯。

4. 制动系统测试点位置

制动系统测试点位置如图 3-16 所示。

5. 制动系统测试点说明

制动系统测试点说明见表 3-5。

6. CLG856H 装载机湿式制动系统工作原理

湿式制动系统通常包括行车制动系统和驻车制动系统（图 3-17）。

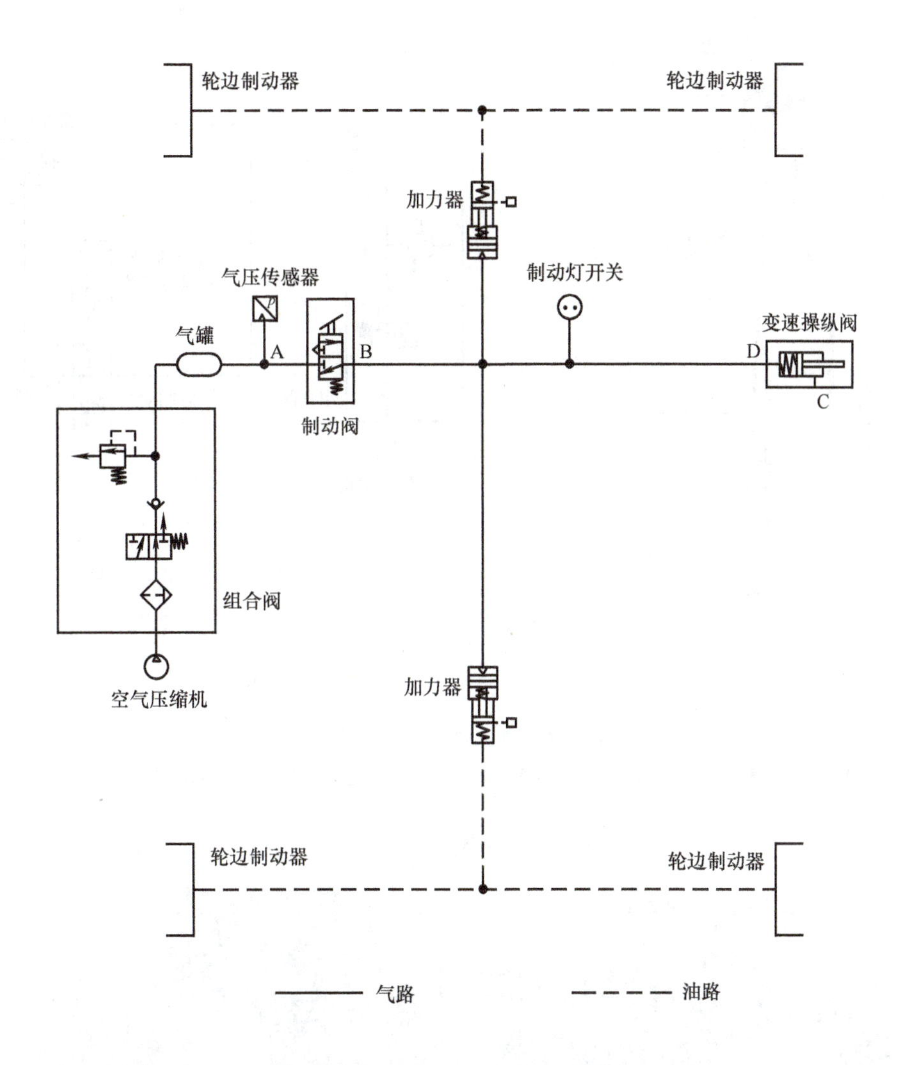

图3-14　行车制动系统的工作原理

行车制动系统用于经常性的一般行驶中的速度控制及停车。

驻车制动系统用于停车后的制动，或在行车制动系统失效时作为紧急制动。另外，当系统压力低于安全压力时，该系统自动使装载机停车，以确保安全。

液压元件包括制动泵、制动阀、充液阀、制动阀块、蓄能器、湿式制动器（传动桥内部）、停车制动液压缸、驻车制动器（箍带）、压力开关及管路等。

采用湿式轮边制动器的特点：全封闭，避免外界不确定因素影响，如沙土侵入、摩擦盘沾油等；浸油冷却，摩擦盘浸泡在油中，避免摩擦盘温升过高使摩擦材料产生热衰退而降低制动能力；降低了对制动用油的要求。

采用湿式制动系统的特点：取消了气路，简化了系统；避免了气路系统中因含水而造成的管路及制动元件的锈蚀，从而提高了系统的可靠性；提高了响应速度。

采用双回路制动的特点：后桥制动管路独立，更安全可靠。

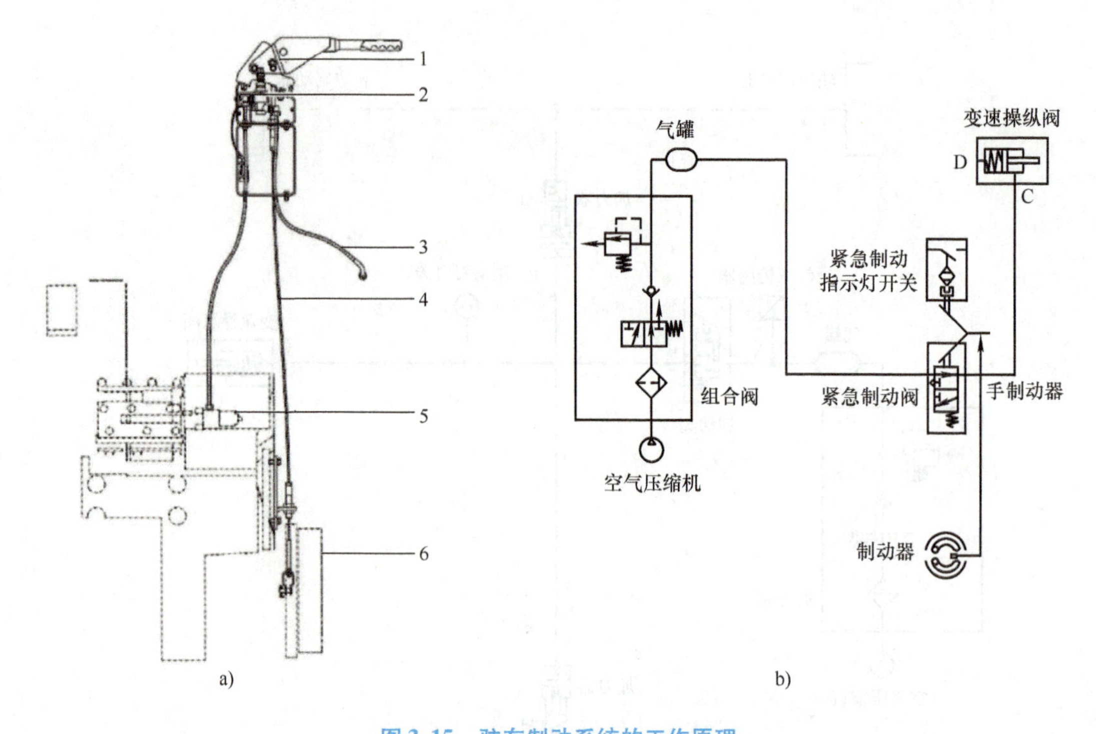

图 3-15 驻车制动系统的工作原理

1—手制动器 2—按钮阀 3—尼龙管 4—软轴 5—变速操纵阀 6—制动器总成

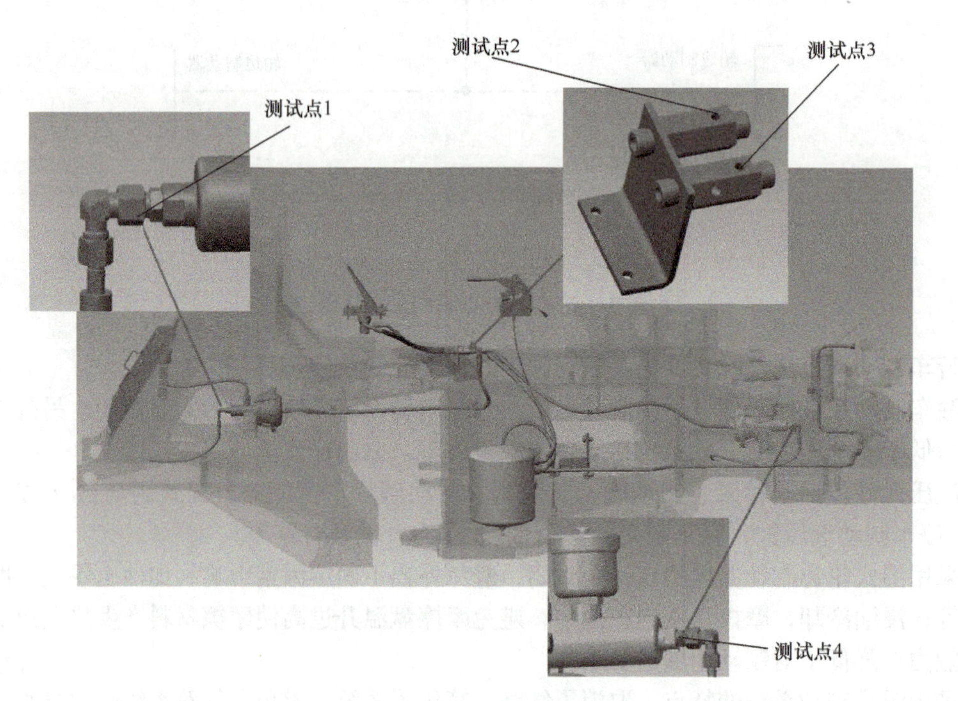

图 3-16 制动系统测试点位置

表 3-5 制动系统测试点说明

测试点	测试点说明	理论压力值/MPa	接口尺寸
测试点 1	前加力器出口油液压力	机型给出的值	M18 × 1.5 – 6g 测压接头
测试点 2	行车制动系统压力	机型给出的值	NPT 1/8
测试点 3	加力器进口压力	机型给出的值	NPT 1/8
测试点 4	后加力器出口油液压力	机型给出的值	M18 × 1.5 – 6g 测压接头

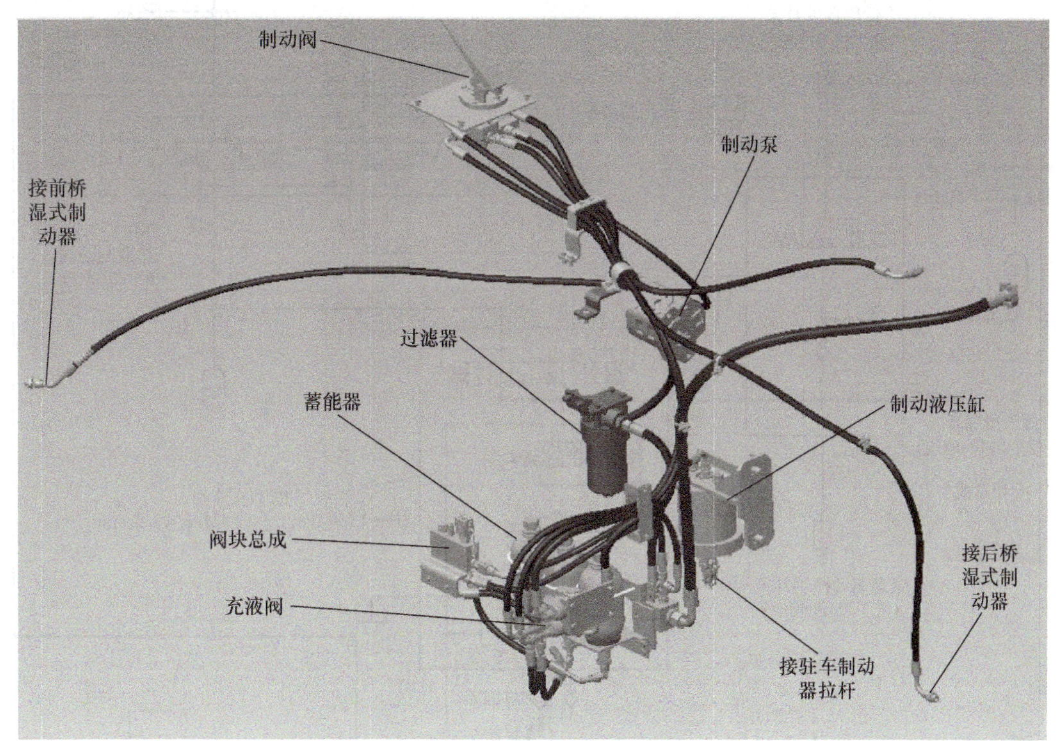

图 3-17 湿式制动系统结构图

如图 3-18 所示，踏板无动作时，无压力反馈，蓄能器储存的能量在 P 口上，制动器口与 T 口相接，P 口到 T 口可能会有内泄。

踏板有动作，即处于调节状态时，压力稳定在最大值上，P 口和 T 口是封闭的，压力反馈在最大值上，踏板力最大，受力平衡。

任务实施

干式制动系统压力的测量：

（1）测量 1 不连接前桥，单独测量加力器出口油液压力。将接往前桥的软管和直角接头拆除，装上测压接头，踩制动踏板，读出的压力表数值应符合要求。

（2）测量 2 连接前桥的同时测试加力器出口压力。将接往前桥的软管和直角接头拆除，装上三通接头和测压接头，再重新将软管和直角接头接到三通位置，踩制动踏板，读出的压力表数值应符合要求。

以上如不符合，则需要更换加力器。

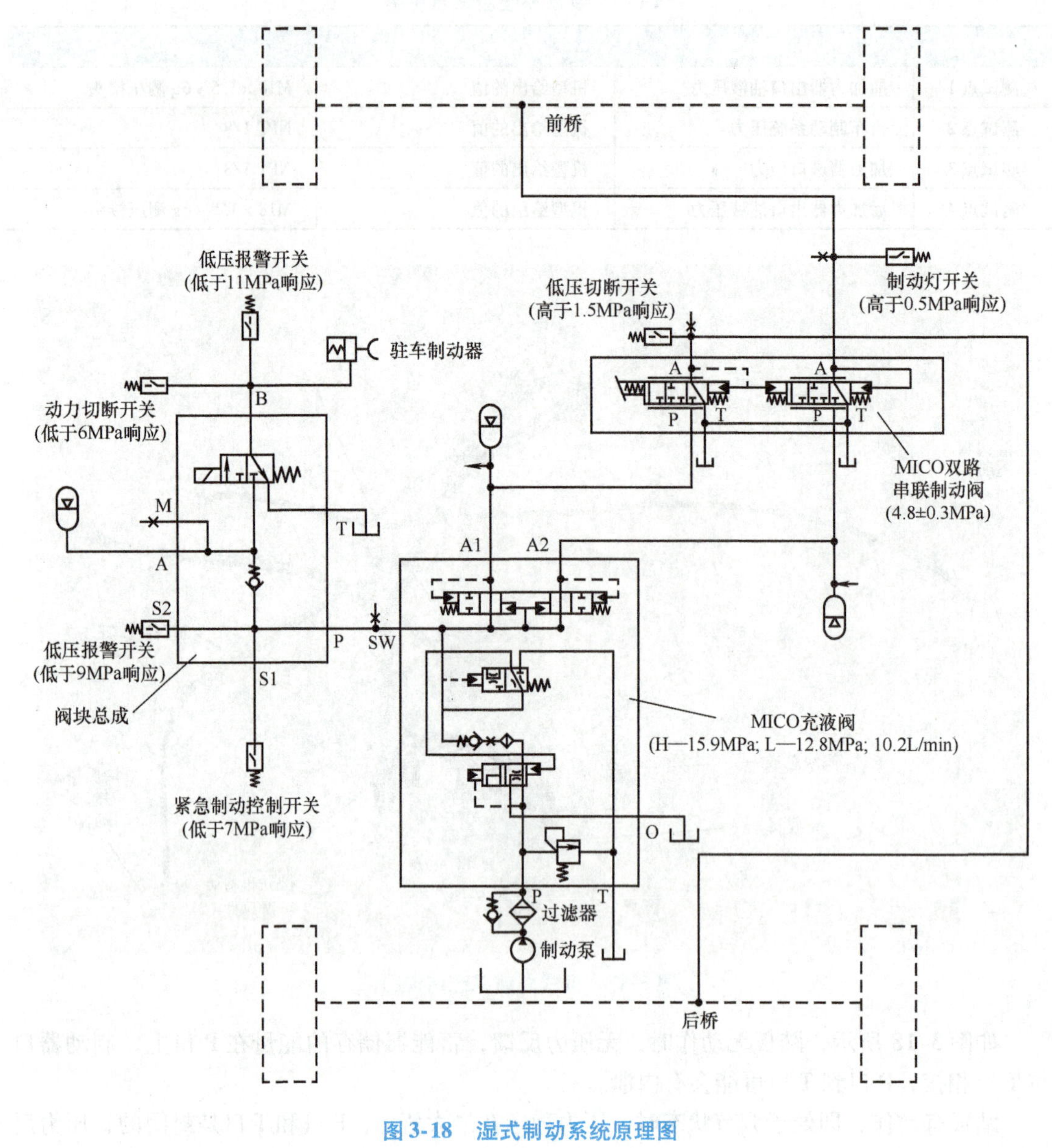

图 3-18　湿式制动系统原理图

任务 3.2　液压系统性能检测

子任务 3.2.1　液压缸沉降量测量

☞ 学习目标

1）能制订测量装载机液压缸沉降量的技术方案。

2）能按规范要求测量装载机液压缸沉降量。

☞ 工作任务

一台装载机的铲斗提升到最高位置后出现掉斗现象，需要对液压缸的沉降量进行测试，以便查找原因。

☞ 相关知识

装载机在工作过程中不允许出现掉臂、掉斗等液压缸自动伸缩现象，特别是一些特殊工况，如吊装作业时，要求液压缸在长时间内保持静止不动，或只能在允许范围内动作，这就需要对液压缸的沉降量进行测量。

影响液压缸沉降量的主要因素有：

1）铲斗液压缸、铲斗阀芯、油口溢流阀、液压缸密封件。

2）动臂液压缸、油口溢流阀、液压缸密封件。

液压缸的沉降量是影响装载机工作装置性能的一个重要指标。

测试条件：装载机按规定注满冷却液、燃油、润滑油、液压油，准备好工具、备件、一名驾驶员（75kg）和其他附件，轮胎压力应达到使用维护说明书的规定；初始测定时，液压系统的油温为（50±3）℃。

静态测试法：铲斗后翻并承载额定载重量（工作载荷），操纵提升液压缸使铲斗升至最高位置，发动机熄火，操纵分配阀处于封闭位置。按表3-6所列项目每15min测量一次提升液压缸和转斗液压缸的活塞杆的外伸长度，测量时间为3h。

表3-6　液压缸沉降量测定记录表

序号	载重量/t	测量时间/h	活塞杆伸出长度/mm				小时沉降量/（mm/h）	
			动臂液压缸		转斗液压缸			
			左	右	左	右	动臂液压缸	转斗液压缸
1								
2								
3								

测量值应符合表3-7所列范围。

表3-7　液压缸沉降量　（单位：mm/h）

	动臂液压缸	转斗液压缸
静态测量3h平均值	≤50	≤20

☞ 任务实施

装载机按规定注满冷却液、燃油、润滑油、液压油，准备好工具、备件、一名驾驶员（75kg）和其他附件，轮胎压力应达到使用维护说明书的规定；初始测定时，液压系统的油温为（50±3）℃。

铲斗后翻并承载额定载重量（工作载荷），操纵提升液压缸使铲斗升至最高位置，发动机熄火，操纵分配阀处封闭位置。按表3-6所列项目每15min测量一次提升液压缸和转斗液压缸的活塞杆的外伸长度，测量时间为3h。

子任务 3.2.2　液压缸循环时间测量

☞ 学习目标

1）能制订液压缸循环时间测量的技术方案。
2）能按规范要求来测量液压缸循环时间。

☞ 工作任务

一用户反馈装载机提升慢，需要对该装载机的液压缸循环时间进行测试，以便查找原因。

☞ 相关知识

测量液压缸循环时间的目的：液压缸伸缩的速度与流量有关，流量越大，液压缸运动速度越快，流量越小，液压缸运动速度越慢。液压缸运动速度可以体现出整机的工作效率。当怀疑液压缸动作慢时，可通过测量液压缸的循环时间来判断液压缸动作是否正常。

测量步骤：

（1）测试条件　装载机按规定注满冷却液、燃油、润滑油、液压油，并包括工具、备件、一名驾驶员（75kg）和其他附件，轮胎压力应达到使用维护说明书的规定；初始测定时，液压系统的油温为（50±3）℃。

（2）提升时间的测定　先将铲斗从基准地平面向后翻转，然后在铲斗内加载至额定载重量（工作载荷），操纵提升液压缸使铲斗提升到最高位置，按表 3-8 测量铲斗提升时间，同时测定铲斗铰销垂直提升高度，并计算铲斗提升速度。

表 3-8　工作装置动作时间测定记录表

项目	铲斗铰销高度/mm	移动距离/mm	测定时间/s	实测速度/(mm/s)	备注
提升					斗内加额定载重量
下降					空载
卸载					空载

（3）卸载时间的测定　测试时，提升铲斗至最高卸载位置，起动转斗液压缸，使空铲斗从最高位置转动到卸载位置，按表 3-8 测量此过程的连续时间（即为卸载时间）。

（4）下降时间的测定　测试时，把空铲斗从最高位置下降到基准地平面，按表 3-8 测量此过程的连续时间，同时测定铲斗铰销垂直下降高度，并计算铲斗下降速度。

☞ 任务实施

（1）提升时间的测定　先将铲斗从基准地平面向后翻转，然后在铲斗内加载至额定载重量（工作载荷），操纵提升液压缸使铲斗提升到最高位置，按表 3-8 测量铲斗提升时间。

（2）卸载时间的测定　测试时，提升铲斗至最高卸载位置，起动转斗液压缸，使空铲斗从最高位置转动到卸载位置，按表 3-8 测量此过程的连续时间（即为卸载时间）。

（3）下降时间的测定　测试时，把空铲斗从最高位置下降到基准地平面，按表 3-8 测量此过程的连续时间，同时测定铲斗铰销垂直下降高度，并计算铲斗下降速度。

子任务 3.2.3 液压油温度测量

☞ 学习目标

1）能够利用测温枪对液压油温度进行测量。
2）能够判断液压油是否变质。

☞ 工作任务

一台装载机液压油温度过高，需要对液压油的温度进行测试。

☞ 相关知识

1. 液压油温度测量方法

通常所说的液压系统温度是指液压油箱内液压油的温度。检测装载机液压系统油温的传感器一般安装在主泵吸油管路上，所测温度值在驾驶室液晶显示屏幕上可以查看。如果没有安装温度传感器，可以打开液压油箱上方的检查盖，然后使用温度计来测量油箱内的液压油温度。

液压油合理的工作温度范围是 50 ~ 70℃。导致液压油温度升高的原因可以分为外部原因和内部原因。外部环境热辐射和液压系统内部的零件摩擦、空穴现象都有可能引起液压油温升过快。液压油长期在高温条件下工作，很容易导致变质。变质的液压油黏性增大，会导致流动摩擦阻力增大，增加发热量。

2. 液压油变质判断方法

1）颜色辨识法。观察液压油的颜色，如果液压油呈黑褐色，表明液压油已经高温氧化，发生了变质。

2）手搓法。把液压油滴到大拇指上反复搓摩，如果感觉手指之间有较大的颗粒物，表明液压油里面的杂质较多。好的液压油有润滑性，基本无摩擦。

3）对光观察法。用螺钉旋具撩起液压油，抬起与水平面呈成 45°左右。在充足的光线下可以观察到液压油的油滴是否存在磨屑。

4）触摸法。取一张白纸，将液压油滴在白纸表面，在液压油渗漏后，会发现原来滴油的表面有一层黑色粉末的颜色。用手去触摸这些黑色粉末，如果感觉光滑、没有粉末滞涩感，表明液压油没有变质。

5）滴渗法。将液压油滴在滤纸上，观察斑点的变化情况，如果液压油迅速扩散，中间无沉积物，表明油品正常。如果液压油扩散慢，中间出现沉积物，表明油已变坏。

☞ 任务实施

使用测温枪对液压油温度进行测量：

现阶段的装载机液压系统都安装有液压油温度传感器，可通过显示器直接读取液压油温度，如果电气系统（如传感器、传感器线束等）出现问题将导致液压油温度显示数值不正确，这时需要通过测温仪器对液压油温度进行测量。

常用的测温仪器为测温枪，使用测温枪测量液压油散热器的进出口温度，如果温差范围在 5 ~ 8℃，说明散热器散热正常。

项目 4

装载机电气系统分析与维护

最初的装载机（如 ZL50 铰接式装载机）电气系统是非常简单的，只有照明与起动功能是通过电来实现的，监控仪表是机械式的。

随着传感器与检测技术的不断进步，电磁式仪表逐步取代了机械式仪表。同时，随着用户需求的不断提高，用于提高用户舒适性的电器部件，如刮水器、风扇、收放机、空调等，也逐步进入了装载机的电气系统。电气系统在整机中发挥着越来越重要的作用。

装载机的品种规格繁多。有的机器电气系统非常简单，只有简单的起动、照明、仪表监控等功能；有的机器电气系统则非常复杂，集机、电、液、信于一身。不管怎样，所有的装载机电气系统都具有下述两个特点：

1）大部分为 24V 标称电压（极少数为 12V）。

2）采用单线制与负极搭铁方式。

蓄电池的正极一般与用电设备相连，蓄电池的负极与机体相连，即利用机体的金属体代替电路中的负极导线，这种方法形成的电路称为单线制。将负极与机体连接在一起的方式称为负极搭铁方式。所有装载机的电气系统都可以归纳为图 4-1 所示的电路模型。

作为电路模型，图 4-1 着重表述的是电源与负载的关系，省略了实际电路中的控制元器件，如开关、继电器、熔断器等。

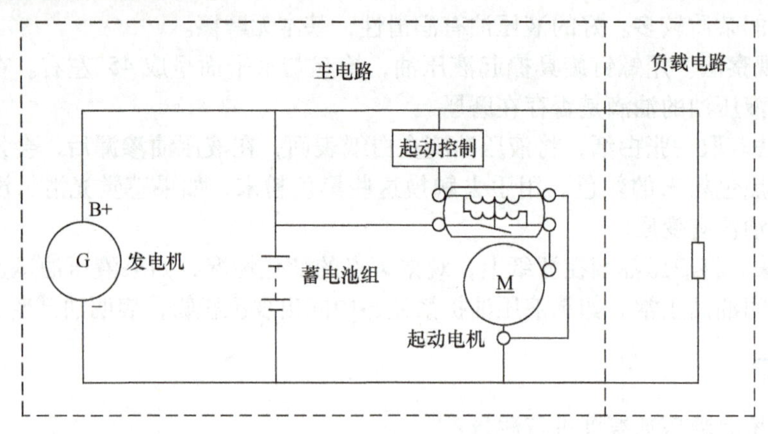

图 4-1　装载机的电路模型

从图 4-1 中可见，装载机的电气系统可分为主电路与负载电路。主电路包括电源系统与起动系统，用来起动发动机并为全车电气提供电源，是电气系统的核心。负载电路一般包括仪表系统，照明系统，辅助电器（如刮水器、电风扇、空调电器、电喇叭、音响等），较高档次装载机的负载电路还包括电子监控系统、工作装置自动复位系统及电液变速操纵控制系统等。

任务 4.1　主电路工作原理分析

学习目标

1）能正确描述装载机主电路的工作原理。

2）能按规范要求对电源回路进行检修。

工作任务

一台装载机发电机有阻滞现象和碰刮响声，需要对发电机进行故障排查和分析，并给出解决办法。

相关知识

1. 主电路的结构与功能

主电路是装载机电气系统的核心，是全车能否正常工作的基础。其主要由电源总开关、蓄电池、发电机、起动电机、起动控制电路、电锁（钥匙开关）、电源继电器等组成。其中蓄电池、发电机、起动电机是主电路的核心元器件。

2. 蓄电池

见本书子任务 2.5.1。

3. 发电机

（1）发电机的作用　发电机（图 4-2）是在发动机的带动下将机械能转化成电能的装置。发电机与蓄电池并联组成整机供电系统，在柴油机未发动之前，由蓄电池向整机用电器进行供电，在柴油机发动之后，主要由发电机对用电器进行供电，同时发电机对蓄电池进行充电，在发电机提供的能量不能满足整机用电器的需要时，发电机与蓄电池共同供电。

（2）发电机的结构　装载机使用发电机的三个接线端口，它们分别是 B +、Ground、D +。B +：发电机正极，为蓄电池充电，为车上的用电器供电；Ground：发电机负极，与整车蓄电池负极端子相连；D +：励磁端，接充电指示灯、起动机保护继电器等，输出电流不超过 1A。

4. 起动电动机

（1）起动电动机的作用　装载机上的起动电动机是将蓄电池电能转化为机械能并起动发动机的装置。起动电机由直流电动机、传动机构、控制装置三部分组成。直流电动机的作用是将电能转化成机械能，产生电磁转矩。传动机构的作用是在发动机起动时，将电磁转矩传递给飞轮，驱动发动机运转并起动，在发动机起动后，使起动电动机驱动齿轮自动打滑，以免发动机反拖起动电动机电枢，并最终与飞轮齿圈脱离啮合。控制装置用来控制直流电动机与蓄电池连接电路的通断，同时控制传动机构与飞轮的啮合与脱离。

目前，装载机上的起动电动机绝大部分是串励式电磁操纵直驱柔性啮合起动电动机。所

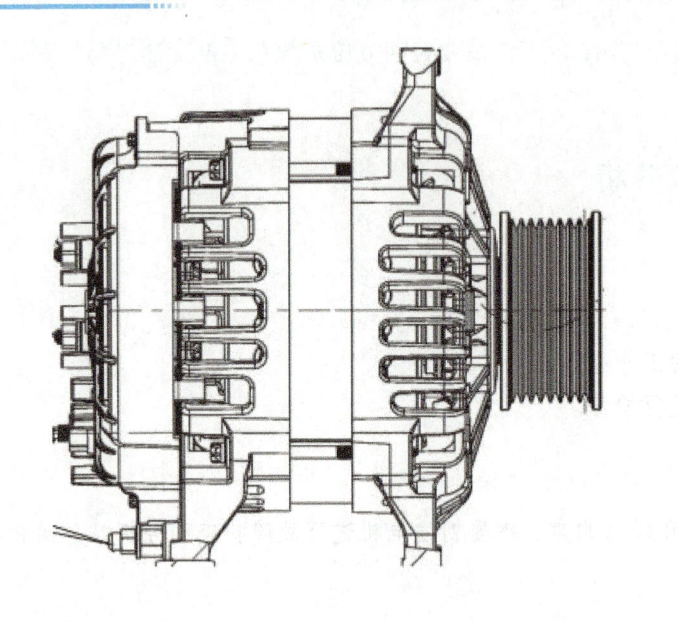

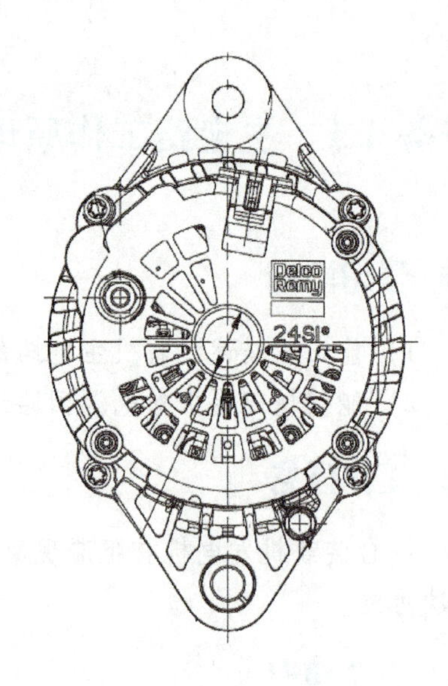

图 4-2　发电机

谓串励式，是指磁场绕组和电枢绕组串联。另外，减速式起动电动机与柔性啮合起动电动机由于优点明显，也在逐步应用于装载机。

（2）起动过程　如图 4-3a 所示为起动前及起动后起动电动机与起动线路的状态，右图 4-3b所示为起动过程中起动电动机与起动线路的状态。

起动时，接通起动开关，起动电动机控制装置的吸引线圈与保持线圈通电，两者产生的电磁力方向相同，相互叠加，吸引控制装置的衔铁克服弹簧力右移，并带动拨叉绕其销轴转动，使驱动齿轮左移；同时，由于吸引线圈的电流流过直流电动机的绕组，电枢开始转动，通过单向器使驱动齿轮旋转。因此，驱动齿轮边旋转边左移。当左移出一定距离后，驱动齿轮齿端与发动机飞轮齿圈齿端相对，不能马上啮合，弹簧被压缩，当驱动齿轮转过一定角度后，两齿轮的齿端错开，在弹簧力的作用下，驱动齿轮迅速左移与飞轮啮合，同时，控制装置的衔铁迅速右移，使控制装置的触点开关迅速闭合。触点开关闭合后，大电流从蓄电池正极通过触点开关流经直流电动机的绕组后回至蓄电池负极，直流电动机便产生较大的电磁转矩驱动发动机旋转并起动（注意：触点开关闭合后，吸引线圈两端电势相等，不再有电流流过，由保持线圈产生的电磁力维持衔铁的位置）。

发动机起动后，其转速迅速上升到怠速，飞轮变成主动齿轮，带动驱动齿轮旋转，但由于单向器的“打滑”作用，发动机的转矩不会传递给电枢，防止了电枢超速运转的危险。

起动后，松开起动开关，起动控制回路断电，电流除从蓄电池正极通过触点开关流经直流电动机的绕组回至蓄电池负极外，还从蓄电池正极通过触点开关流经控制装置的吸引线圈后经保持线圈回至蓄电池负极。很明显，此时吸引线圈与保持线圈是串联关系，流经两者的电流相等。由于两者的匝数相等，因此两者产生的电磁力大小相等，但方向相反，相互抵消。控制装置的衔铁在弹簧力的作用下迅速左移，使触点开关断开，直流电动机的绕组与控制装置的吸引线圈与保持线圈断电；衔铁左移带动拨叉绕其销轴转动，使驱动齿轮右移，脱开驱动齿轮与飞轮的啮合。

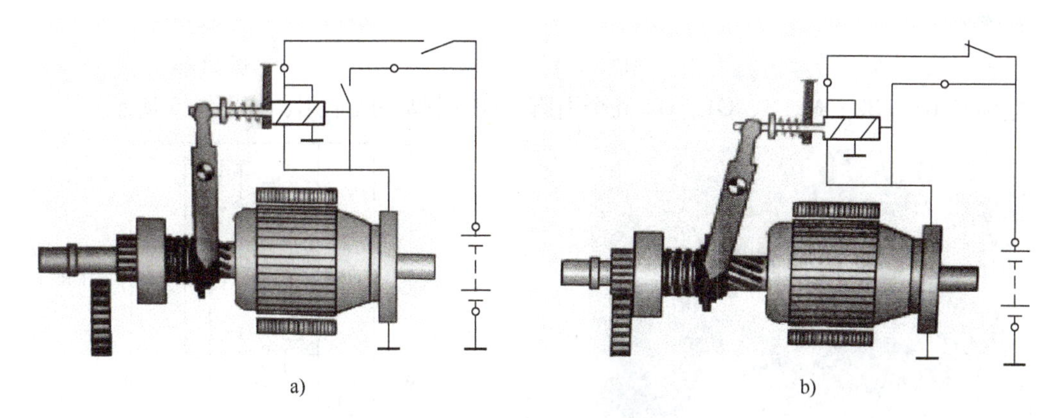

图4-3 起动构成示意

(3) 起动电动机的使用与维护 起动电动机使用与维护的注意事项如下：

1) 任何原因引起的起动电动机正极或者负极电缆上的电压降，都会降低起动性能，导致起动困难甚至无法起动。

① 要保证蓄电池线路中所有的接线柱清洁、连接牢固，以减少接触电阻。

② 由于绝大部分起动电动机是外壳接地的，应检查发动机外壳与蓄电池负极搭铁是否良好。

③ 蓄电池线路中电缆的截面积、材料应符合要求，电缆的总长度应尽量短，以减少导线电阻。

2) 起动电动机的防尘罩、密封垫等密封元件一定要装好，防止变速器润滑油、尘土等窜入起动电动机内部。

3) 发动机起动后，起动电动机应立即停止工作，以减少起动电动机不必要的运转所造成的磨损与蓄电池电能消耗。此外，如起动电动机连续运转时间过长，会导致内部直流电动机绕组温升过高而烧毁，同时，使蓄电池过度放电，影响蓄电池的寿命。一般来说，每次起动时间不超过5s，如一次未能起动成功，应间歇15s以上再进行第二次起动。连续三次起动不成功，应查明原因，排除故障后再起动。

4) 起动前，应关闭所有与起动无关的用电设备，同时将装载机挂空档，动臂与铲斗操纵杆置中位，以增加起动电动机的起动能力，减少发动机的阻力矩。

5) 当环境温度过低导致起动困难时，在起动前，要对发动机进行充分预热，以降低发动机润滑油的黏度，减少发动机的阻力矩。

(4) 起动电动机的常见故障

1) 控制装置故障，主要有吸引线圈或保持线圈的短路、断路、搭铁，触点与接线柱烧蚀等。

2) 直流电动机故障，主要有换向器严重脏污或烧蚀，电刷严重磨损，电刷架内卡死，电枢绕组或磁场绕组短路、断路、搭铁等。

3) 传动机构故障，主要有单向器打滑，单向器弹簧折断，驱动齿轮或飞轮齿圈严重磨损或损坏，电枢轴衬套磨损严重，拨叉折断或脱离位置，驱动齿轮与限位环之间间隙过大等。

5. 电锁

电锁（图 4-4）俗称钥匙开关，用来控制全车通电/断电、起动、熄火等功能。

电锁有 B1 – B2、M、S、G1、G2 五个引脚，G1 引脚一般不用，如图 4-5 所示。

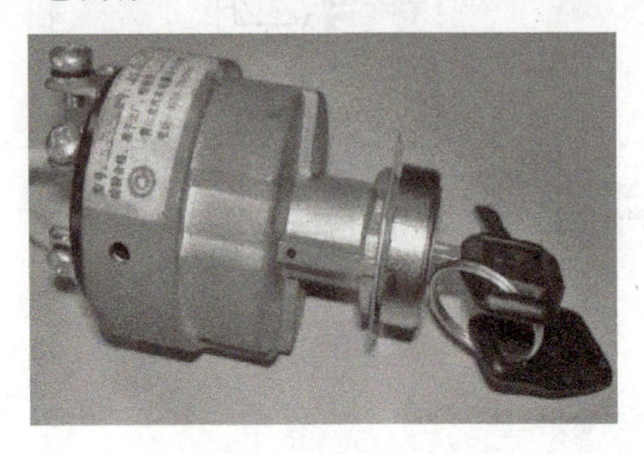

图 4-4　电锁

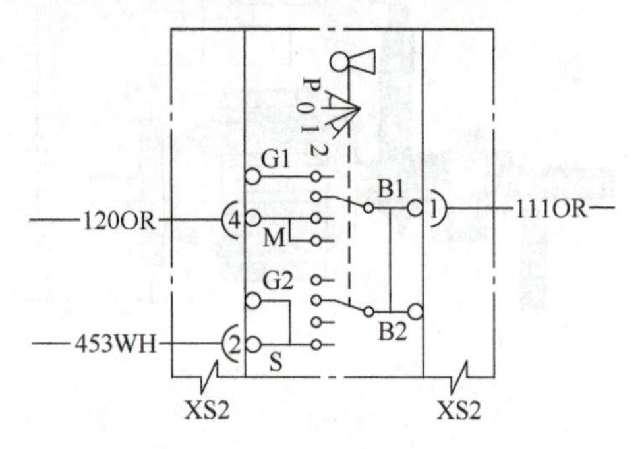

图 4-5　电锁原理图

B1 ~ B2 为电源引脚，B1 接 111 号导线，M 为点火引脚，接 120 号导线，S/G2 为起动引脚，接 453 号导线。电锁的功能与档位关系见表 4-1。

表 4-1　电锁的功能与档位

	B1	B2	M	S	G1	G2
OFF	●	●				
ON	●	●	●			
START	●	●	●	●		●
辅助	●	●			●	

6. 熔断器

熔断器在电路中主要起短路和过电流保护的作用。CLG856H 电源系统中使用了各种系列的熔断器、插片式熔断器和螺栓式熔断器。插片式熔断器使用的规格为 20A、15A、10A、7.5A、5A，用于各分支回路。螺栓式熔断器规格为 30A、50A、60A、80A、150A，用于主回路。

插片式熔断器的颜色与规格有关，表 4-2 列出了装载机使用的插片式熔断器的规格与颜色。

表 4-2　插片式熔断器规格与颜色

规格	颜色
5A	褐色
7.5A	棕色
10A	红色
15A	蓝色
20A	黄色

7. 负极开关

负极开关（图 4-6）控制蓄电池负极与整机车架之间的连接。当负极开关处于“O”

（关闭）位置时，整机电源负极被切断。即使打开电锁，整机用电器也不能工作。当负极开关处于“I”（打开）位置时，整机电源负极接通，此时可以通过电锁控制整机用电器与电源的接通，并可起动柴油机。

注意事项：

1）在每次作业或行驶结束后都必须关闭负极开关，否则会造成漏电的严重后果。

2）严禁在机器运转的过程中关闭负极开关。这种错误会对整机的电气系统造成非常严重的伤害；每次停机时应先关闭电锁，再关闭负极开关。

3）每次开机时应先打开负极开关，再打开电锁；在连接蓄电池电缆或紧固蓄电池电缆桩头或拆卸蓄电池电缆时，必须关闭负极开关；在对整机进行焊接作业时，必须关闭负极开关。

图 4-6　负极开关

8. 继电器

继电器有30、87、87a、86、85五个接线柱。86与85之间为线圈，电阻值约为300Ω。30与87之间为常开触点，30与87a之间为常闭触点。继电器内部带有续流二极管。

继电器的工作原理是线圈通电后，30与87接通，与87a断开；断电后，30与87断开，与87a接通。

继电器是否损坏的判断方法：用万用表的电阻档测量，86、85之间的电阻约为300Ω；30、87之间的电阻为无穷大，30、87a之间的电阻为0。将86接至直流24V电源的正极，85接至负极，30与87应导通，30与87a断开。

由于继电器内部带有续流二极管，故86端必须接电源正极，85端必须接地，不能接反，如图4-7所示。

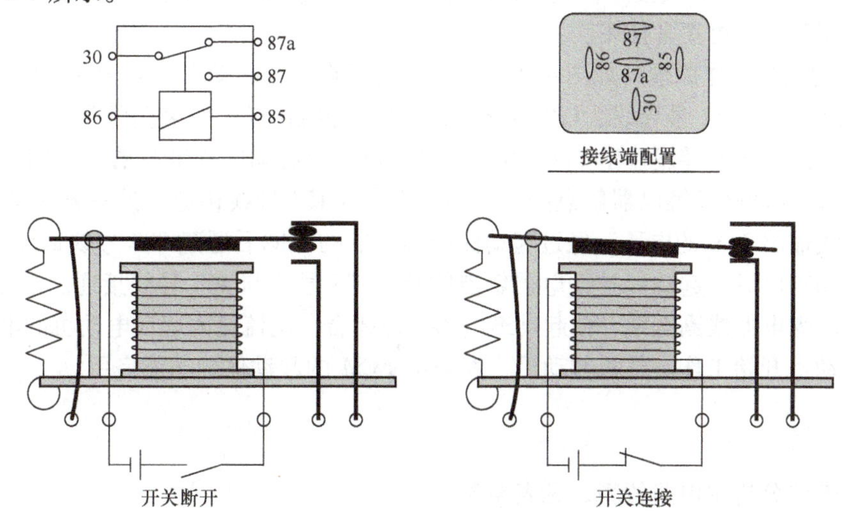

图 4-7　继电器

9. 主电路工作原理与故障分析

不同型号的装载机的主电路是不完全一样的，然而基本原理相同，因此，只要掌握了一种装载机的主电路，其余的都可以触类旁通。

（1）电源系统工作原理　图 4-8 所示为 CLG856H 装载机的主电路电源系统。

负极开关闭合后，蓄电池的电流一路通过 50A 电器集中控制盒熔断器、100 号导线，到达非过电锁电源总线熔断器。此时可正常工作的电器部件为壁灯、旋转警示灯、驻车灯、喇叭等；同时电锁熔断器通过 111 号线供电给电锁电源端（B1～B2）。另一路通过 60A 主电源熔断器、176 号导线，到达电源接触器。

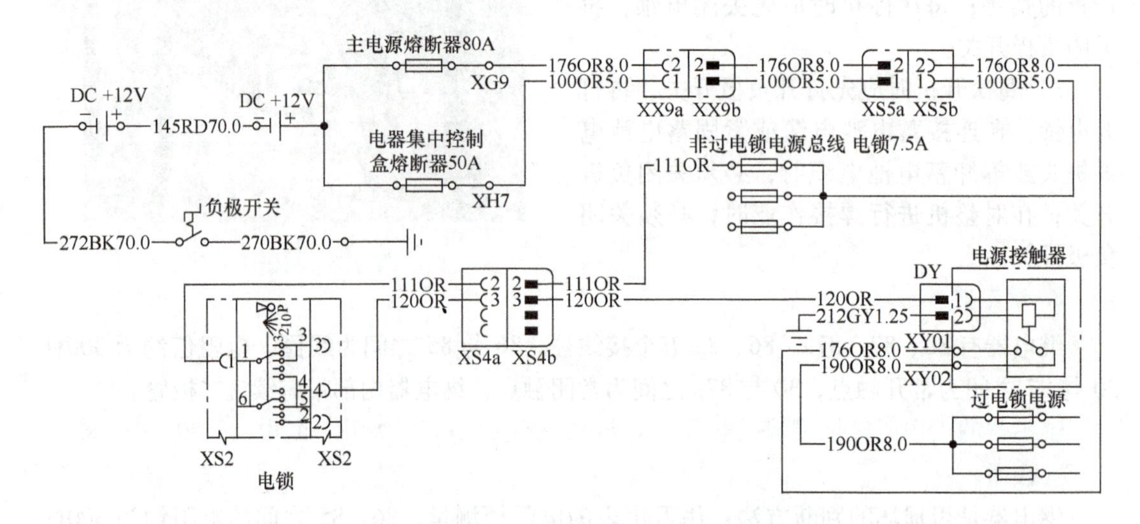

图 4-8　CLG856H 装载机主电路电源系统原理图

电锁旋转至 ON 档后，此时电锁 B1～B2 端便与 M 端接通，111 号导线与 120 号导线接通，电流通过 120 号线、电源接触器的线圈、212 号线至地。故电源接触器触点开关闭合，176 号导线便与 190 号导线接通，过电锁电源熔断器得电。此时除倒车警报与空调功能模块外，其他电器部件均可正常使用。

（2）起动系统工作原理　如图 4-9 所示，将电锁旋转至起动 START 档，B1～B2 端、M 端、S 端互相接通，111 号导线、120 号导线、453 号导线接通。如果换档手柄挂在空档，则变速控制器通过 584 号导线输出 24V 的电压，通过档位/起动联锁继电器的线圈至地，线圈得电后，档位/起动联锁继电器触点闭合，453 号线与 454 号线接通。另一方面发动机 ECM 通过 982 号线输出 24V 的电压，经过起动保护继电器的线圈后通过 983 号线回流至 ECM 内部接地，线圈得电后，起动保护继电器触点闭合，454 号线与 460 号线接通。电流通过 460 号导线、起动继电器线圈至地，使起动继电器触点闭合，电流流入起动电动机的电磁开关线圈，起动电动机开始工作。整机起动后，发动机 ECM 的起动保护功能启用。

☞ 任务实施

采用列表法分析主电路故障，见表 4-3。

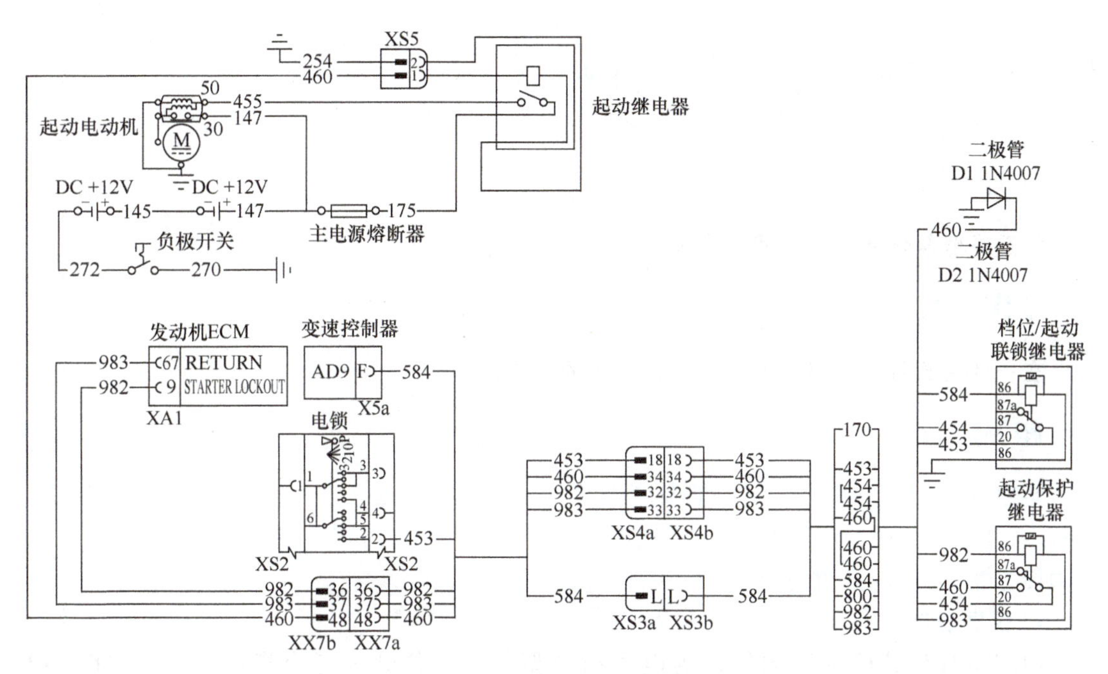

图 4-9　起动系统工作原理

表 4-3　列表法故障诊断

故障	现象	原因	维护与预防
定子绕组短路、断路	短路可分为相间短路、匝间短路和绕组搭铁等。相间短路和匝间短路可从外观检查，一般可见到线圈烧焦、变色	定子浸漆处理不好，绝缘漆没能填满槽内空隙，使导线固定作用减弱；轴承因缺油松动或烧蚀，致使轴承径向间隙过大使转子轴偏心，造成转子与定子碰刮（扫膛），使定子绕组局部温升过高而短路或机械性断路	定期保养，对轴承松旷严重的要及时更换。发现轴承外圈与轴承室有磨损时，应立即修复或更换壳体
定子铁心损坏	发电机有阻滞现象和碰刮响声	发电机长时间过载（匹配不合理）造成定子烧毁，前后端盖紧固螺钉紧度不一或部分失效，造成碰刮（扫膛）使定子损坏；绝缘纸磨破，造成对地短路	检查发电机前后盖紧固螺栓是否松动或丢失，并及时拧紧
转子绕组断路	绕组引线脱焊或断线；电刷磨损严重或松动；电刷弹簧卡死、折断；集电环烧蚀	线路与集电环焊接处因焊接品质和其他原因脱焊；集电环与转子轴间配合松动，引线断；集电环与电刷接触面接触不良，造成发电机不能产生磁场	对于磨损大于 50% 的电刷和压力失效的弹簧应给予更换；对于表面烧蚀严重和跳动大于 0.05mm 的集电环，用车床车圆和抛光处理；用酒精等溶剂擦拭集电环表面
整流组件短路、断路	发电机的输出电压过低或电压不能输出	电压调节器损坏导致二极管击穿；二极管的正向或反向工作电压、电流过大而击穿	避免在装载机上加装其他大功率电气设备

任务 4.2　仪表系统工作原理分析

☞ 学习目标

能正确理解仪表上各区间的含义。

☞ 工作任务

在一台装载机上找出压力开关及传感器的位置。

☞ 相关知识

装载机仪表监控系统一般包括温度表（如发动机水温表、变矩器油温表等）、压力表（制动气压表、变速器油压表等）、燃油油位表、电压表、计时器等指示仪表和温度传感器、压力传感器等，以及其他采用压力开关驱动的报警指示灯。

1. 仪表总成

CLG856H 装载机配置的仪表为电子组合型仪表，共分为三个指示区域，如图 4-10 所示。

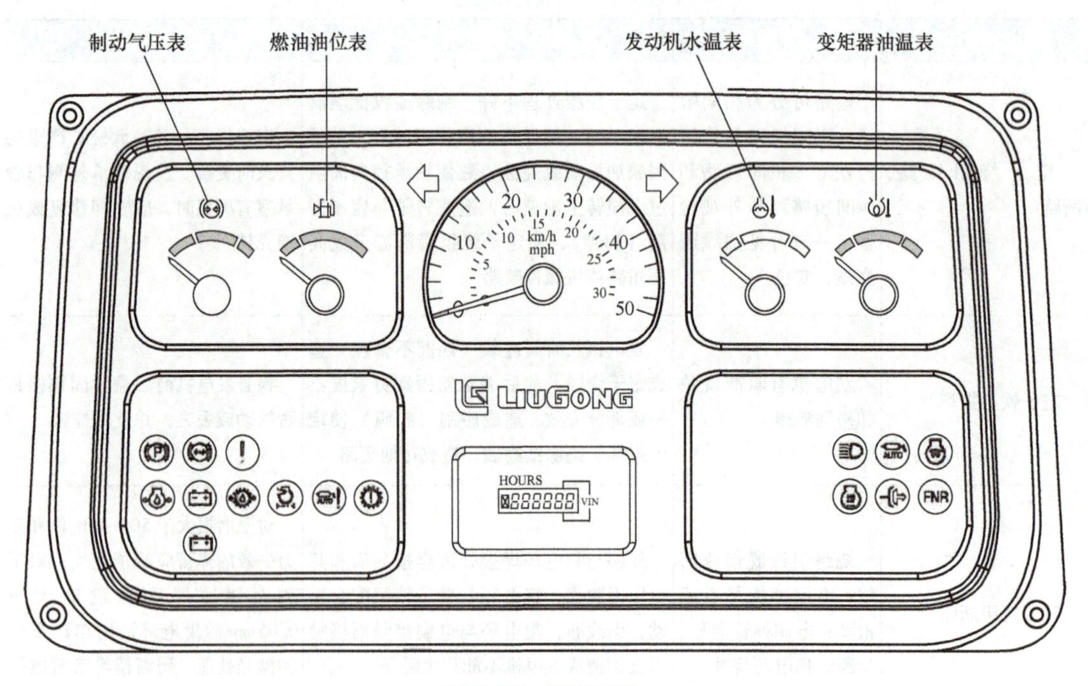

图 4-10　仪表总成

1）指针式区域，见表 4-4。

表 4-4 指针式区域

项目	指示区域		
	第一区域	第二区域（绿色）	第三区域（红色）
发动机水温表	40～55 ℃（黄色）	55～101 ℃	101～120 ℃
变矩器油温表	40～60 ℃（黄色）	60～116 ℃	116～140 ℃
燃油油位表	0～0.2（红色）	0.2～1	—
制动气压表	0～0.4MPa（红色）	0.4～0.8MPa	0.8～1.0MPa

2）指示灯区域，如图 4-11 所示。

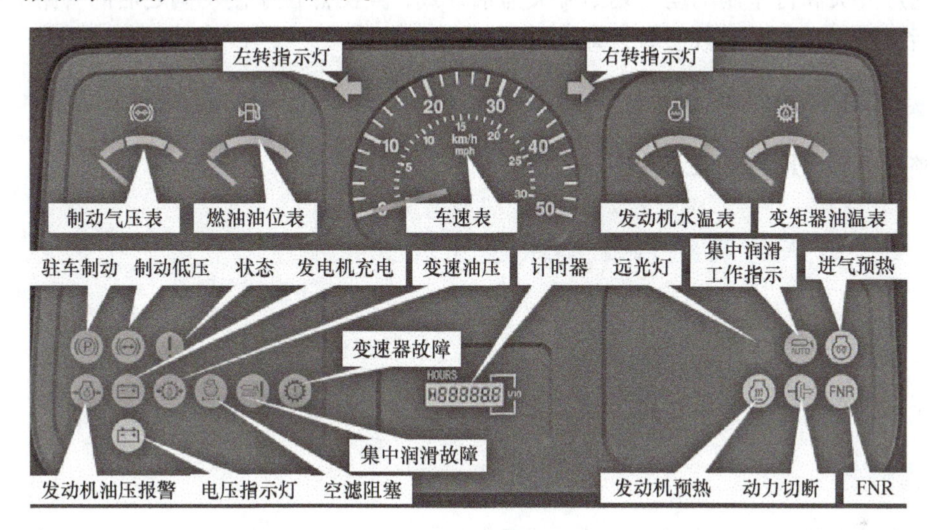

图 4-11 仪表指示灯

3）液晶显示屏区域。液晶显示屏可显示车速、系统电压、发动机故障代码、ZF 故障代码等内容，如图 4-12 所示。车速、系统电压、发动机故障代码等屏幕显示项目可通过方向机柱右侧上的屏幕切换开关进行切换。

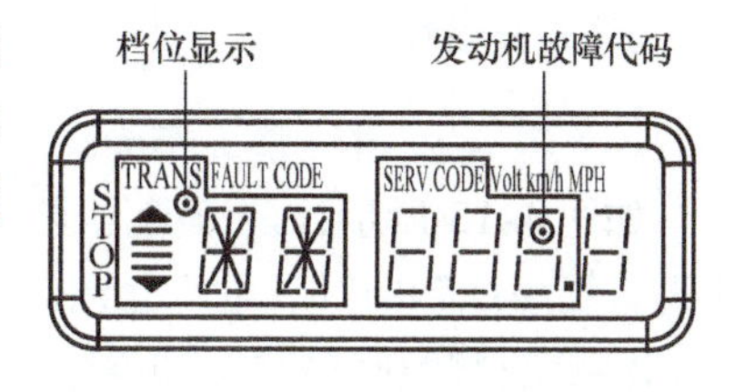

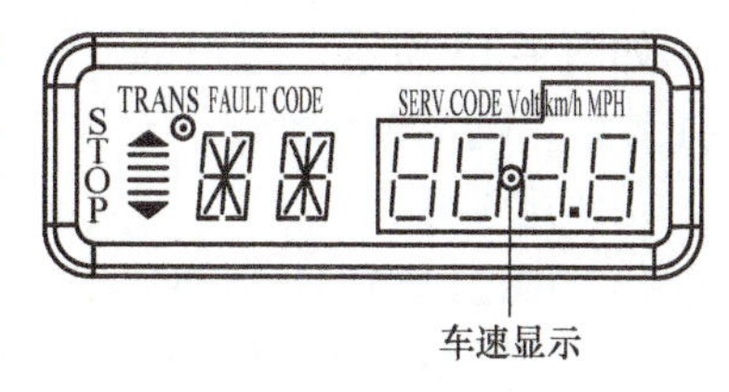

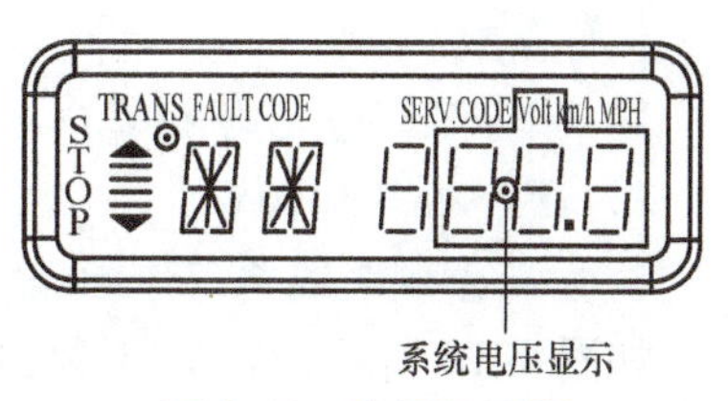

图 4-12 液晶显示屏

2. 传感器

1）温度传感器。装载机用温度传感器一般采用传热性能很好的铜外壳将负温度系数的热敏电阻封装而成。热敏电阻是用陶瓷半导体材料掺入适量氧化物在高温下烧结而成的。所谓“负温度系数”，就是在热敏电阻的工作范围内，当温度升高时，热敏电阻的电导率会随着温度的升高而增加，即它的电阻值随着温度的升高而减少。

2）压力传感器。压力传感器主要由膜盒腔、传动机构与滑线电阻组成。膜盒腔的膜片是一种弹性敏感元件，用来感受介质（如油、气等）的压力变化并将其转换为机械位移。当介质压力变化时，膜片产生机械位移，传动机构将机械位移放大并传递给滑线电阻的滑动触片，从而改变传感器的输出电阻值。

3. 压力开关

压力开关分常开式和常闭式，装载机上的压力开关一般包括机油压力、变速油压、动力切断、制动灯等压力开关。

机油压力报警开关为常闭式，其作用是对机油压力进行监测，当机油压力过低时，压力开关触点闭合，机油压力指示灯闪烁报警，同时蜂鸣器报警。

变速油压报警开关为常闭式，其作用是对变速器油压力进行监测，当压力过低时，压力开关触点闭合，变速油压指示灯闪烁报警，同时蜂鸣器报警。

制动灯开关和行车制动动力切断开关为常开式，监测点位于制动阀之后的管路上，当进行制动后，压力达到一定数值时，开关触点闭合。

☞ 任务实施

在整机上找出压力开关及传感器的位置。

任务 4.3　开关与灯组件工作原理分析

☞ 学习目标

能正确描述前车架照明线路、驾驶室照明线路、后车架照明线路的工作原理。

☞ 工作任务

对一台装载机，判断其制动灯开关是否损坏。

☞ 相关知识

1. 前车架照明线路工作原理

如图 4-13 所示，前车架照明线路包括小灯及大灯开关、组合开关和左、右前组合灯。前组合灯集成转向灯、远/ 近光灯及小灯功能。

2. 驾驶室照明线路工作原理

如图 4-14 所示，驾驶室照明线路包括壁灯、工作灯开关和工作灯。该装载机配置了四个工作灯，前后各两个。

工作灯参数：

1） 额定工作电压为 24V。

2） 额定功率为 70W。

壁灯参数：

1） 额定工作电压为 24V。

2） 额定功率为 10W。

3. 后车架照明线路工作原理

如图 4-15 所示，后车架照明线路包括制动灯压力开关和后组合灯。后组合灯集成转向灯、行车/ 制动灯与倒车灯功能。

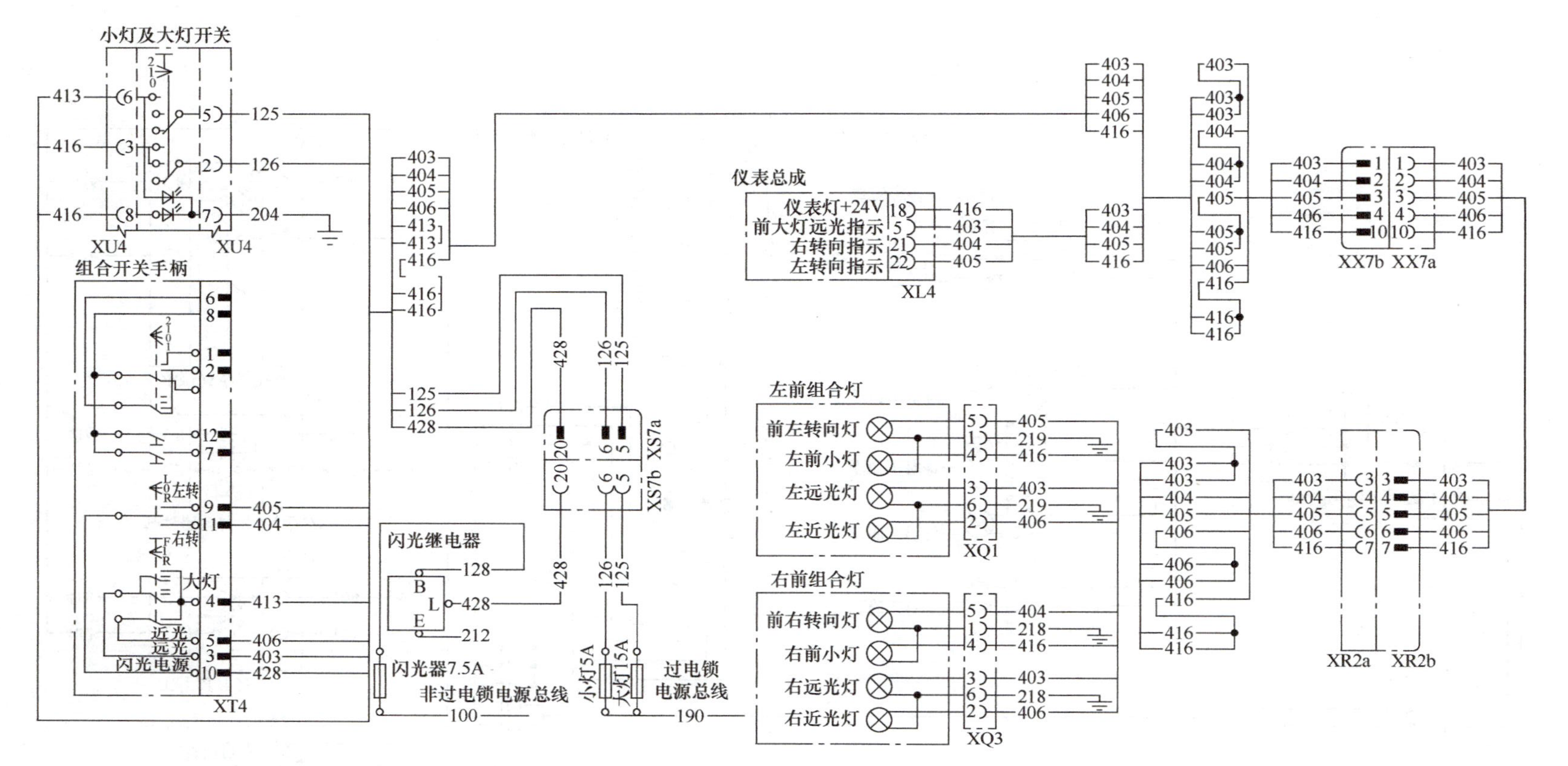

图4-13 前车架照明线路工作原理图

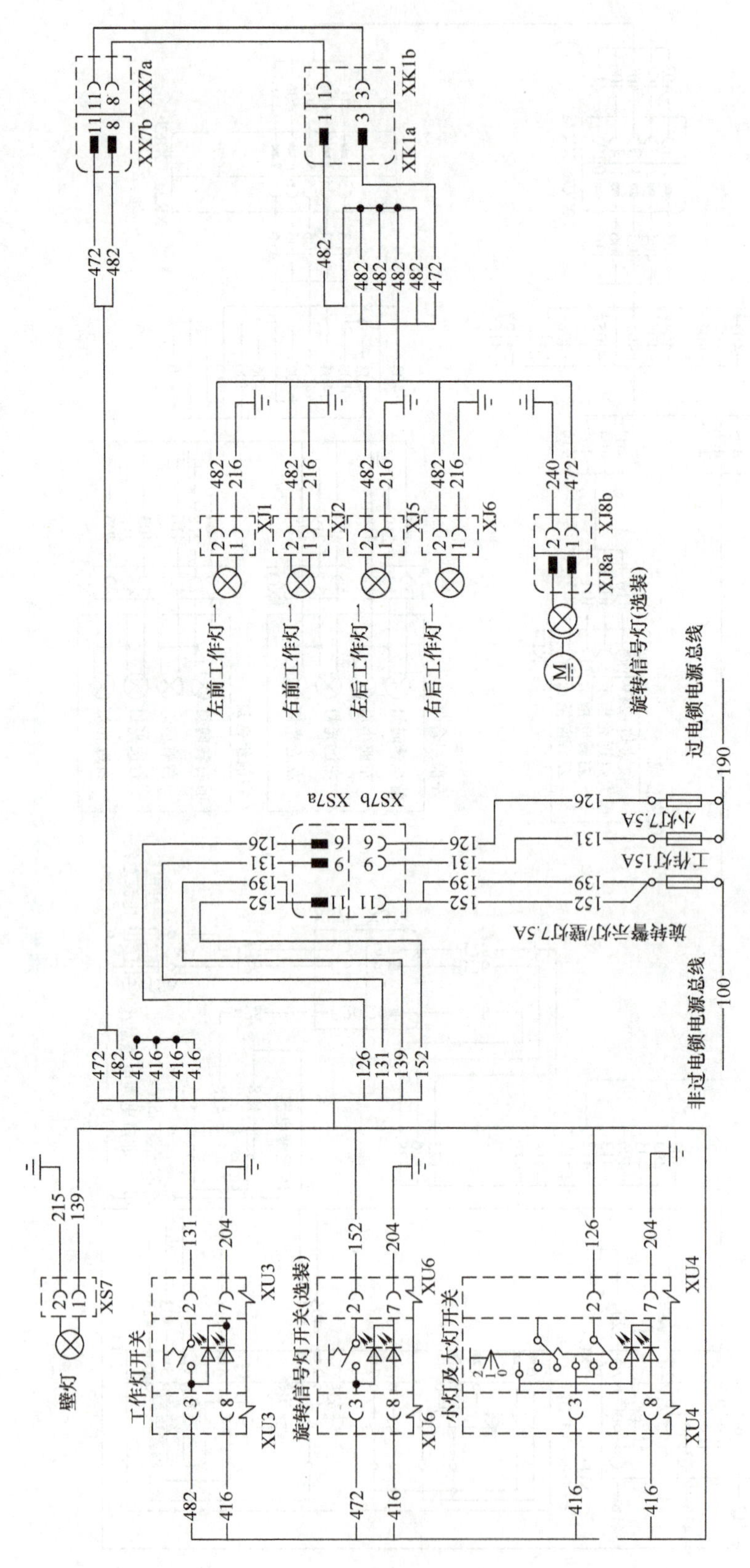

图 4-14 驾驶室照明线路工作原理图

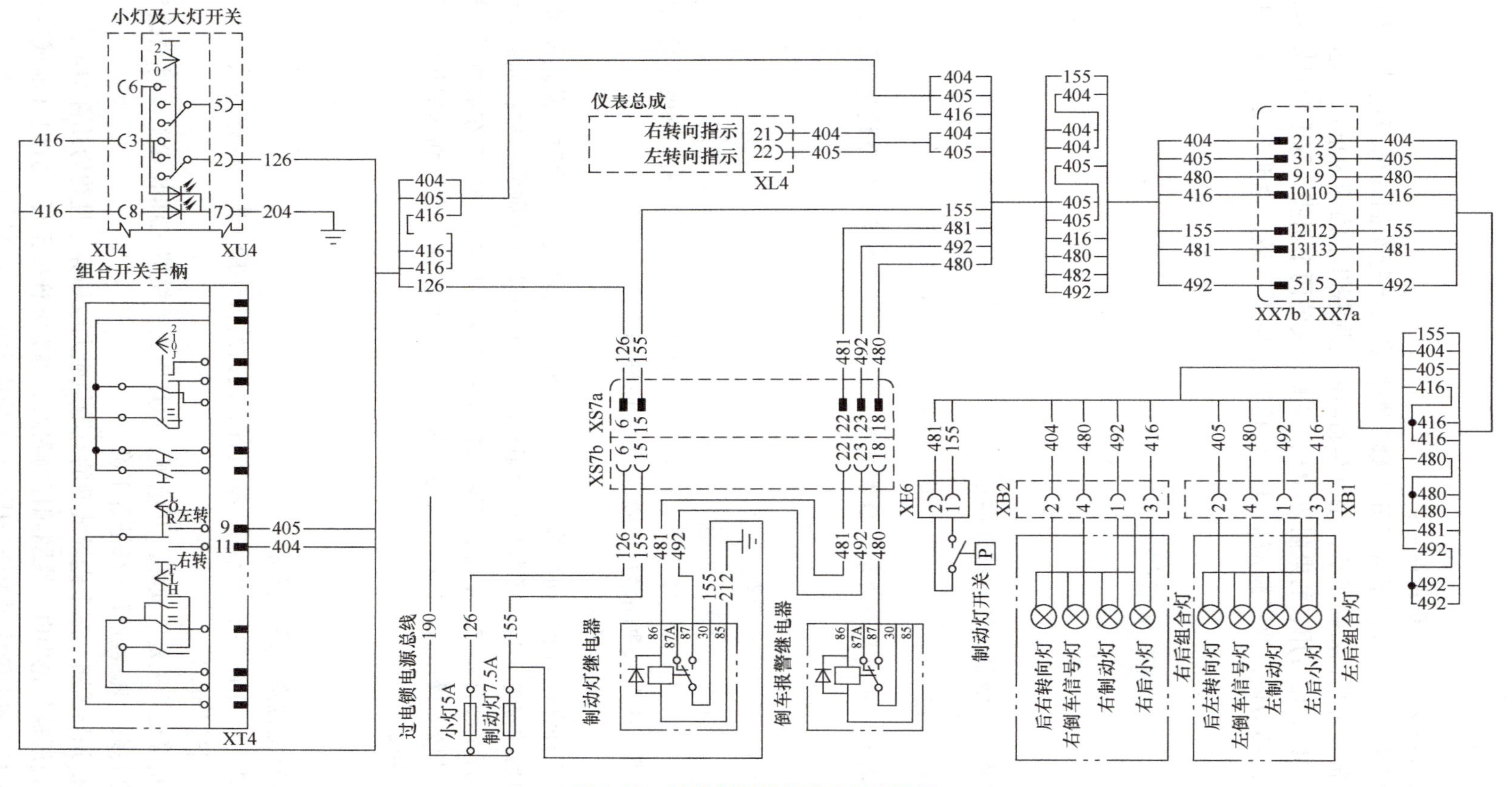

图 4-15　后车架照明线路工作原理图

☞ 任务实施

如何判断制动灯开关是否损坏：

首先确定制动压力是否正常（开电锁，如仪表板上行车制动低压报警灯不闪烁，说明制动压力正常，否则，发动车子，至行车制动低压报警灯不闪烁为止），如正常，拔下制动灯开关处的导线，用万用表的电阻200Ω档检测开关的两个引脚。踩下制动踏板，两引脚应导通；松开制动踏板，两引脚应断开。如检测结果不一致，说明压力开关已损坏，需要更换。

任务 4.4　自动复位系统工作原理分析

☞ 学习目标

能正确描述自动复位系统的工作原理。

☞ 工作任务

一台装载机，用户反映自动复位系统不工作，请检查自动复位系统是否正常。

☞ 相关知识

1. 自动复位系统的作用及组成

自动复位系统主要用来实现动臂限位、动臂浮动、铲斗收平功能，以达到减少全车冲击、减轻操作劳动强度、提高工作效率的目的。该系统由接近开关、继电器和先导阀上的三个电磁铁组成，图 4-16 所示为自动复位系统原理图。

2. 自动复位系统操作方法

操作者将操纵杆向前或向后推，进行动臂下降或动臂提升作业时，操纵杆都会自动保持在向前或向后的位置，当动臂提升到达限位时，操纵杆会自动弹回中位；而动臂浮动下降到底时，操纵杆不会自动弹回中位，需要操作者将其扳回中位。操作者在卸料后将操纵杆向左推，进行铲斗收斗操作时，操纵杆会自动保持在向左的位置，直至到达铲斗放平位置，操纵杆自动弹回中位。

☞ 任务实施

检查自动复位系统是否正常：

1）检查 10A 熔断器是否熔断。检查各插接头是否连接良好；检查磁铁与接近开关的间隙（一般不超过 8mm）。

2）检查接近开关是否损坏。开电锁，绿灯应亮；模拟工作装置工作时磁铁与接近开关的相对运动关系，观察红灯状态是否正确。

3）检查先导线圈。三个先导线圈的电阻值应大致相等，且约为几百 Ω。

4）检查压板与先导电磁线圈阀杆的间隙。将操纵杆扳至任一方向（前或后）的极限位置，在相反方向的电磁线圈阀杆与压板的间隙应为 0.5 ~ 1.27mm。

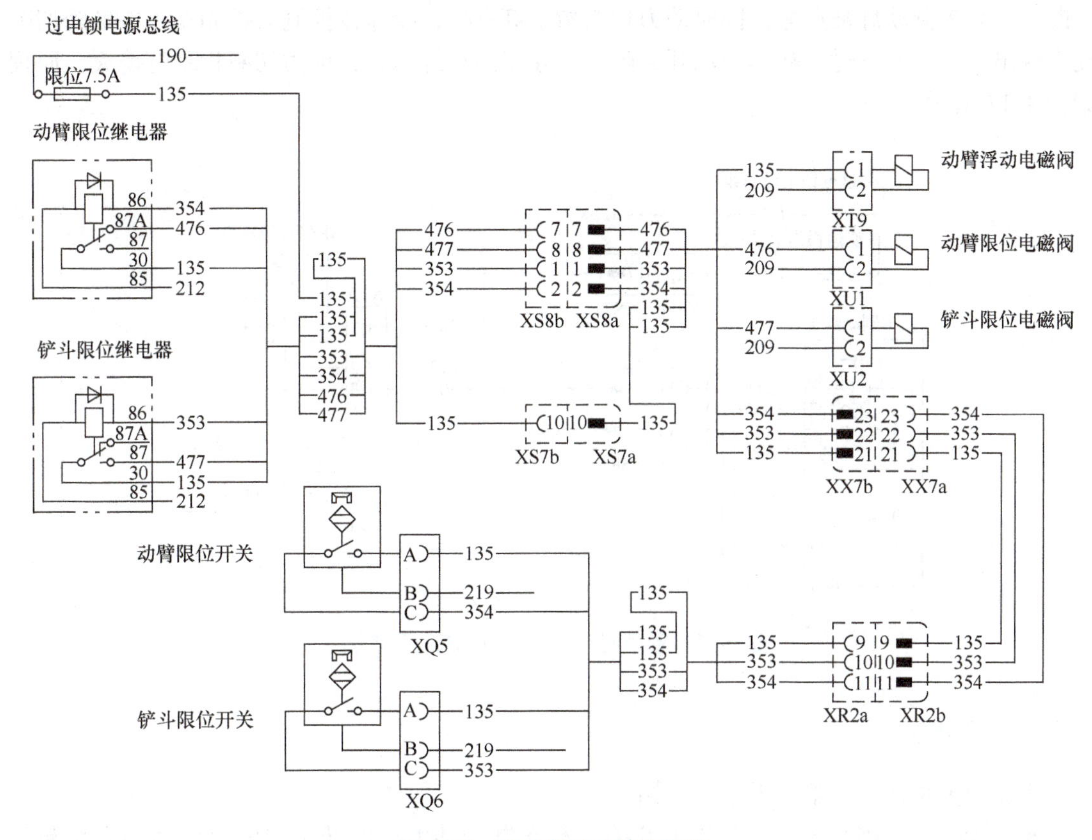

图 4-16 自动复位系统原理图

任务 4.5 紧急制动与动力切断系统工作原理分析

☞ 学习目标

能正确描述紧急制动与动力切断系统的工作原理。

☞ 工作任务

一台装载机，用户反馈整车无一、二档，需要对故障现象进行排除，并检查动力切断系统。

☞ 相关知识

1. 紧急和驻车制动系统

紧急和驻车制动系统用于停车后的制动，或在行车制动系统失效时作为应急制动。另外，当系统压力低于安全压力时，该系统将自动使装载机停车，以确保安全。

2. 工作原理分析

紧急制动与动力切断系统用于监测整车的制动与非制动状态。手刹拉起时，驻车制动开关断开，驻车制动灯被点亮，同时动力切断指示灯点亮，变速器控制盒控制换档电磁阀切断动力输出；手刹放下时，驻车制动开关闭合，驻车制动灯熄灭，动力切断指示灯熄灭，原理如图 4-17 所示。

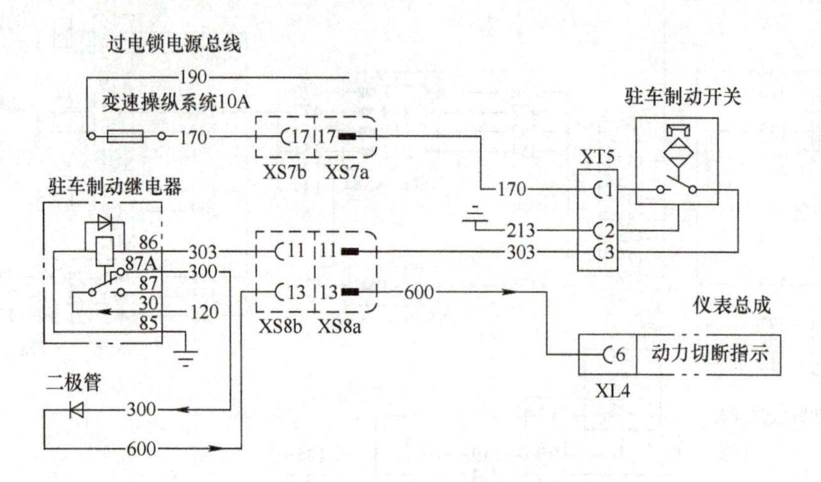

图 4-17 紧急制动与动力切断系统原理图

☞ 任务实施

常见故障检修——整车无一、二档：

整车无一、二档一般是由于压力开关（行车制动动力切断开关与紧急制动动力切断开关）损坏原因导致600 号导线总有 24V 电压输入控制单元 EST－17T，从而切断变速器一、二档动力输出所致。可通过检测 600 号导线的电压判定。具体步骤如下：

1）检查插接器接线是否正确。

2）拔下紧急制动动力切断开关与行车制动动力切断开关的插接器，如有一、二档，则可能为动力切断开关与行车制动动力切断开关损坏。

3）插上紧急制动动力切断开关处插接头，试车，如一、二档没有，则为紧急制动动力切断开关损坏。

4）插上行车制动动力切断开关处插接头，试车，如一、二档没有，则为行车制动动力切断开关损坏。

任务 4.6 变速操纵系统工作原理分析

☞ 学习目标

能正确描述变速操纵系统的工作原理。

☞ 工作任务

一台装载机不能起动，需对不能起动的故障进行排除。

☞ 相关知识

1. 工作原理分析

变速操纵系统的 EST117 控制单元接收来自变速操纵手柄、速度传感器、KD 键等的外部信号，并对这些外部信号进行处理，送入 CPU 中进行计算，CPU 最终通过输出电路驱动位于变速器上的变速操纵阀电磁阀组合动作，切换到操作者选定的档位，如图 4-18 所示。

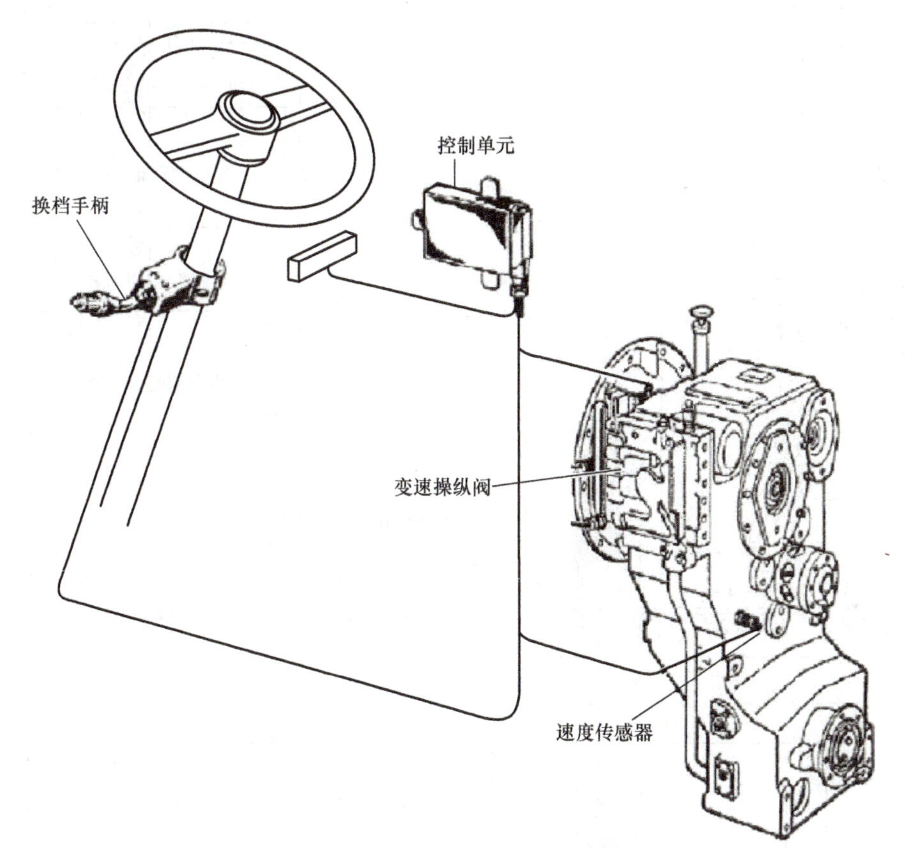

图 4-18　变速操纵系统结构示意

2. 系统主要功能（图 4-19）

1）空档起动联锁功能。当 DW－3 变速操纵手柄拨至空档时，电控单元 584 号导线输出 24V 电压，用来驱动空档联锁继电器。因此，只有当换档手柄挂空档时整机才能起动。

2）动力切断功能。控制单元检测到来自 600 号导线的动力切断输入信号后决定是否给变速操纵阀发出切断动力的指令。一般要求制动时（包括行车制动和驻车制动），动力切断信号有效。

动力切断功能在前进或后退 1、2 档（低速档）中发生作用，当装载机处于 3、4 档（高速档）时，为保证行车安全，控制单元不会切断变速器动力输出，这是由装载机的行驶特性决定的。

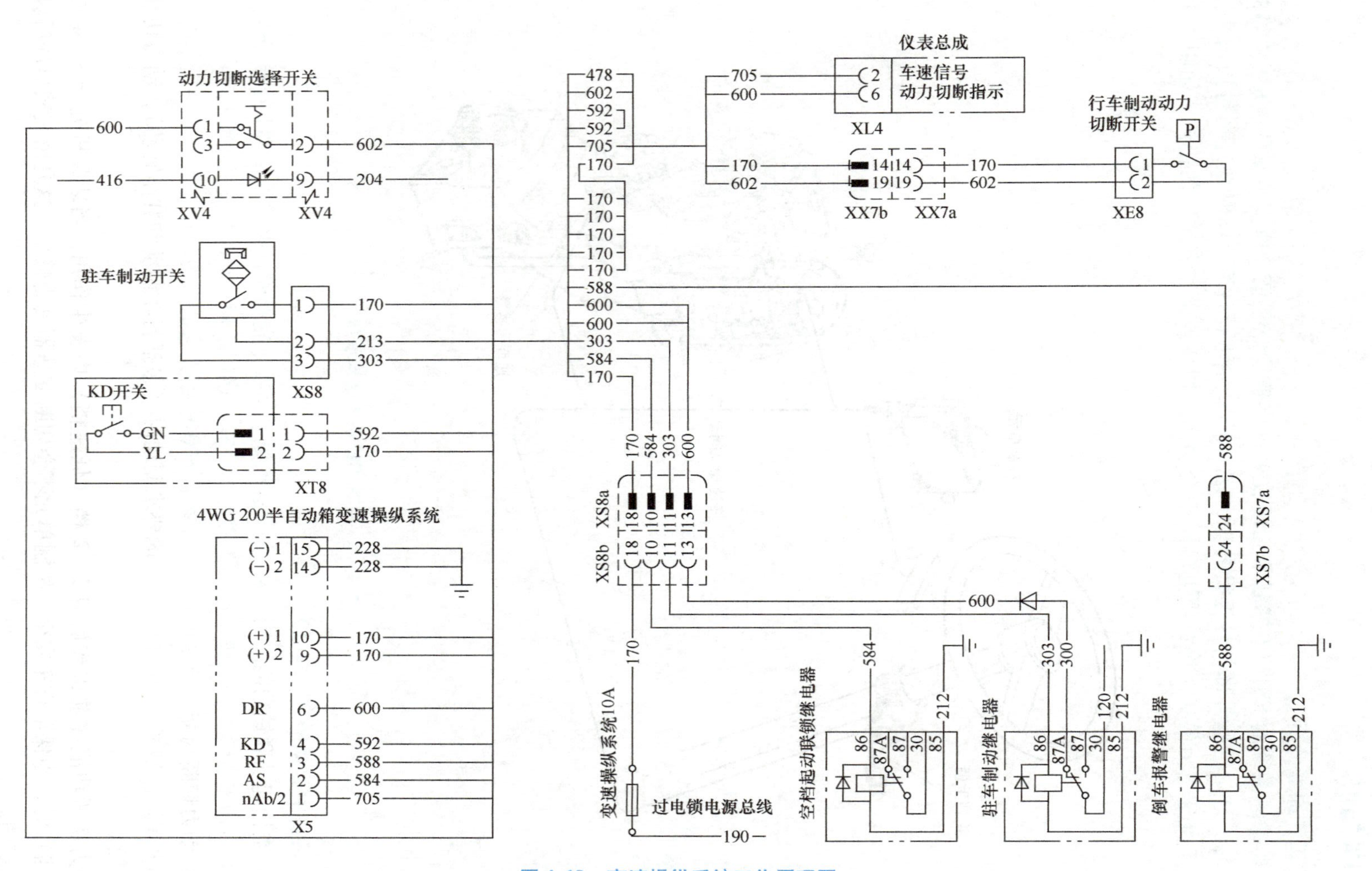

图 4-19 变速操纵系统工作原理图

3）强制换低档功能（KD 功能）。通过变速操纵手柄上的强制换低档 KD 键，当档位设置在前进 2 档或后退 2 档时，按一下 KD 键，变速器档位可自动切换到相应的 1 档。

装载机以前进 2 档的速度行走，接近料堆时，按 KD 键，自动降为前进 1 档；在装好料后，再挂倒档，则自动升为后退 2 档，装载机直接以 2 档速度退出铲装作业区，从而节省了装载机传统的从前进 2 档换至前进 1 档、空档、后退 1 档、后退 2 档所花费的时间，提高了工作效率。

4）失效时系统自我保护功能。控制单元连续监控所有来自换档手柄、速度传感器的输入信号和电磁阀的输出信号，当出现异常信息组合（如线路断开、控制单元地线断路）时，控制单元立即转换至空档状态锁止所有输出信号，电压超过规定限值或发生断路时也如此。因此当装载机发生无法挂档故障时，需仔细检查控制单元外围电路，以判断是否出现元件或线路故障。

输出速度传感器发生故障时，控制单元将只允许变速器挂 1、2 档。

3. 系统主要部件

1）控制单元（图 4-20）。EST117 控制单元是变速操纵系统的核心元件，操作者选择的档位信号、速度传感器信号、KD 信号、动力切断信号，均送入控制单元进行处理及运算，处理完毕后，控制单元输出控制信号驱动变速操纵阀内的五个电磁阀组进行组合动作，最终完成变速器档位的选择切换。控制单元同时还输出空档信号，以驱动档位/空档联锁继电器在 DW－3 换档手柄处于空档时动作。输出倒车警报信号驱动倒车警报继电器工作。

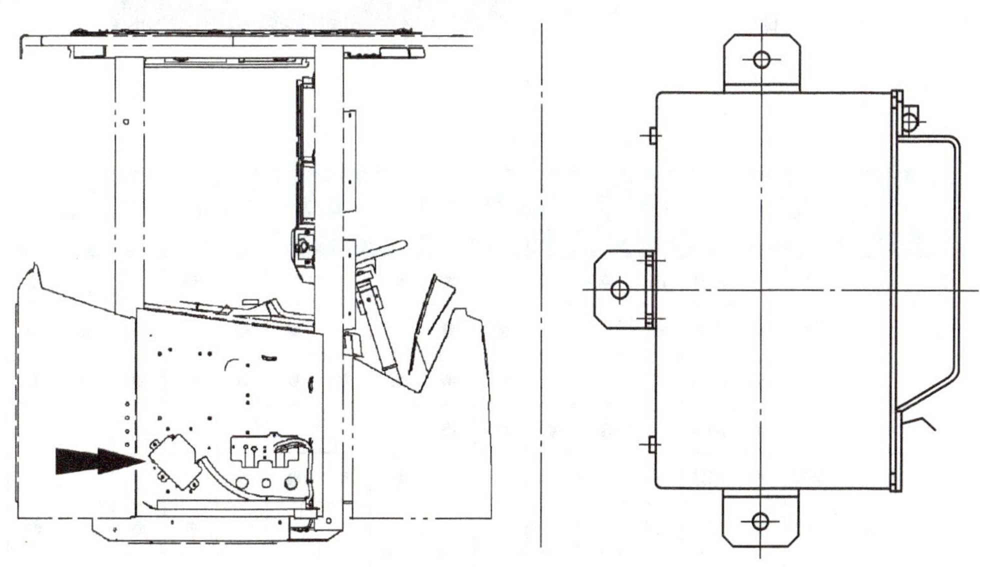

图 4-20　控制单元

2）换档手柄。换档手柄内部由许多微型开关组成。操作者对档位及方向进行选择时，内部的微动开关发生动作，并通过线束的连接，这些变化的档位及方向信号最终进入 EST117 控制单元内部，如图 4-21 所示。

换档手柄检测逻辑见表 4-5。

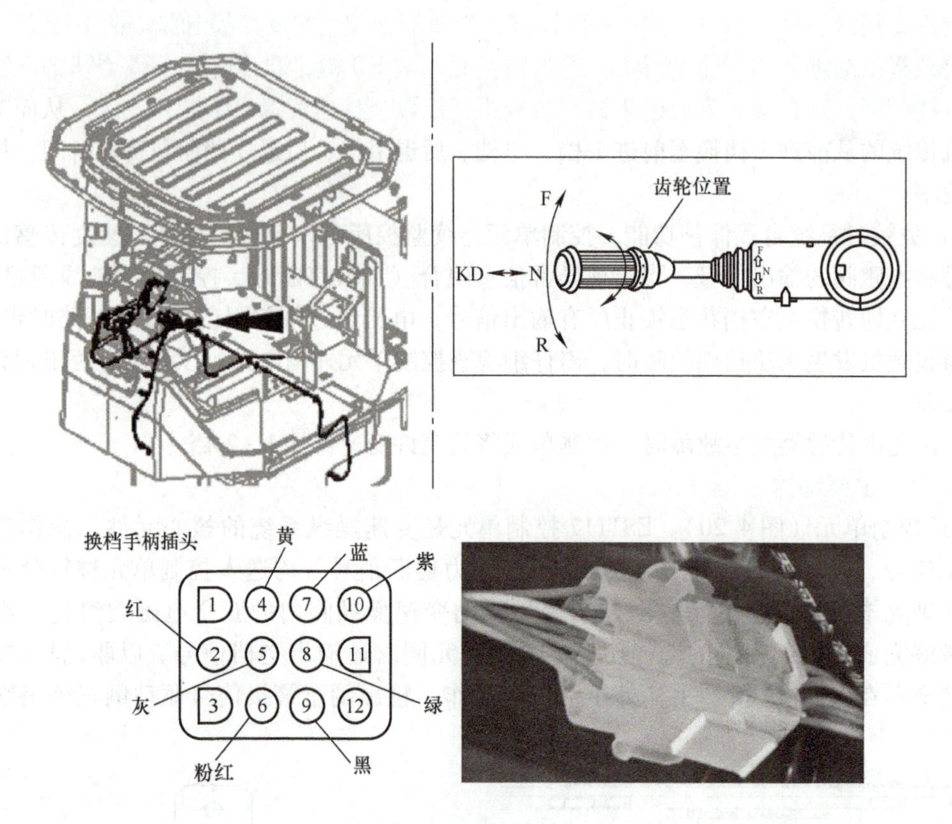

图 4-21 换档手柄

表 4-5 换档手柄检测逻辑

电脑盒处电缆接头	换档手柄		换档手柄		前进档（V）				倒退档（R）				空档（N）				KD
	线色	接头			1	2	3	4	1	2	3	4	1	2	3	4	
26	蓝	7	AD1	B1	●			●	●			●	●			●	
8	绿	8	AD2	B2			●	●			●	●			●	●	
25	黑	9	AD3	B3	●	●	●	●	●	●	●	●	●	●	●	●	
23	黄	4	AD4	V	●	●	●	●									
5	粉红	6	AD5	R					●	●	●	●					
29	灰	5	AD6	AS									●	●	●	●	
24	紫	10	AD7														●
19	红	2	ED1		（+）												

注：●表示得电。

3）换档电磁阀组件。位于变速操纵阀上的五个电磁阀，作为系统的执行元件，接收控制单元发出的换档指令，通过控制变速操纵阀内部油路来控制变速器内的档位离合器，从而使装载机处于某一档位。五个电磁阀的电阻均为 90 ~ 110Ω（图 4-22）。

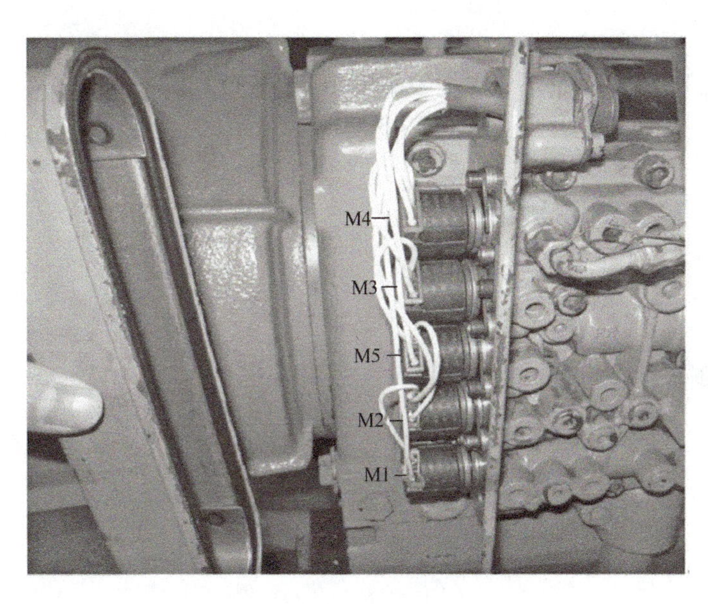

图 4-22 电磁阀组件

电磁阀检测逻辑见表 4-6，用于检测电阻值的逻辑。

表 4-6 电磁阀检测逻辑

电磁阀： 电阻值：90 ~ 110 欧姆 顺序（从上到下）：M4—M3—M5—M2—M1

电脑盒处电缆接头	变速器处电缆接头	电磁阀	前进档（V）				倒退档（R）				空档（N）				KD
			1	2	3	4	1	2	3	4	1	2	3	4	
33	B	M1					●	●	●						
15	C	M2	●				●				●				
32	D	M3	●	●	●										
14	E	M4	●	●			●	●			●	●			
31	A	M5				●									
35	F	（—）													
离合器			KV	KV	KV	K4	KR	KR	KR						
			K1	K2	K3	K3	K1	K2	K3						

注：●表示得电。

4）速度传感器。速度传感器检测变速器输出齿轮的转速，控制单元采集此转速信号后，再综合换档手柄的档位指令，决定是否将变速器挂高档（由 2 档至 3、4 档）。因此，如果速度传感器损坏，装载机将无 3、4 档。此速度传感器为磁电式传感器，电阻值为（1050 ± 100）Ω。

☞ 任务实施

常见不能起动故障的检修：

1）未挂空档，挂空档，重新起动。

2）7.5A 熔断器烧断，更换 7.5A 熔断器，如熔断器仍然熔断，需仔细检查电路，查明原因后再更换。

3）控制单元 EST－17T 损坏，更换控制单元 EST－17T。

4）手柄 DW－3 损坏，更换手柄 DW－3。

任务 4.7　倒车报警器更换

学习目标

能正确描述倒车报警的工作原理。

工作任务

一台新的装载机，确定其倒车报警是否有问题。

相关知识

1. 倒车报警原理

倒车雷达由超声波传感器（俗称探头）、控制器和显示器（或蜂鸣器）等部分组成。倒车雷达一般采用超声波测距原理，在控制器的控制下，由传感器发射超声波信号，当遇到障碍物时，产生回波信号，传感器接收到回波信号后经控制器进行数据处理、判断出障碍物的位置，由显示器显示距离并发出其他警示信号，得到及时警示，从而使驾驶员倒车时做到心中有数，使倒车变得更轻松。

整车发动后，档位选择器拨至 R 档，倒车报警继电器使能，倒车报警器鸣叫，倒车灯点亮，如图 4-23 所示。

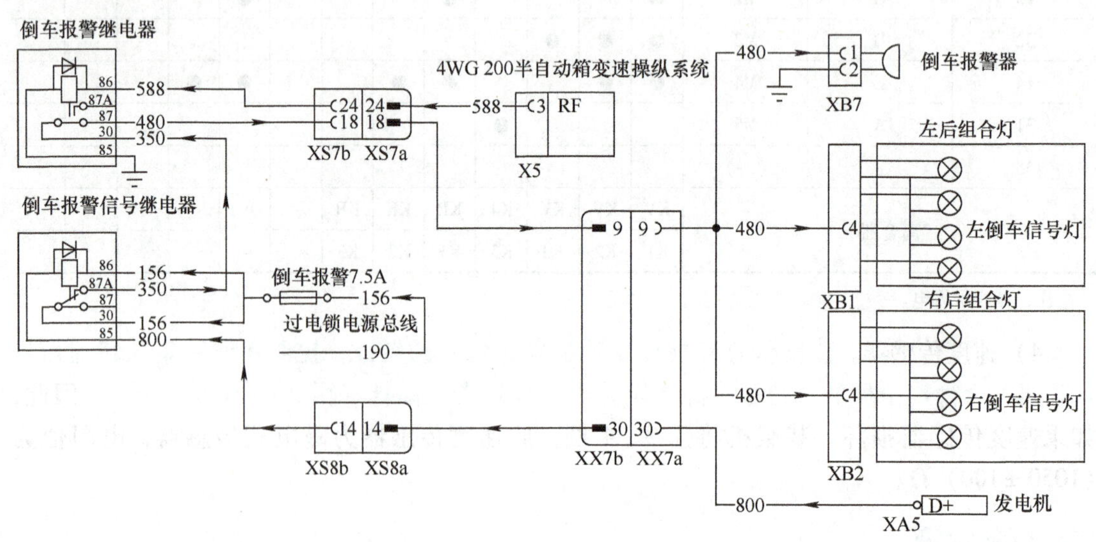

图 4-23　倒车报警原理图

2. 倒车警报器参数

倒车警报器是一个间歇鸣叫的蜂鸣器，在驾驶员进行倒车作业时起作用，与倒车灯同时工作。参数：

1）额定电压：24V。

2）工作电流：3A。

3）基本频率：960/1440Hz。

4）声级：（107 ±4）dB（A）范围内。

☞ 任务实施

倒车报警的测试：

1）将发动机起动开关转到“Ⅰ”或“ON”位置接通整车电源，以便进行试验。

2）将驻车制动按钮（手柄）提起，实施驻车制动。

3）将发动机起动开关转到“Ⅱ”或“START”位置，起动柴油机。

4）将变速操纵手柄置于后退位置，倒车报警器应该立即开始鸣响。

5）将变速操纵手柄放到中位或前进位置，倒车报警器应停止鸣响。

项目 5

装载机空调系统维护

空调即空气调节器，是一种用于给空间区域（一般为密闭）提供处理空气温度变化的机组，它的功能是对一定空间区域内的空气温度、湿度、洁净度和空气流速等参数进行调节，以满足人体舒适或工艺过程的要求。

为了使乘车人员感到舒适，现代工程车辆空调有四种基本功能。

1）空调能控制车厢内的气温，既能加热空气，也能冷却空气，以便把车厢内温度控制到舒适的水平。

2）空调能够排除空气中的湿气，营造舒适的环境。

3）空调可吸入新风，具有通风功能。

4）空调可过滤空气，排除空气中的灰尘和花粉。

车用空调的特点：工程车辆内部空间有限，工作环境温度高、振动大；车辆一起动，空调就要对车内空间进行快速冷却，要求空调压缩机效率高、体积小、性能可靠；对空调的其他部件也有同样要求。因此，工程车辆空调必须采取相应的技术措施，以适应上述特点。

空调基本工作原理：冷却和取暖可以简单地解释为从空气中取走热量或对空气加热。冷却意味着是通过从车内空气中取走热量来降低温度。取暖则是通过供热给车内空气来加热车辆内部，即加热空气。

当酒精涂在人的皮肤上时人会感到凉快，这是因为酒精从皮肤上蒸发吸收了潜热。同样，夏天在花园里浇水会感到凉快，这是因为浇在土壤中的水蒸发时从周围空气中吸收了潜热。这些自然现象体现了制冷的基本原理。

空调工作需要经历四个过程的变化（图 5-1）：

1）压缩过程。制冷剂在蒸发器中吸收热量后变成低温低压的气态制冷剂，经压缩机吸入压缩后，制冷剂被压缩为高温高压的气态制冷剂，排入冷凝器。

2）冷凝过程。高温高压的气态制冷剂进入冷凝器后，在发动机散热器风扇的驱动下，强制空气通过冷凝器表面将制冷剂热量带走，制冷剂被冷凝为中温高压的液态制冷剂。

3）节流过程。中温高压的液态制冷剂通过干燥器过滤后经膨胀阀节流（制冷剂从膨胀阀毛细孔中喷出，体积突然膨胀），变成低温低压的液态制冷剂雾进入蒸发器。

4）蒸发过程。经膨胀阀节流成为低温低压液态的制冷剂在蒸发器中汽化，车厢内空气在蒸发器风机的驱动下流过蒸发器表面，制冷剂吸收车厢内空气热量而使车厢内空气降温，同时析出冷凝水。吸收热量后的制冷剂蒸发成低温低压气态制冷剂，经压缩机吸入再进行压缩，完成一次制冷循环。

压缩机不停转动，上述制冷过程连续不断地进行循环，驾驶室内热量不断被蒸发器内制冷剂带走，从而完成整车的降温除湿。

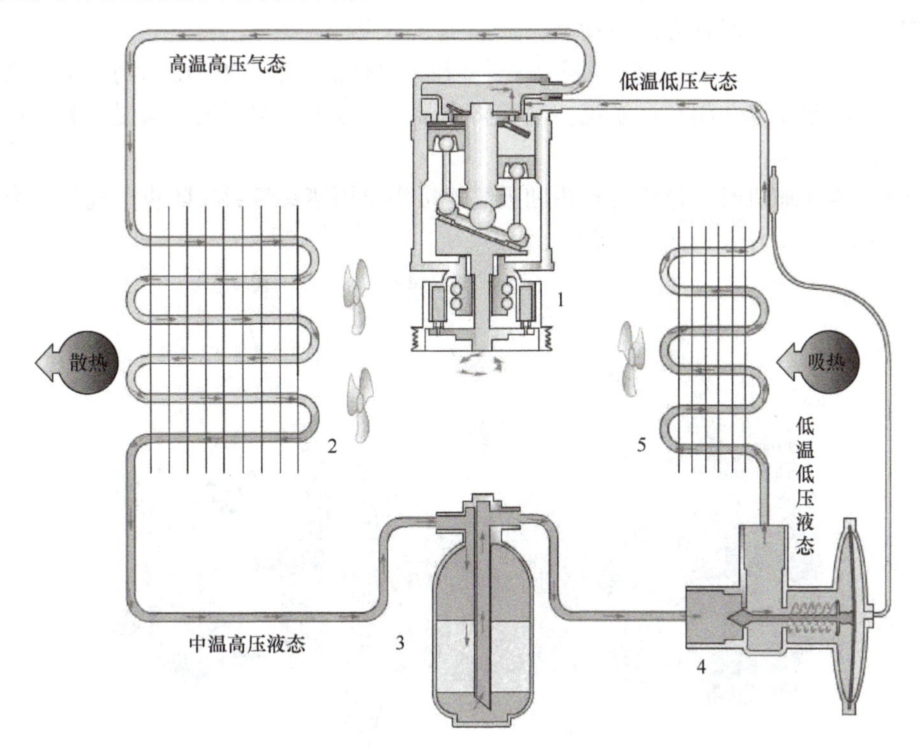

图 5-1　空调工作过程

1—压缩机　2—冷凝器　3—储液干燥器　4—膨胀阀　5—蒸发器

任务 5.1　压缩机的拆卸

☞ 学习目标

能正确描述压缩机的工作原理。

☞ 工作任务

一台装载机，用户反馈空调不制冷，经检查是压缩机损坏了，需要对压缩机进行拆卸。

☞ 相关知识

1. 压缩机作用

空调系统中的压缩机是系统的“心脏”部件，它使制冷剂可以在空调系统中反复使用，它由发动机曲轴通过传动带轮和传动带来驱动，将制冷剂压缩到高温高压状态。

空调系统中的压缩机大都采用往复活塞式压缩机，压缩机工作介质为 R134a 制冷剂，冷冻机油牌号为 PAG100。

2. 压缩机结构（图 5-2）

装载机空调系统的压缩机选用斜盘翘板式压缩机，它采用独特的斜盘翘板式驱动系统。

其结构精巧、体积小、安装运动零件布置在压缩机的纵轴周围，提高了机械效率。而且压缩机可以顺时针方向或逆时针方向转动。

电磁离合器是按需要驱动或停止压缩机的装置，例如，当车内温度达到预定的热敏感电阻温度时，它将起动或关闭空调系统。电磁离合器的工作由空调开关、水温开关、双重压力开关来控制。

警告：存放压缩机时，排气管接头和吸气管接头要用堵头密封，防止湿气与脏物进入压缩机。压缩机为铝质外壳，应轻拿轻放。

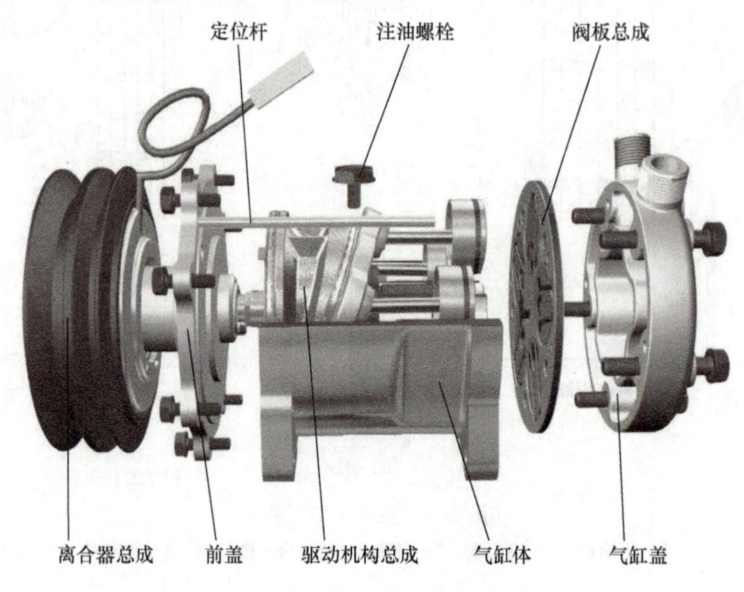

图 5-2 压缩机结构

☞ 任务实施

压缩机的拆卸：

1）停机，或关闭空调系统 1h 以上后，关闭负极开关，挂“禁止操作”警示牌。

2）将歧管表的仪表处高低压阀门关闭，如图 5-3 所示。

3）将高低压接头的阀门关闭。

4）打开后罩，将后罩打开至充分暴露压缩机及其上的管路。

5）将压缩机高低压充注口上的堵帽旋下来，将歧管表的高低压阀分别连接到高、低压管路上的充注口。连接好之后再将接头的阀门打开。注意，管路较粗，接口部分较细的为低压管；管路较细，但接口部分较粗的为高压管。此处接口为防错设计，错接无法接上。

6）缓慢松开歧管表上的低压阀门，使制冷剂慢慢喷出。注意控制阀的开度大小，流速过大容易导致冷冻油随制冷剂喷出来，控制好没有冷冻油喷出来即可（冷媒放出后为气态，冷冻机油为液态，可将冷媒出口对准一张白纸，观察是否有液滴喷出）。10min 后可缓慢打开高压阀门，同时仍然需要控制流速，避免冷冻机油流失。整个释放过程应持续13 ~ 30min。

7）用扳手松开压缩机管路的 M6 固定螺栓，取下管路，如图 5-4 所示。

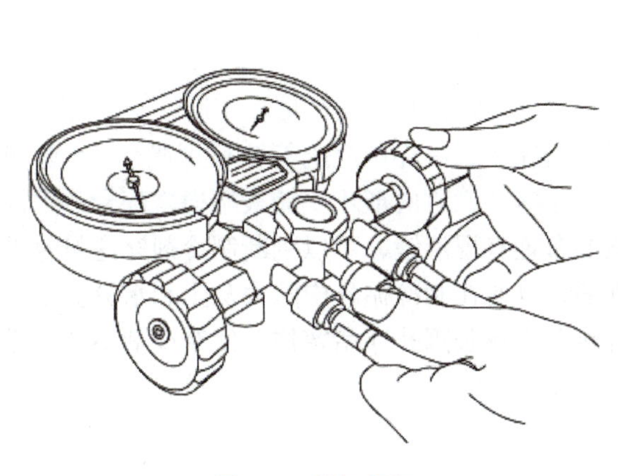

图5-3　关闭阀门

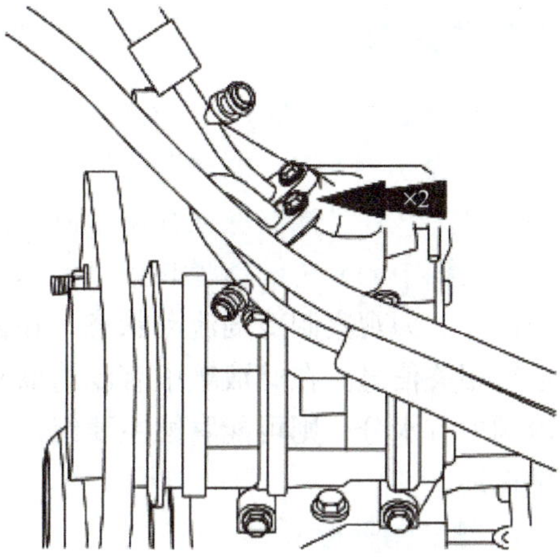

图5-4　管路及螺栓

8）用干净、无破损的塑料袋包住拆下的两根管路接头，并用胶布扎紧，防止空气和异物进入管路中。

9）拆开压缩机离合器线束插接件，松张紧轮，拆传动带，如图5-5所示。

10）拆下固定压缩机的4颗M8螺栓，即可取下压缩机，如图5-6所示。

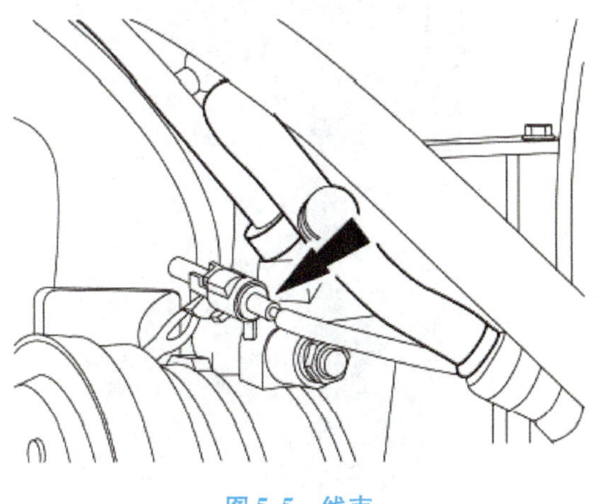

图5-5　线束

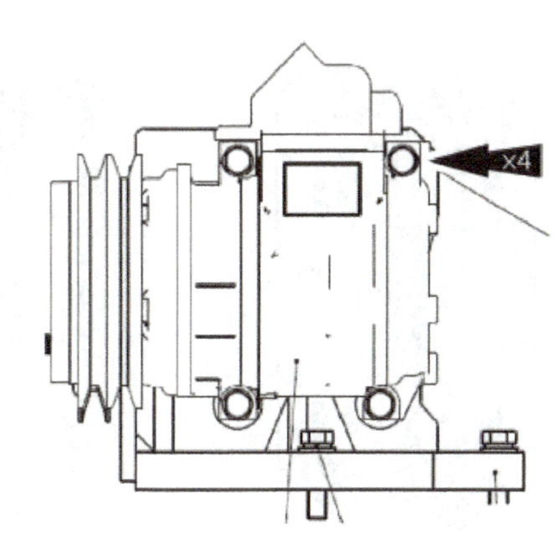

图5-6　螺栓位置

任务5.2　储液罐检测

☞ 学习目标

能按照规范要求对储液罐进行检测。

☞ 工作任务

一台装载机空调不制冷，需对储液罐进行检测。

☞ 相关知识

1. 储液罐的作用

储液罐（图 5-7）也称为储液干燥器，用于过滤和储存制冷剂，可在工况变化时储存多余的制冷剂，并且能够保证在微量制冷剂泄漏的情况下，制冷系统仍能够有效工作，更重要的是储液罐中的分子筛能吸收少量的水蒸汽，防止酸性物质的形成。储液罐通常配合热力膨胀阀使用。为观察制冷剂流动状态，在储液罐上布置有视液镜，可以判断制冷剂充注量及系统运行相关信息。在储液罐中有滤清器和干燥剂，以便在制冷循环中除去灰尘和水分。如果制冷剂中有水分，则可能腐蚀主要组件，并可能造成膨胀阀孔冰堵塞住，使制冷剂不能在系统中循环。

2. 储液罐外形

储液罐外形如图 5-8 所示。

注意：储液罐存放时，出口和入口要用堵头密封，防止湿气进入储液罐。储液罐出口和入口不能接反，否则制冷系统无法工作。

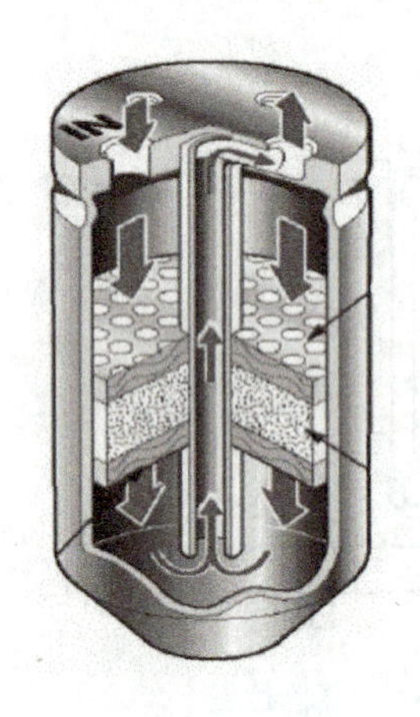

图 5-7　储液罐

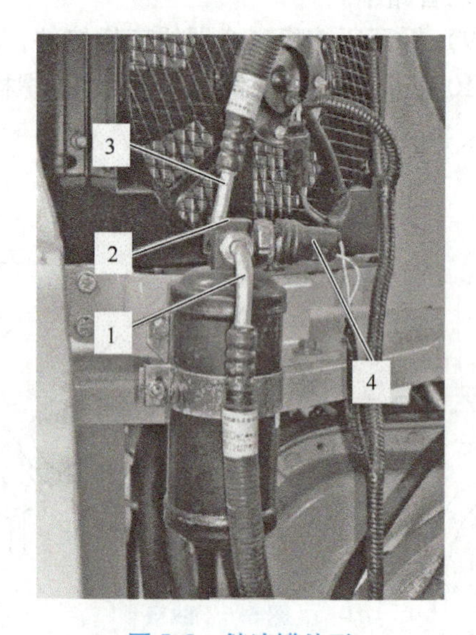

图 5-8　储液罐外形

1—出口　2—视液镜　3—入口　4—高低压开关

☞ 任务实施

储液罐的检测：

观察储液罐旁边的视液镜（图 5-9），并用手摸储液罐下端，如果储液罐下端是冷的而视液镜内起泡，说明储液罐内滤网堵塞。正常情况下储液罐各处的温度与冷凝器出口温度基本相同。

检查步骤：

1）在下列工况下进行检查，见表 5-1。

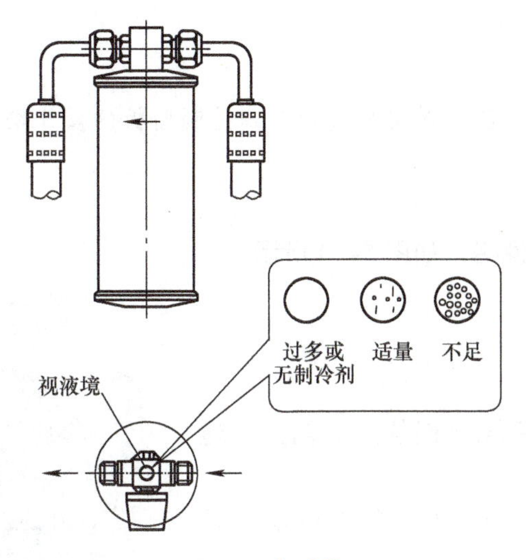

图5-9 视液镜

表5-1 工况检查

空调开关	接通
压缩机转速	1800r/min
鼓风机开关	最大
温度控制	最大冷却
车门	全开
进气阀的位置	再循环

2）用观察窗检查空调系统制冷剂的条件并将它与下述情况做比较，见表5-2。

表5-2 对比比较

序号	条件	原因	改进措施
1	通过观察窗可见到气泡	制冷剂不够	用检漏仪检查漏气
2	通过观察窗看不到气泡	制冷剂不够或过多	参见序号3、5或6
3	压缩机进气侧与输出侧无温差	无制冷剂	空调系统抽真空并重新填充，用检漏仪检查渗漏
4	压缩机进气侧与输出侧的温差很大	无制冷剂适量或过多	参见序号5或6
5	当空调停机时观察窗下面的制冷剂立即变透明	制冷剂过多	排出过多的制冷剂，把制冷剂调整到规定量
6	当空调停机时观察窗下面的制冷剂产生一些气泡，然后很快变成透明的，并保持下去	制冷剂填充适量	无需修理

任务5.3 操作空调控制面板

☞ 学习目标

能对空调控制面板上的各开关进行正确操作。

☞ 工作任务

一台新的装载机交机，需要对用户进行空调控制面板的操作培训。

☞ 相关知识

空调控制面板的操作说明，如图 5-10 所示。

☞ 任务实施

（1）制冷操作

1）发动机起动后，将风速档位开关调至合适档位。

2）将模式切换开关顺时针方向旋至最右端（制冷档）。

3）将制冷温度调节开关旋至“COOL”端（此时绿指示灯亮），制冷系统开始工作，冷气开始从风口送出。

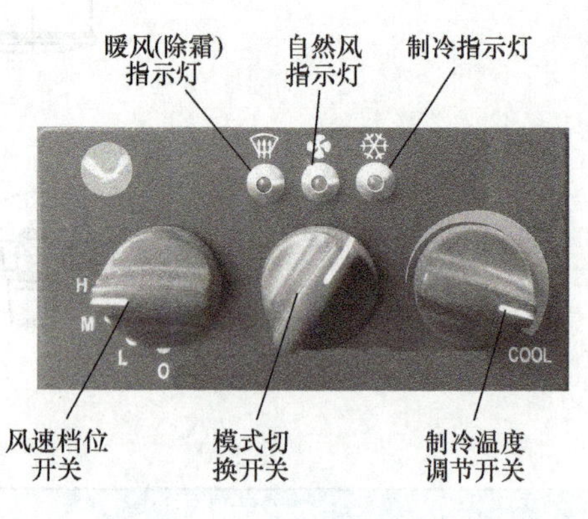

图 5-10　空调控制面板

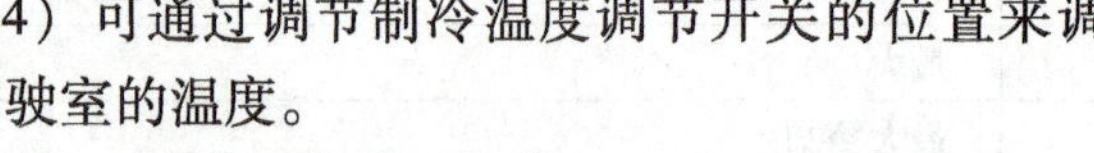

4）可通过调节制冷温度调节开关的位置来调节驾驶室的温度。

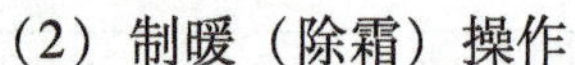

（2）制暖（除霜）操作

1）发动机起动一段时间后（冷却水温度高于 70℃），将风速档位开关调至合适的档位。

2）将模式切换开关逆时针方向旋至最左端（即暖风档，此时红色指示灯亮），系统开始工作，暖风开始从出风口送出。

（3）自然风操作　在春、秋凉爽季节时，当只想吹自然风并借此来调节驾驶室内的空气流动时，可打开风速开关至合适档位，并将模式转换开关调至中间位置（此时黄色指示灯亮）。

任务 5.4　冷凝器检测

☞ 学习目标

能对冷凝器进行故障检测。

☞ 工作任务

用户反馈某台装载机的空调不制冷，需要对冷凝器进行检测。

☞ 相关知识

1. 冷凝器的作用

冷凝器是热交换设备，其作用是将压缩机排出的高温高压的制冷剂气体的热量，通过发动机冷却风扇强制冷却散发到车外的空气中，使制冷剂气体在其中发生状态变化，成为液态制冷剂。冷凝器结构为管带式，散热效率比管片式高，一般安装在散热器附近。

2. 冷凝器外形

冷凝器外形如图5-11所示。

注意：冷凝器存放时，排气管接头和进气管接头要用堵头密封，防止湿气与脏物进入冷凝器。不要损伤冷凝器散热片，如发现散热片弯曲，可用旋具或手钳加以矫正。冷凝器导管及翅片外表不能有污垢、残渣，以免造成散热不良。

☞ 任务实施

冷凝器的检测：

分别用手触摸冷凝器上部靠近进气口部分和下部靠近出气口部分，如图5-12所示，正常情况下，进气口部分应该感觉有些烫手，出气口部分应该感觉接近环境气温。进气口附近温度不高，说明冷媒不足，或压缩机未正常工作；如果常温区域温度与高温区域没有明显差异，说明冷凝器散热不良，应清理冷凝器表面的污物。

图5-11 冷凝器外形

1—进气管接头 2—排气管接头

图5-12 冷凝器

任务5.5 传感器检测

☞ 学习目标

能正确描述空调系统各传感器的工作原理，并对传感器进行检测。

☞ 工作任务

一台装载机仪表上水温报警，需要对传感器进行检测。

☞ 相关知识

1. 传感器的作用

传感器是一种检测装置，能感受到被测量对象的信息，并能将感受到的信息，按一定规律变换成为电信号或其他所需形式的信息输出，以满足信息的传输、处理、存储、显示、记录和控制等要求。

2. 传感器的种类（图5-13）

装载机上使用的传感器主要有温度传感器、速度传感器、压力传感器、液位传感器等。

温度传感器主要有发动机水温传感器、变矩器油温传感器、液压油温传感器、环境温度传感器、发动机进气温度传感器，安装在不同的被测物体上，分别检测发动机水温、变矩器油温、液压油温、环境温度、发动机进气温度。

速度传感器主要有变速器转速输出传感器、发动机转速传感器。变速器转速传感器安装在变速器输出齿轮上，用于检测变速器转速，同时经处理后可转换成整机的行驶速度显示在仪表上。发动机转速传感器主要用于检测发动机的转速，安装在飞轮壳上。

压力传感器用于检测压力的变化，主要有制动压力传感器、发动机机油压力传感器、进气压力传感器等。

液位传感器有燃油油位传感器、冷却液水位传感器、机油液位传感器等，主要用于监测液位的变化。

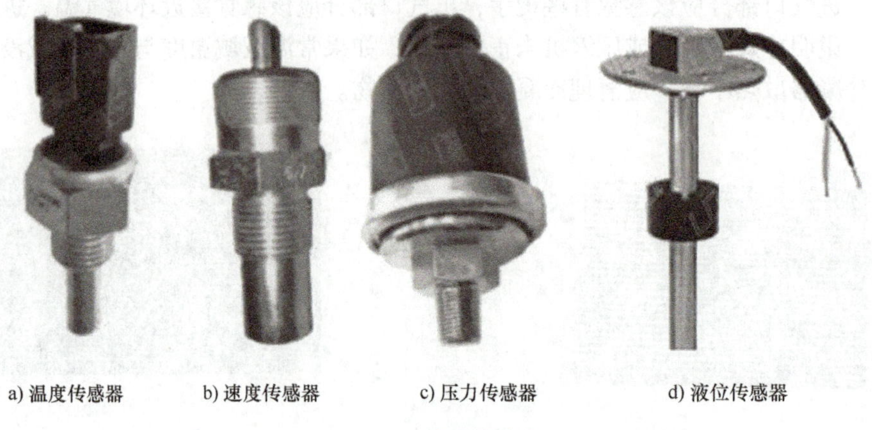

a) 温度传感器　　b) 速度传感器　　c) 压力传感器　　d) 液位传感器

图 5-13　传感器种类

☞ 任务实施

发动机温度传感器的检测：

1）将后车架线束中与温度传感器所连接的插接器拔出。

2）将温度传感器从整机上拆除。

3）将温度传感器测温探头置于液体中（水/变矩器油），给液体加温，根据以下参数对温度传感器性能进行判定，见表 5-3。

表 5-3　温度判定

温度/℃	电阻/Ω	误差（±%）	最小值	最大值
0	33650	8.8	30700	36600
25	10000	7	9304	10700
90	865.5	2.1	847.3	883.7

项目 6

装载机传动系统分析

装载机动力装置和行走装置（驱动轮）之间的传动部件总称为传动系统。

传动系统的作用是将动力装置输出的动力按需要传给驱动轮和其他机构（如工作液压泵、转向液压泵等），并解决动力装置功率输出特性和行走装置动力需求之间的各种矛盾。动力传递基本路线是：发动机飞轮→弹性板→变矩器→变速器→前后输出法兰→前后传动轴→前后驱动桥→轮胎。

（1）传动系统构成 装载机传动系统主要由变矩器、变速器、后桥传动轴、前桥传动轴、前驱动桥、后驱动桥、轮胎等组成。主要功能有：

1）降低转速，增大转矩。

2）实现装载机倒退行驶。

3）必要时中断传动。

4）差速作用。

（2）传动系统的分类 传动系统按结构和传动介质的不同可分为机械式传动、液压机械传动、全液压传动和电力式传动四种形式。

轮式装载机液压机械传动分类：

1）行星式液压机械传动系统。

2）定轴式液压机械传动系统。

（3）液压传动的概念 在传动系统中，以液体（矿物质油）为介质进行能量传递与控制的装置称为液压传动装置，简称液压传动。

任务 6.1 变矩器工作原理分析

☞ 学习目标

能正确描述变矩器的工作原理。

☞ 工作任务

用户想了解装载机变矩器，请在整机上指出变矩器位置，并为用户介绍变矩器结构原理。

☞ 相关知识

1. 变矩器的结构及原理

变矩器（图 6-1）由泵轮、涡轮、导轮组成一个液体流动循环的空间，通过变速泵不断将油液冲入变矩器内，当进入的油液从泵轮流入涡轮时，流动的方向与冲力的大小被改变，根据负载的大小产生不同的涡轮力矩，紧接着液流从涡轮流入导轮，再一次改变流动方向后流入泵轮，同时变矩器产生的热量通过循环油被带走；当涡轮转速为泵轮转速的80%时，变矩比为1，即涡轮转矩等于泵轮转矩，当涡轮不转动时，转矩比最大。

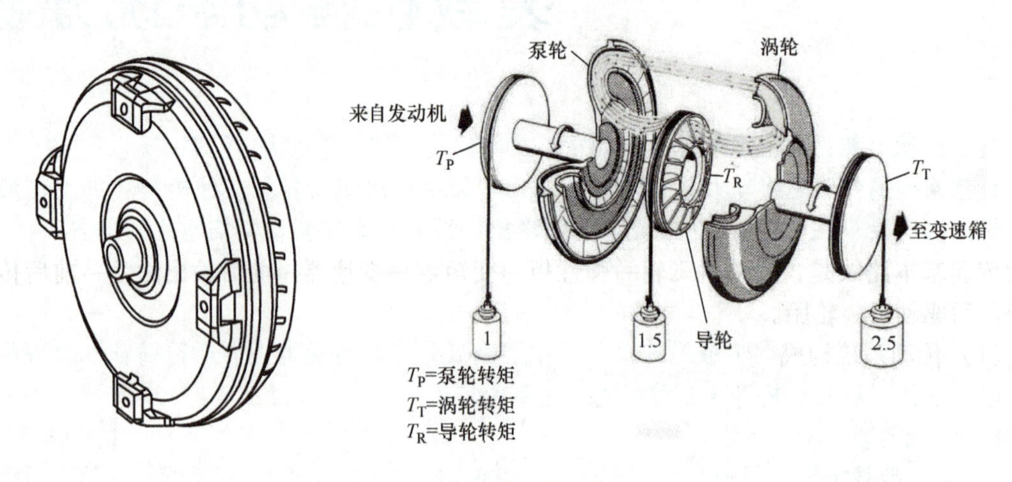

图6-1 变矩器结构

2. 变矩器的特点

1）变矩器可以自动调节输出的转速和转矩，使车辆根据路面状况和阻力大小自动改变速度和牵引力的大小以适应工况的变化。

2）变矩器高效区宽、效率高，可充分利用发动机功率，获得更好的经济性。

3）采用了以油为介质的液压传动，对车辆起到了缓冲、减振作用，提高了车辆的舒适性。

变矩器的缺点是传动效率低；被拖车时转向泵不工作，整机没转向（装有应急转向泵除外）；以油液作为传递介质，有泄漏的可能，污染环境。

任务实施

指出变矩器位置，并说明其结构原理。

任务6.2 变速器工作原理分析

学习目标

能正确描述变速器的工作原理。

工作任务

一台新装载机，在交车时，需要检查变速器的油位并确定变速器油是否合适。

☞ 相关知识

1. 变速器的工作原理

变速器可以分为行星式变速器及定轴式变速器。

BS305 变速器（图 6-2）为行星式变速器，采用双涡轮，四元件变矩器，包含泵轮、Ⅰ级涡轮、Ⅱ级涡轮、导轮，且变矩器可拆分，具有如下特点：

1）自动调节输出转矩和转速；自动实现低速重载和高速轻载的转换；变矩比大，高效区域宽。

2）以油液为传动介质，吸收和消除外来振动和冲击，保护柴油机和传动系统，当外载荷突然增大或不可克服时，发动机也不会熄火；大大减轻驾驶员操作的劳动强度，从而提高舒适性。

BS305 变速器通过机械 - 液压动力换档，可实现 2 前 1 后。因为变速器内部有超越离合器，可实现自动结合和分离（根据外负载的需要），所以变速器实际档位为 4 前 2 后。

2. 变速器特点

1）改变发动机和驱动轮之间的传动比，目的是改变机器行驶的速度和牵引力，满足机器作业和行驶的需要。

2）使机器能倒档行驶（挂倒档）。

3）具有动力切断功能，可自动切断传递给行走机构的动力。

4）降轴距，解决发动机输出和驱动桥输出不同轴的问题。

5）结构简单，维修成本低。

图 6-2　行星式变速器

3. 定轴式变速器——采埃孚（ZF）4WG - 200 变速器

如图 6-3 所示，ZF 系统主要用于装载机的传动件，为装载机提供了先进、安全、可靠的传动技术，使驾驶员可以更轻松、更便捷地操作车辆，极大地提高了装载机的工作效率，ZF 技术的应用，全方位体现出现代科技给产品带来的优势。

目前，大部分装载机都采用采埃孚（ZF）4WG - 200 动力换档变速箱，它的变速操纵系统为微电脑集成控制的电液换档，驾驶员通过按钮即可完成变速器操作，并配装先导液压操纵系统，大大地减轻了驾驶员劳动强度，提高了工作效率。

1）装载机配备的采埃孚（ZF）4WG - 200 变速器为定轴式，采用半自动电液控制，确保换档迅速、准确、平稳；液压变矩器可使车辆根据路面状况和阻力自动改变速度及牵引力，适应多种不同的工况。

2）采用油液为介质的液压传动，使车辆起到缓冲、减振作用，提高了车辆的舒适性；采用三元件变矩器，将发动机的输出动力传递到变速器上。

图 6-3　定轴式变速器

4. （ZF）4WG200 动力换档变速器的优点

具有 4 个前进档，3 个倒退档，独特的 KD 按钮（强制换档功能，仅用于 1、2 档）工作效率高，维修率低，超长的工作寿命耗油低，噪声低。

5. 电控系统的组成

EST－17T 变速器换档电控盒，通过选择不同的电控盒可实现全自动或半自动功能；WG200 系列动力换档变速器；DW－2 旋转式档位选择器；整车电路；变速器控制换档操纵电缆；输出速度传感器电缆。

6. 电控系统的主要功能

1）空档/起动联锁保护功能，确保车辆操作安全性；动力切断功能（制动脱档功能）只在 1、2 档（低速档）起作用，有效保护车辆传动系统。

2）专用强制换低档功能（KD 档功能），提升工作效率；起步限速功能；直接换向功能；系统保护功能，确保在工作中电控系统发生故障时电控盒自行转换成空档并锁止所有信号输出，避免引起严重后果。

7. 采埃孚 WG－200 变速器操纵系统的组成

变速器操纵系统主要由电控盒、档位选择器、速度传感器、电磁阀、操纵阀及线缆等组成；电控盒接收到档位选择器发出的档位信号后，转换成系统内设定的程序向电磁阀发出指令，电磁阀根据信号打开相应的油路，使不同的两组离合器结合，实现档位并向前后驱动桥输出转矩驱动车辆行走。

注意：在发动机起动后，不得关闭电锁，否则可能会造成半自动换档系统及整车线束和仪表的损坏。

☞ 任务实施

检查变速箱油位：

发动机怠速运转，油温应在正常工作温度；当油温为 40℃时，油位应在油尺中间刻度线和下刻度线之间；当油温为 80℃时，油位应在油尺中间刻度线和上刻度线之间。

任务 6.3　驱动桥工作原理分析

☞ 学习目标

1）能够向客户介绍驱动桥结构和工作原理。

2）能够向客户推荐合适的驱动桥。

☞ 工作任务

客户来购买机器，如何更好地向客户介绍驱动桥，以体现专业水准？请了解驱动桥的相关知识，以便给客户推荐合适的产品。

☞ 相关知识

1. 驱动桥的作用

驱动桥是指变速器或传动轴之后，驱动轮（轮胎轮辋）之前的所有传动机构的总称。功能如下：

1）承载。承载机器负荷重量。

2）驱动。吸收变速器功率输入，通过传动减速放大输入转矩，驱动机器作业。

3）转向。差速器提供驱动桥左右轮胎灵敏差速功能需求，实现机器转向灵活性。

4）制动。装在驱动桥上的制动器是机器行车制动的执行元件。

通过主传动锥齿轮改变传力方向；通过主传动和终传动（轮边减速器）将变速器输出轴的转速降低，转矩增大；通过差速器解决左右差速问题；通过差速器将动力分传给驱动轮；通过制动器使装载机实现减速或停车。

2. 驱动桥分类

驱动桥按照制动器的结构形式的不同可分为两类，一类是干式外置钳盘式制动器的驱动桥，即干式桥（图6-4）；另一类是制动器在驱动桥壳体的内部，浸在油里面的驱动桥，即湿式桥（图6-5）。

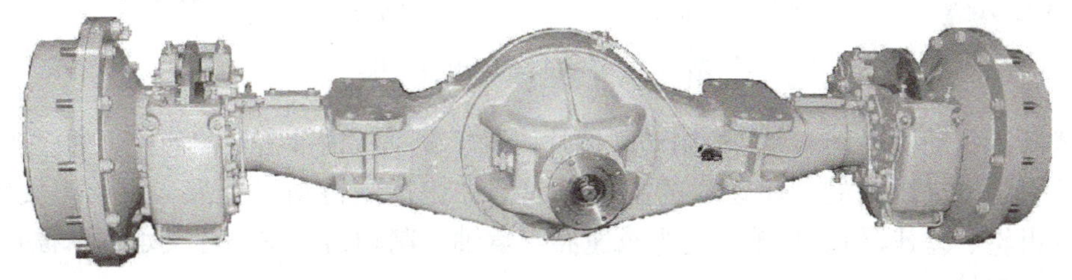

图6-4　干式桥（前桥）

图6-5　湿式桥（前桥）

驱动桥按照安装部位的不同可分为两类，一类是与前车架刚性连接，即前桥；另一类是与后车架摆动连接，即后桥，如图6-6所示。

3. 驱动桥的结构及工作原理

干式制动驱动桥（图6-7）的制动盘和摩擦片露在空气中，制动盘与摩擦块之间是干摩擦。干式制动驱动桥的主要组成部件包括主传动（托架、主动螺旋锥齿轮、大螺旋锥齿轮、差速器）、半轴、轮边减速器支承轴、夹钳、制动盘、制动蹄片、轮边减速器、桥壳等。

图 6-6　后桥

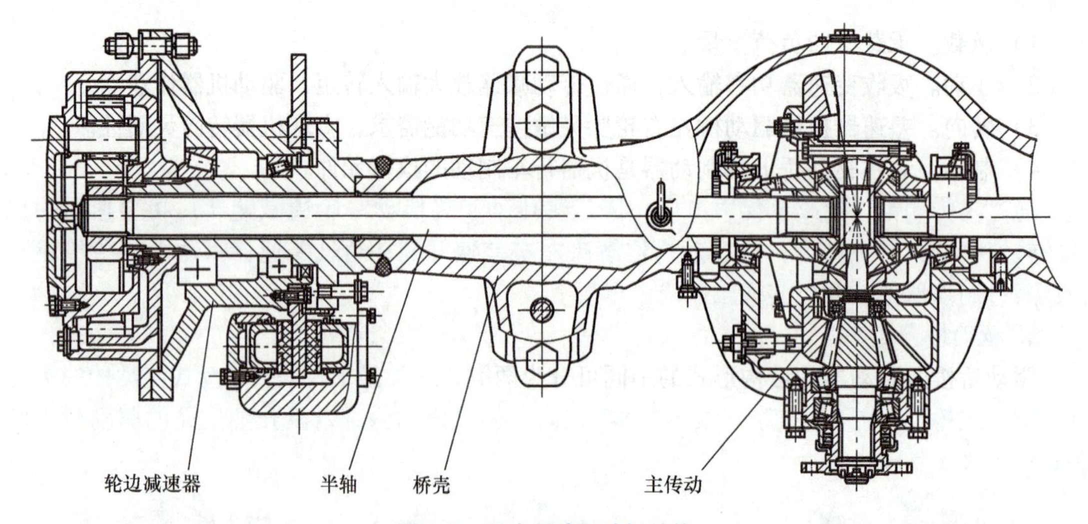

图 6-7　干式制动驱动桥结构

湿式制动驱动桥（图 6-8）的制动摩擦片内置于驱动桥内部密封的油中，通过多片摩擦片的摩擦来制动。湿式制动驱动桥的主要组成部件包括主传动（托架、主动螺旋锥齿轮、大螺旋锥齿轮、差速器）、半轴、轮边减速器支承轴、制动器、摩擦片支承、摩擦片、轮边减速器、桥壳等。

驱动桥传动路线：

输入法兰→主动螺旋锥齿轮→大螺旋锥齿轮→差速器壳体→十字轴→行星锥齿轮→半轴齿轮 →半轴→太阳轮轴→行星齿轮→内齿圈→行星轮架→轮毂/轮辋/轮胎，如图 6-9 所示。

☞ 任务实施

编写一份介绍装载机驱动桥的 PPT，需要包含以下内容：

1）驱动桥的作用。

2）驱动桥的构成。

3）驱动桥的分类。

4）驱动桥的动力传递路径。

5）驱动桥减速增扭的主要部件。

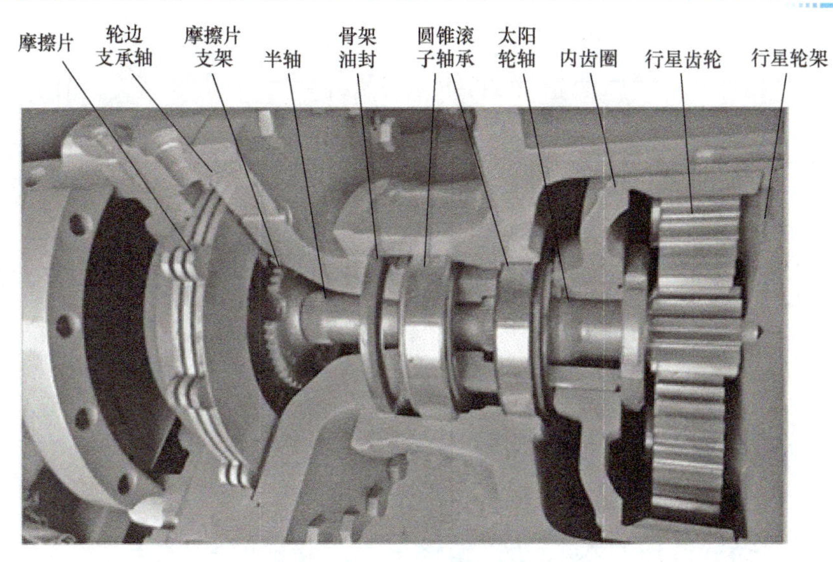

图 6-8　湿式制动驱动桥结构

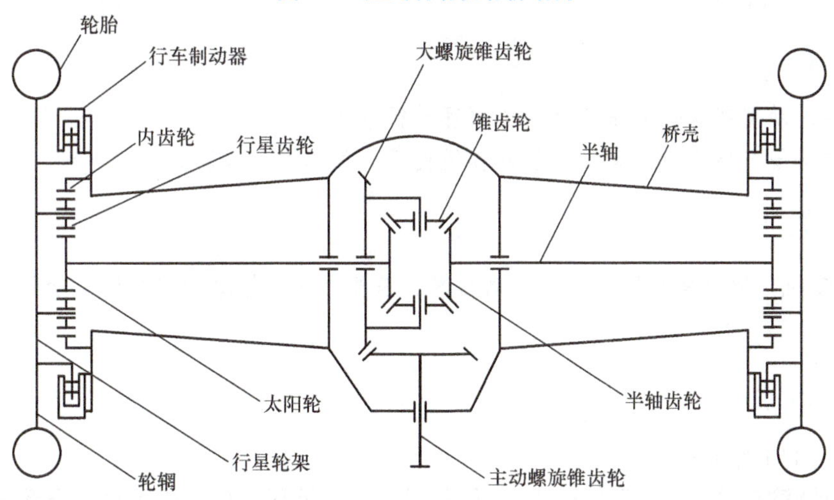

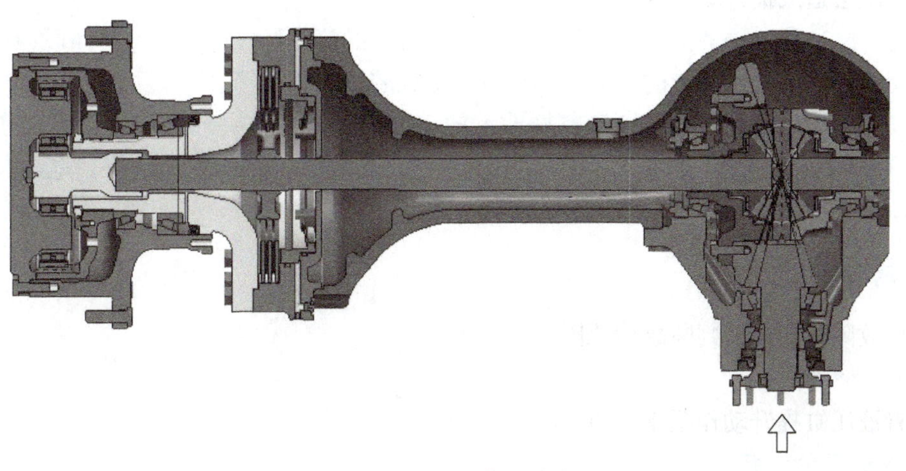

图 6-9　驱动桥传动路线

项目 7

装载机故障诊断与排除

任务 7.1　液压系统故障诊断与排除

子任务 7.1.1　动臂液压缸提升动作缓慢无力故障诊断与排除

☞ 学习目标

1）能够对动臂液压缸液压油路工作原理进行分析。

2）能够按规范要求对动臂液压缸动作缓慢无力故障进行检测。

☞ 工作任务

一台装载机，用户反馈提升无力、动作慢，需要诊断与排除动臂液压缸动作缓慢无力的故障。

☞ 相关知识

动臂提升的工作原理：

装载机动臂液压缸活塞杆伸出时，动臂提升。如果液压元件发生了故障导致液压系统压力变小，就会造成提升缓慢或无力。

动臂下降，先导手柄前推，工作泵与转向泵合流的流量经过分配阀的转斗联中位和动臂联的下降位进入动臂液压缸的小腔，动臂液压缸大腔回液压油箱。

如图 7-1 所示，动臂液压缸活塞杆伸出的主进油路为：P1→转斗滑阀中位→负载单向阀→动臂滑阀右位→铲斗液压缸大腔；主回油路为动臂液压缸小腔→动臂滑阀右位→过滤器→液压油箱。

☞ 任务实施

采用故障树法分析判断故障原因。

1. 故障现象

动臂液压缸提升动作缓慢无力。

2. 故障原因分析

故障树如图 7-2 所示。根据故障现象，初步判断故障发生在动臂液压缸活塞杆伸出的主

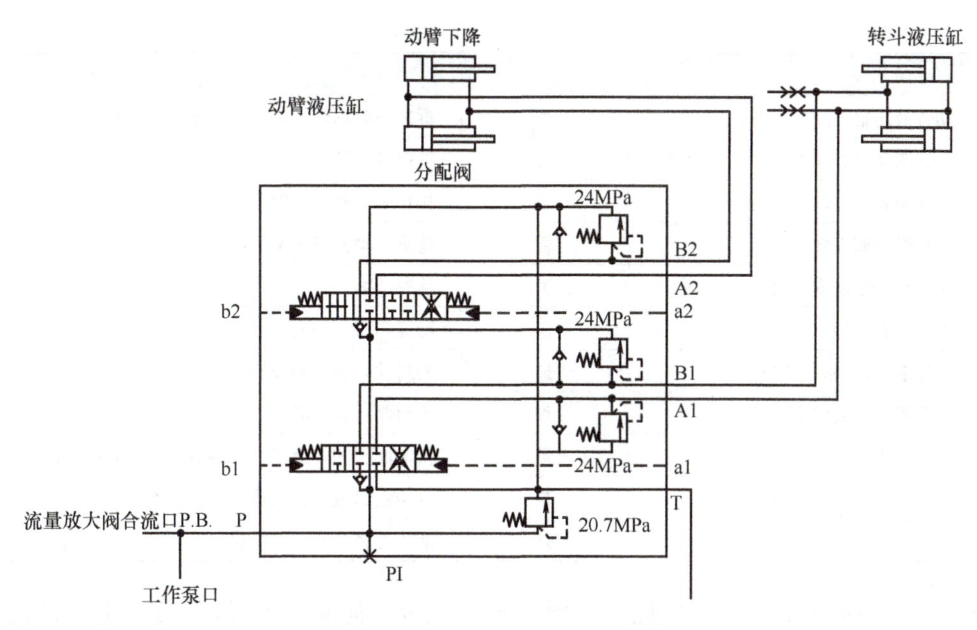

图7-1 动臂提升的工作原理

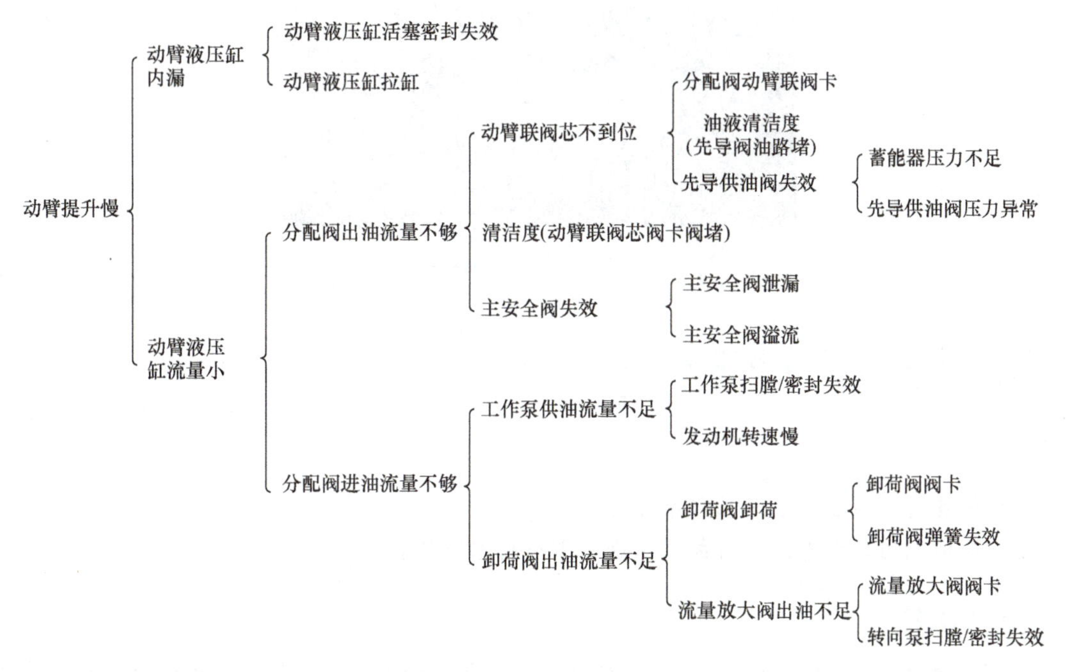

图7-2 故障树

进油路上，可能原因是油路上的元件发生泄漏，或者是动力元件损坏导致，可从压力、流量判断。

3. 故障排除方法

见表7-1。

表 7-1 故障排除方法

序号	引起失效的可能原因描述	是否可能原因	判断和说明
1	动臂液压缸内漏	否	拆掉小腔钢管，基本无油流出
2	分配阀动臂联阀卡或阀芯不到位	否	拆解动臂联阀芯无卡滞
3	先导阀弹簧不回位	否	操作先导阀无异常
4	先导供油阀失效	否	熄火，动臂可正常下降
5	主安全阀泄漏或溢流	否	检查压力
6	发动机转速慢	否	转速 2230r/min，符合要求
7	工作泵扫膛/密封失效	是	密封件碎裂，侧板有缺口
8	卸荷阀阀卡或弹簧失效	否	拆解阀芯无卡滞
9	流量放大阀阀卡	是	换向阀杆的油道里，发现有骨架油封断裂的碎片堵塞
10	转向泵扫膛/密封失效	是	骨架油封破碎，泵体发烫，侧板烧伤
11	液压油污染，清洁度不达标	是	液压油箱内有大量铁屑、泵的破碎骨架油封

通过对元件拆检，发现侧板烧伤（图 7-3），造成内泄漏，从而影响了流量，故此更换泵后故障得到解决。

图 7-3 侧板烧伤

子任务 7.1.2 转向沉重故障诊断与排除

☞ 学习目标

1）能够对转向液压油路工作原理进行分析。

2）能够利用鱼刺图法分析故障原因和叙述排除方法。

☞ 工作任务

一台装载机，用户反映使用过程中转向沉重、驾驶费力，请利用鱼刺图法诊断与排除转向沉重故障。

☞ 相关知识

转向液压系统采用流量放大系统，系统油路由控制油路与主油路组成。所谓流量放大，是指通过全液压转向器以及流量放大阀，保证控制油路的流量变化与主油路中进入转向缸的流量变化具有一定的比例，达到低压小流量控制高压大流量的目的。驾驶员操作平稳轻便，系统功率利用充分，可靠性好。

转向器为闭芯无反应型，方向盘不转动时中位断开。此时，流量放大阀主阀杆在复位弹簧作用下保持在中位，转向泵与转向液压缸的油路被断开，主油路经过流量放大阀中的流量控制阀卸荷回油箱。转动方向盘时，转向器排出的油与方向盘的转速成正比，先导油进入流量放大阀后，作用在流量放大阀的主阀杆端，控制主阀杆的位移，通过控制开口的大小，从而控制进入转向液压缸的流量，由于流量放大阀采用了压力补偿，因而进入转向液压缸的流量与负载基本无关，只与阀杆上开口大小有关。停止转向后，进入流量放大阀主阀杆一端的先导压力油通过节流小孔与另一端接通回油箱，阀杆两端油压趋于平衡，在复位弹簧的作用下，阀杆回复到中位，从而切断主油路，装载机停止转向。通过方向盘的连续转动与反馈作用，可保证装载机的转向角度。系统的反馈作用是通过转向器和流量放大阀共同完成的。流量放大阀回油一部分通过节流孔回油箱，一部分经散热器回油箱。

通过方向盘的连续转动与反馈作用，可保证装载机的转向角度。系统的反馈作用是通过转向器和流量放大阀共同完成的。

转向液压系统测试点如图 7-4 所示。

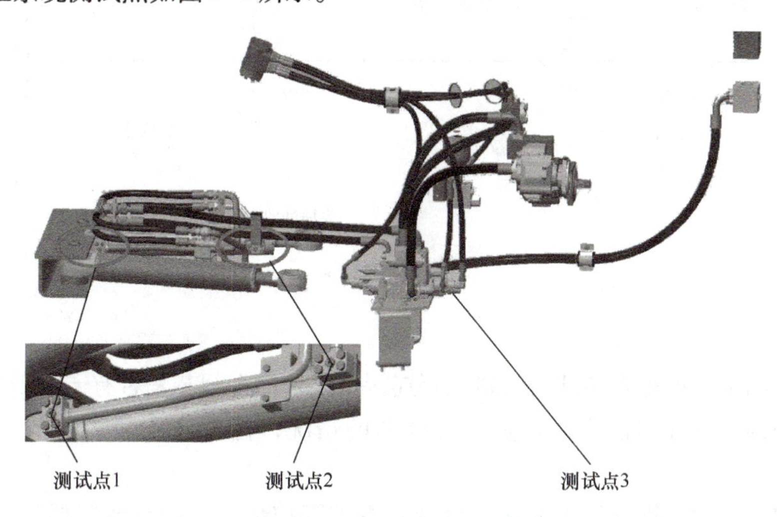

图 7-4　转向液压系统测试点

转向液压系统压力测试点说明见表 7-2。

表 7-2　转向液压系统压力测试点说明

测试点	测试点说明	理论压力值/MPa	接口尺寸
测试点 1	转向液压缸大腔压力	机型给出的值	M14 ×1.5 –6H
测试点 2	转向液压缸小腔压力	机型给出的值	M14 ×1.5 –6H
测试点 3	流量放大阀进口（转向泵出口）压力	机型给出的值	M18 ×1.5 –6g 测压接头

鱼刺图（因果分析）法是以多级箭头线的形式表达故障原因，形成因果关系，将故障的原因以主、次次序分别表示出来，尽可能细化。

（1）箭头

1）主干箭头线：箭头指向右侧，在箭头前端列明故障现象或故障分析的主题。

2）次干箭头线：箭头指向主干箭头线，列明故障主要原因（大原因）。

3）支箭头线：箭头指向次干箭头线，列明故障次要原因（次原因）。

4）次支箭头：箭头指向支箭头线，列明故障下一级原因（小原因）。

（2）故障原因　在除了主干箭头线以外的其他箭头线末端写上故障原因，前一级的故障原因包含下一级的故障原因。

☞ 任务实施

1. 原因分析

用鱼刺图将故障原因表达出来，如图7-5所示。

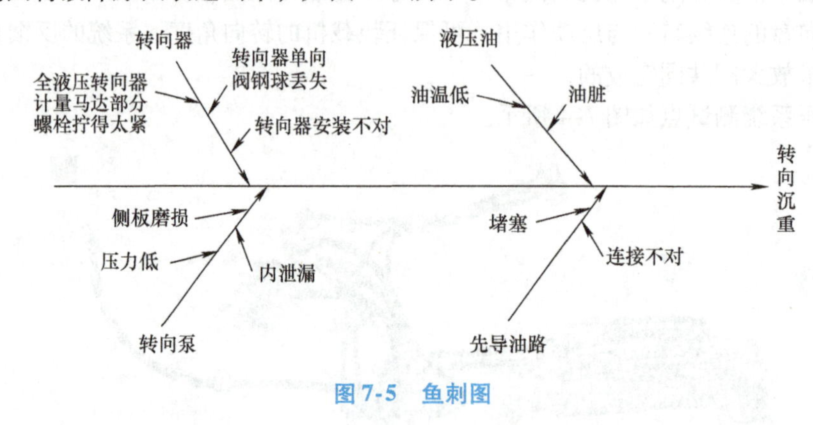

图7-5　鱼刺图

2. 故障排除

（1）先导油路　检查先导油路管路是否连接错误，优先阀芯是否有卡滞现象。

（2）液压油　打开回油室，检查回油滤芯的情况，检查液压油的清洁度以及油温的情况。

（3）转向器　检查钢球是否存在，钢球是否被脏物垫住，检查转向器的花键安装是否正确。

（4）转向泵　检查泵口压力是否达到系统所需压力，触摸泵壳体是否发烫，拆开转向泵检查泵的侧板是否磨损。

子任务7.1.3　制动踏板制动力不足故障诊断与排除

☞ 学习目标

1）能够对制动系统工作原理进行分析。

2）能够利用列表法分析制动踏板制动力不足故障的原因并叙述排除方法。

☞ 工作任务

一台气制动装载机，用户反馈制动软，请诊断并排除制动力不足故障。

☞ 相关知识

1. CLG856H 干式制动系统工作原理

行车制动系统的工作原理如图7-6所示。空气压缩机由发动机驱动将空气转化成压缩空气，压缩空气经过组合阀后存储在气罐中。当气罐内的压力达到制动系统最高工作压力时（一般为0.78MPa），组合阀就打开卸荷口，将空压机输出的压缩空气直接排向大气。当气罐内的压力低于制动系统最低工作压力时（一般为0.71MPa），组合阀就打开通向气罐的出口，关闭卸荷口，补充气罐内的压缩空气，直到气罐内压力达到最高工作压力。在制动时，操作者踩下制动踏板后，压缩空气通过制动阀进入加力器的气缸，推动加力器活塞，活塞将加力器液压缸的制动液送进轮边制动器的夹钳分泵，挤压摩擦片实现制动。

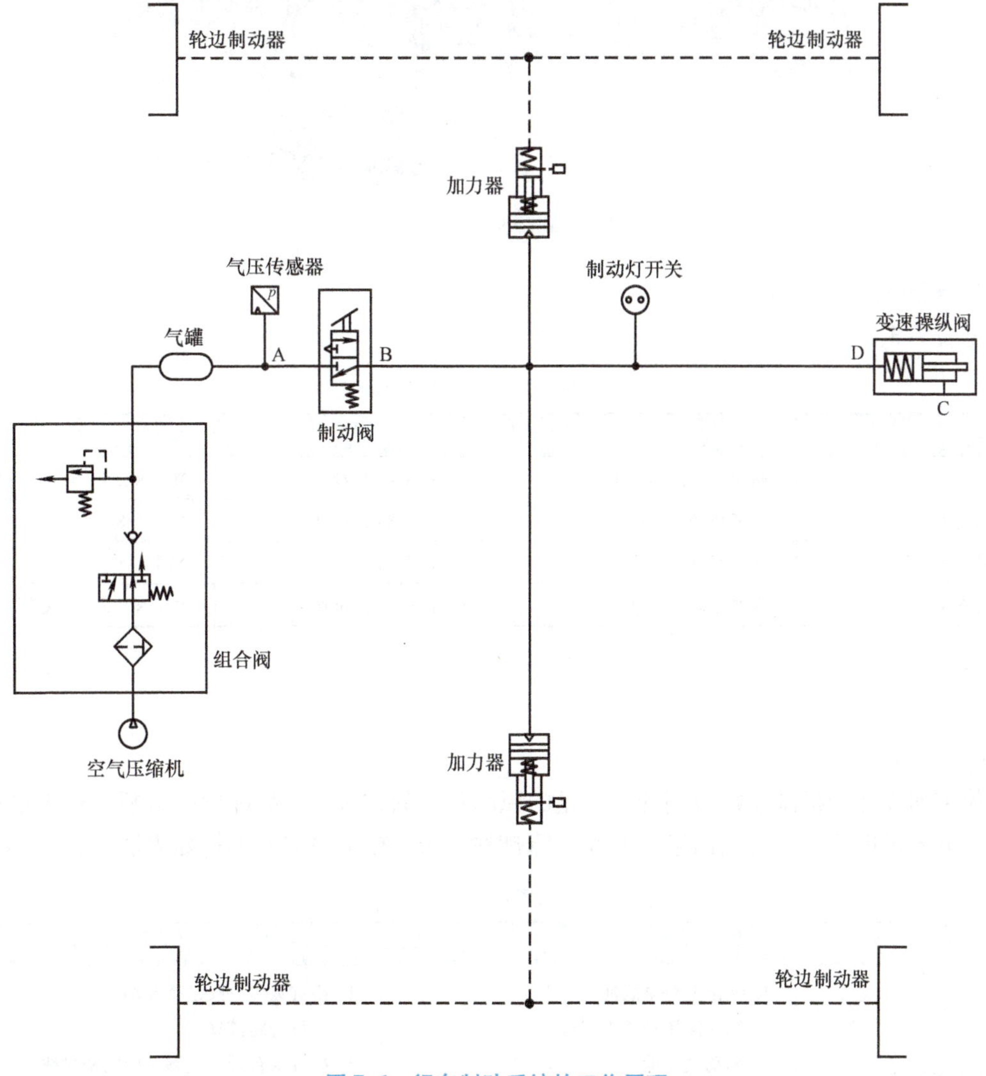

图7-6　行车制动系统的工作原理

2. 制动系统压力测试

制动系统压力测试点如图7-7所示。

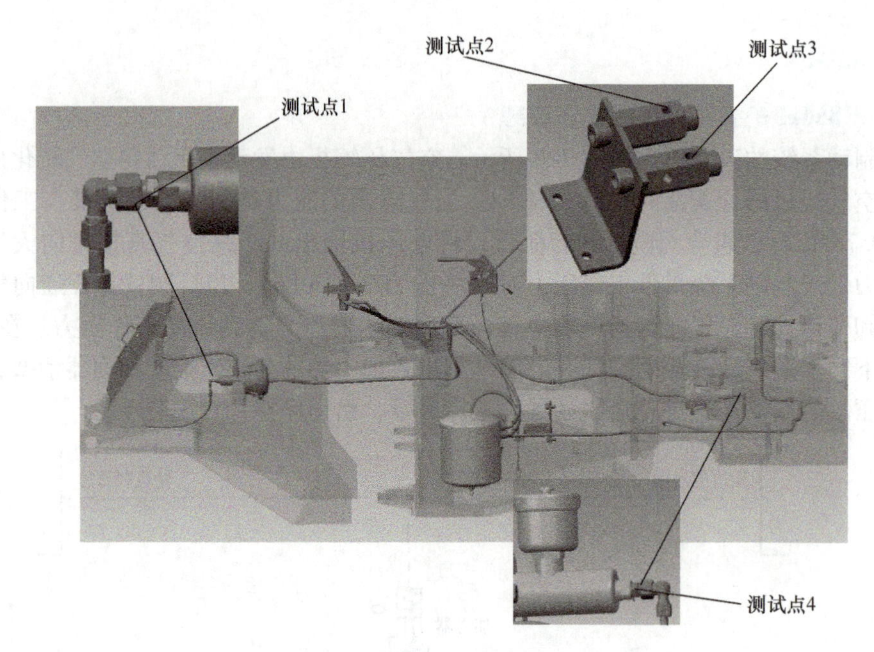

图 7-7 制动系统压力测试点

3. 制动系统压力测试点说明

见表 7-3。

表 7-3 制动系统压力测试点说明

测试点	测试点说明	理论压力值/MPa	接口尺寸
测试点 1	前加力器出口油液压力	机型给出的值	M18 ×1.5 –6g 测压接头
测试点 2	行车制动系统压力	机型给出的值	NPT 1/8
测试点 3	加力器进口压力	机型给出的值	NPT 1/8
测试点 4	后加力器出口油液压力	机型给出的值	M18 ×1.5 –6g 测压接头

任务实施

采用列表法分析故障

使用列表法分析制动踏板制动力不足的原因及排除方法，在故障原因分析一栏中尽量将导致故障的可能原因分类列出来，并在故障排除方法一栏填写对应的排除方法，见表 7-4。

表 7-4 列表法分析故障实例

故障现象	故障原因分析	故障排除方法
制动踏板制动力不足	1. 夹钳上分泵漏油 2. 制动液压管路中有气 3. 制动气压低 4. 加力器皮碗磨损 5. 制动片上沾有油 6. 制动片已到磨损极限	1. 检查夹钳分泵是否漏油 2. 进行管路排气 3. 检查组合阀、气罐及管路密封性 4. 更换皮碗 5. 检查或更换轮毂油封 6. 更换制动片

任务7.2 电气系统故障诊断与排除

子任务7.2.1 整车无电故障诊断与排除

☞ 学习目标

能够对整车电气系统故障进行诊断并排除。

☞ 工作任务

一台装载机，用户反馈整车无电，请对整车无电的故障进行诊断与排除。

☞ 相关知识

1. 电源系统工作原理

图7-8所示为CLG856H装载机的主电路电源系统工作原理。

负极开关闭合后，蓄电池的电流一路通过50A电器集中控制盒熔断器、100号导线，到达非过电锁电源总线熔断器。此时可正常工作的电器部件为壁灯、旋转警示灯、驻车灯、喇叭等；同时电锁熔断器通过111号线供电给电锁电源端（B1－B2）。另一路通过60A主电源熔断器、176号导线，到达电源接触器。

电锁旋转至ON档后，此时电锁B1－B2端便与M端接通，111号导线与120号导线接通，电流通过120号线、电源接触器的线圈、212号线至地。故电源接触器触点开关闭合，176号导线便与190号导线接通，过电锁电源熔断器得电。此时除倒车警报与空调功能模块外，其他电器部件均可正常使用。

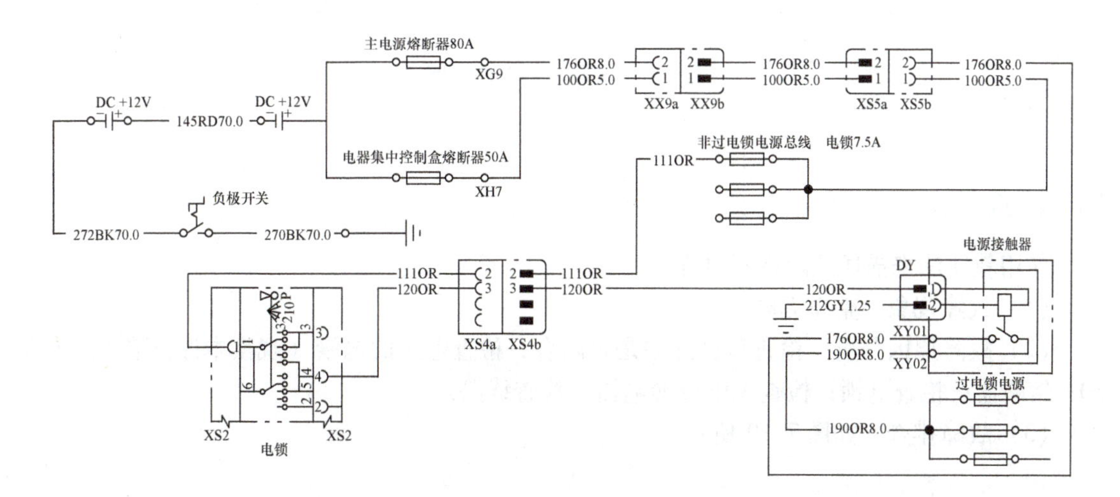

图7-8 主电路电源系统工作原理

2. 起动系统工作原理（图 7-9）

将电锁旋转至 START 挡，B1－B2 端、M 端、S 端互相接通，111 号导线、120 号导线、453 号导线接通。如果换档手柄挂在空档，则变速控制器通过 584 号导线输出 24V 的电压，通过档位/起动联锁继电器的线圈至地，线圈得电后，档位/起动联锁继电器触点闭合，453 号线与 454 号线接通。另一方面发动机 ECM 通过 982 号线输出 24V 的电压，经过起动保护继电器的线圈后通过 983 号线回流至 ECM 内部接地，线圈得电后，起动保护继电器触点闭合，454 号线与 460 号线接通。电流通过 460 号导线、起动继电器线圈至地，使起动继电器触点闭合，电流流入起动电动机的电磁开关线圈，起动电动机开始工作。整机起动后，发动机 ECM 的起动保护功能启用。

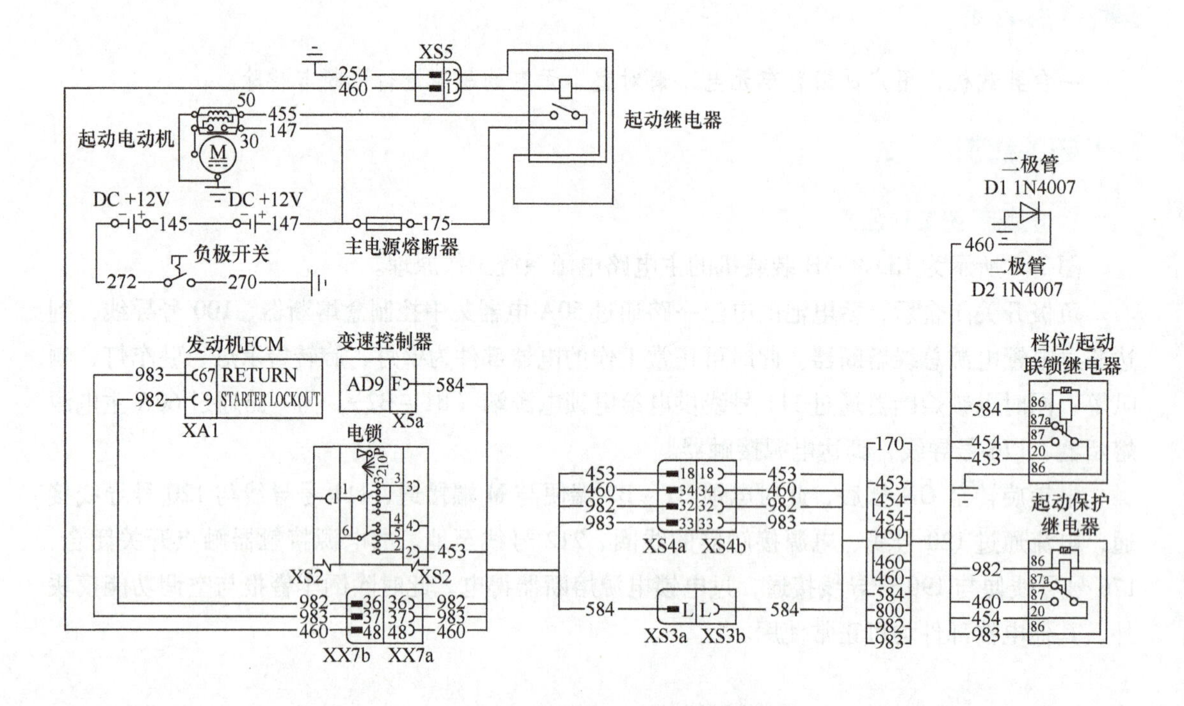

图 7-9 起动系统工作原理

☞ 任务实施

使用叙述法来描述故障诊断过程。

（1）故障现象　整车无电。

（2）故障原因分析　检查负极开关是否闭合；检查电瓶的电量（观察电眼颜色）；检查 60A 熔断器；检查电锁；检查主电源继电器；检查线路。

（3）故障排查　如图 7-10 所示。

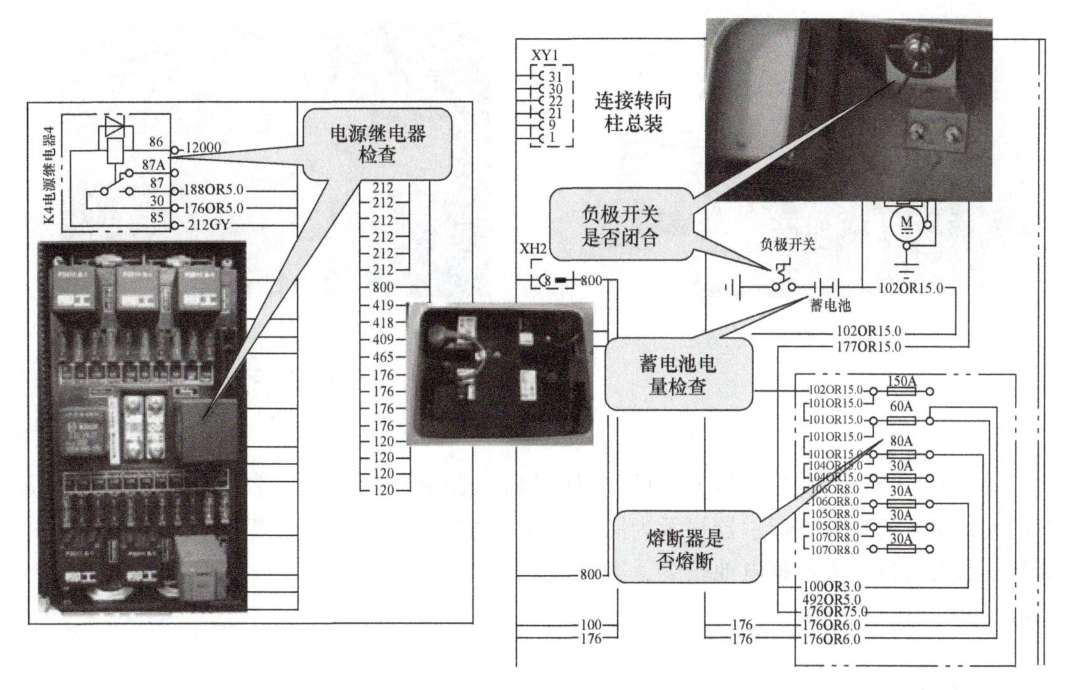

图7-10　故障排查

子任务7.2.2　蓄电池无法充电故障诊断与排除

☞ 学习目标

能够对蓄电池无法充电的故障进行诊断与排除。

☞ 工作任务

一台装载机，用户反馈蓄电池无法充电，请对蓄电池无法充电的故障进行诊断与排除。

☞ 相关知识

1. 蓄电池的类型及规格

装载机一般采用两个标称电压为12V的蓄电池串联。第一个蓄电池的负极经电源总开关搭铁，正极接到第二个蓄电池的负极；第二个蓄电池的正极接至起动电动机的30端子。

2. 蓄电池的作用及组成

见本书子任务2.5.1中的蓄电池相关内容。

3. 装载机蓄电池的检查

1）蓄电池位于整机尾部左侧电瓶箱内，拧开4颗盖板螺栓即可看到蓄电池，如图7-11所示。

2）检查蓄电池的压板螺母、蓄电池端子以及电缆接头是否松动。如果松动请将压板螺母、蓄电池端子以及电缆接头拧紧，如图7-12所示。

图 7-11 蓄电池位置

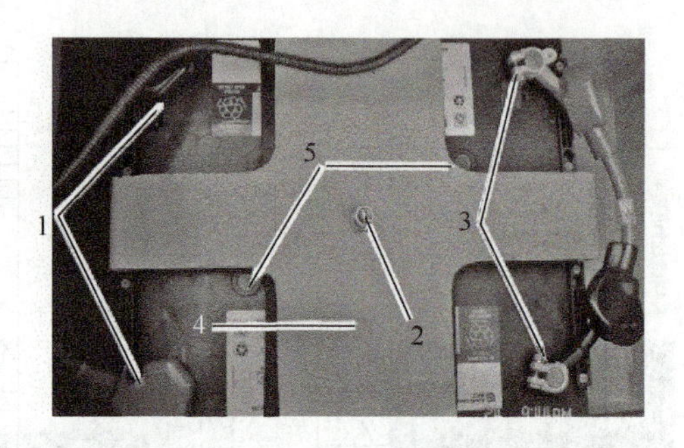

图 7-12 检查蓄电池

1—蓄电池端子 2—压板螺母

3—蓄电池端子（拔开护套的） 4—压板 5—指示器

3）检查蓄电池状态指示器（电眼）。如果指示器显示为绿色，表示蓄电池电量充足，可以正常起动车辆；如果指示器显示为黑色，表示蓄电池电量不足，蓄电池需补充电；如果指示器显示为白色，表示蓄电池报废，需更换。

4）关闭蓄电池箱盖。

☞ 任务实施

使用叙述法来描述故障诊断过程。

（1）故障现象 蓄电池无法充电。

（2）故障原因 线路检查；蓄电池电量检查；发电机 B + 端电压检测；熔断器是否熔断。

（3）故障排查 如图 7-13 所示。

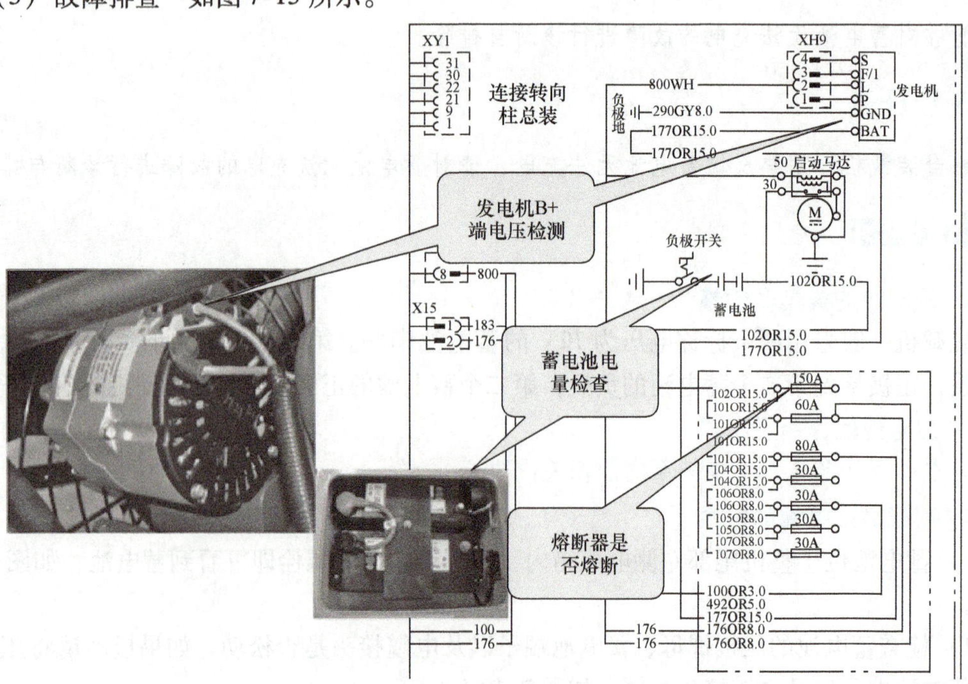

图 7-13 故障排查

子任务7.2.3　起动电动机不转故障诊断与排除

☞ 学习目标

能够对起动电动机不转的故障进行诊断与排除。

☞ 工作任务

一台装载机起动电动机不转，请对其故障进行诊断与排除。

☞ 相关知识

1. 起动过程

图7-14a所示为起动前及起动后起动电动机与起动线路的状态，图7-14b所示为起动过程中起动电动机与起动线路的状态。

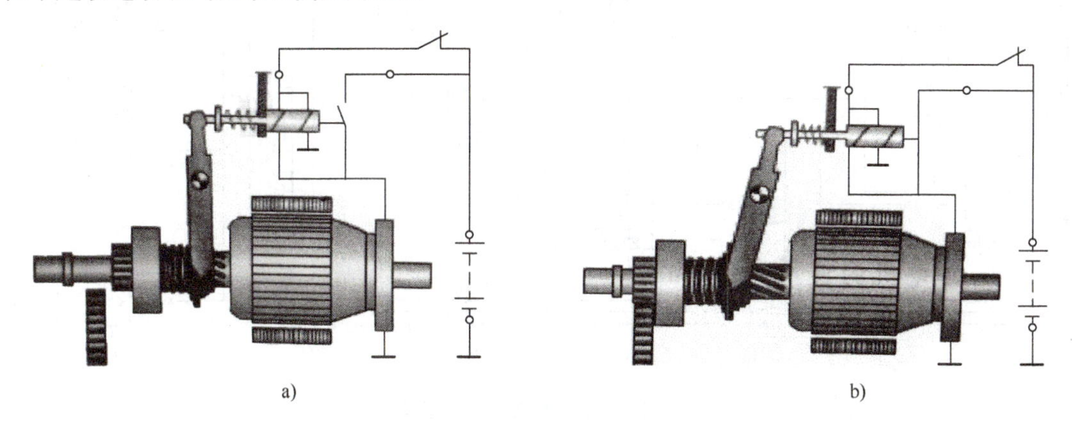

图7-14　起动过程示意

起动时，接通起动开关，起动电动机控制装置的吸引线圈与保持线圈通电，两者产生的电磁力方向相同，相互叠加，吸引控制装置的衔铁克服弹簧力右移，并带动拨叉绕其销轴转动，使驱动齿轮左移；同时，由于吸引线圈的电流流过直流电动机的绕组，电枢开始转动，通过单向器使驱动齿轮旋转。因此，驱动齿轮边旋转边左移。当左移出一定距离后，驱动齿轮齿端与发动机飞轮齿圈齿端相对，不能马上啮合，弹簧被压缩，当驱动齿轮转过一定角度后，两齿轮的齿端错开，在弹簧力的作用下，驱动齿轮迅速左移与飞轮啮合，同时，控制装置的衔铁迅速右移，使控制装置的触点开关迅速闭合。触点开关闭合后，大电流从蓄电池正极通过触点开关流经直流电动机的绕组后回至蓄电池负极，直流电动机便产生较大的电磁转矩驱动发动机旋转并起动（注意：触点开关闭合后，吸引线圈两端电势相等，不再有电流流过，由保持线圈产生的电磁力维持衔铁的位置）。

发动机起动后，其转速迅速上升到怠速，飞轮变成主动齿轮，带动驱动齿轮旋转，但由于单向器的“打滑”作用，发动机的转矩不会传递给电枢，防止了电枢超速运转的危险。

起动后，松开起动开关，起动控制回路断电，电流除从蓄电池正极通过触点开关流经直流电动机的绕组回至蓄电池负极外，还从蓄电池正极通过触点开关流经控制装置的吸引线圈后经保持线圈回至蓄电池负极。很明显，此时吸引线圈与保持线圈是串联关系，流经两者的电流相等。由于两者的匝数相等，因此两者产生的电磁力大小相等，但方向相反，相互抵消。控制装置的衔铁在弹簧力的作用下迅速左移，使触点开关断开，直流电动机的绕组与控制装置的吸引线圈与保持线圈断电；衔铁左移带动拨叉绕其销轴转动，使驱动齿轮右移，脱开驱动齿轮与飞轮的啮合。

2. 起动电路的工作原理

如图 7-15 所示。

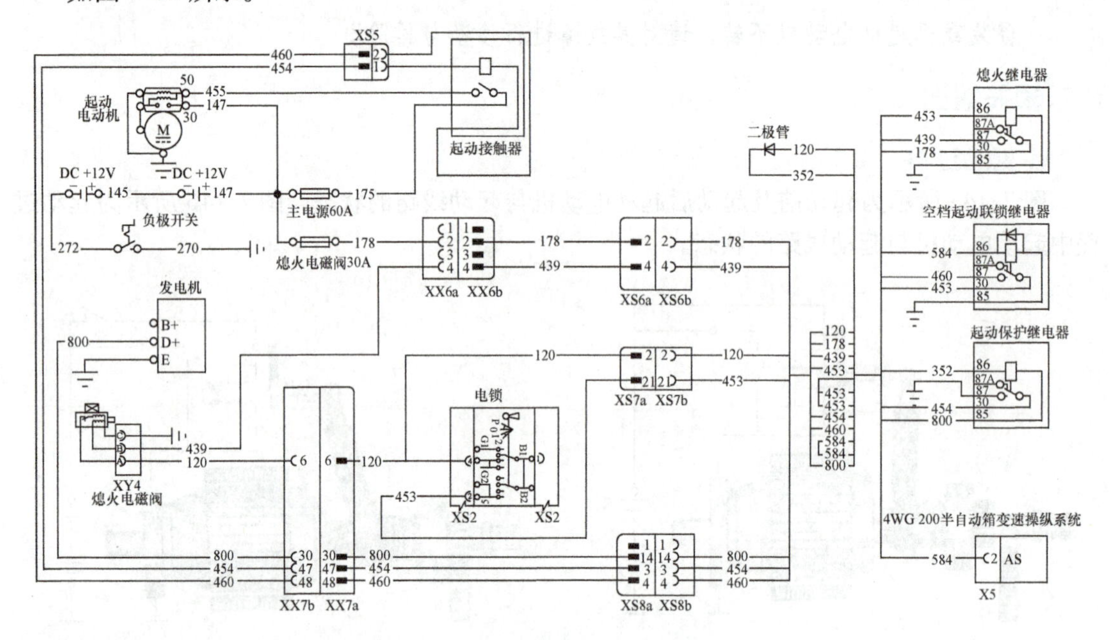

图 7-15 起动电路的工作原理

1）打开负极开关，电锁旋转至 ON 档，此部分工作原理解析参见主电路电源系统。

2）变速操纵手柄拨至中位后，ZF 变速控制器输出高电平信号（24V）“584”；电锁旋转至 START 档，整机起动后，电锁档位由 START 档回到 ON 档，发电机 D+ 输出高电平信号，起动保护功能启用。

3. 框图法

框图法是故障诊断与排除的一种逻辑分析方法。它是利用菱形、矩形、指向箭头和文字按照一定的逻辑关系组合，形成故障原因分析和故障排除流程图。将故障现象放在菱形框内，将故障原因按照概率大小或易难程度依次列在矩形框内。如果故障原因成立，则在横向指引框内写上故障排除方法。

☞ 任务实施

使用框图法分析和排除故障：

（1）故障现象　起动电动机不转。

（2）故障原因　用框图法分析故障原因，并列出排除方法，如图 7-16 所示。

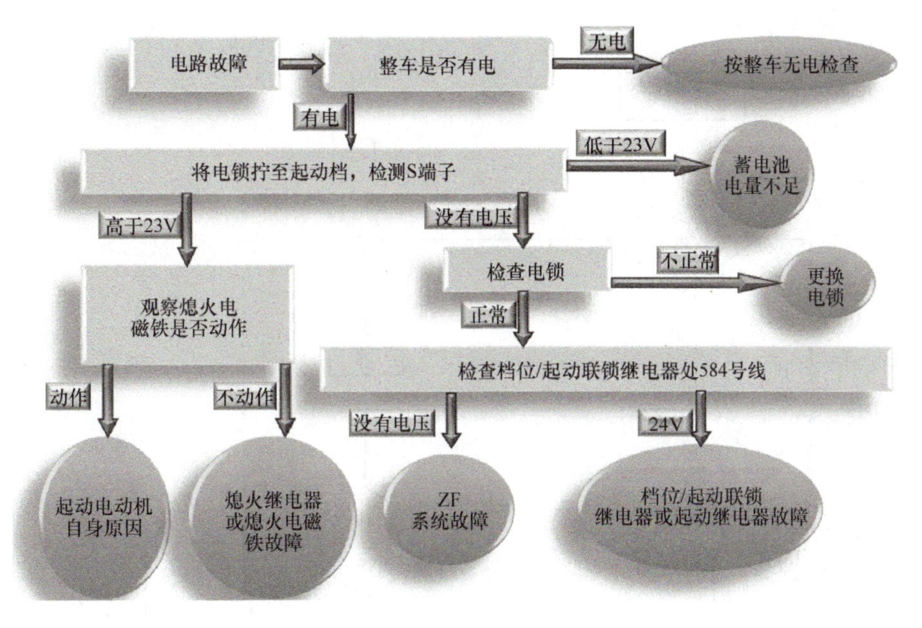

图7-16　框图法分析故障并排除

任务7.3　装载机空调系统故障诊断与排除

☞ 学习目标

能够对送风系统的电气故障进行诊断与排除。

☞ 工作任务

用户反映一台装载机打开空调后无风，请对空调无风的故障进行诊断与排除。

☞ 相关知识

空调的送风系统主要由蒸发器、鼓风机、风管和控制线路等组成，用于将车外新鲜空气或经蒸发器冷却后的空气通过导风管从驾驶室各出风口送出。操作驾驶室面板上相应的开关可以调节空调送风及风量。

如果空调无风，可能是线路出现故障造成鼓风机不运转，或鼓风机自身发生故障，无法工作。

☞ 任务实施

首先将出风量设为最大，听有无风机运转的声音。如鼓风机不运转，按图7-17所示流程进行检修。

如鼓风机正常运转，按图7-18所示流程进行检修。

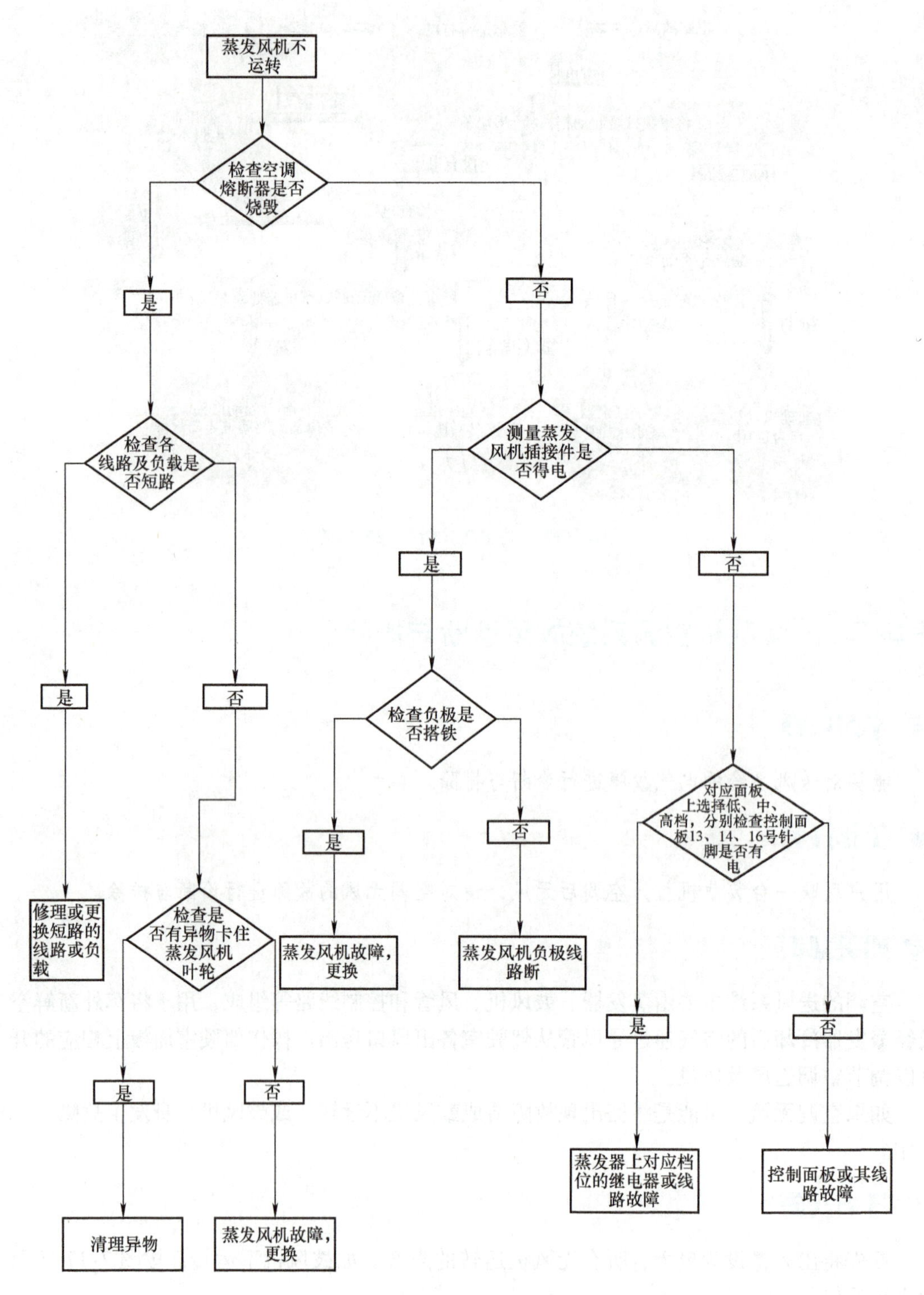

图 7-17　鼓风机不运转检修流程

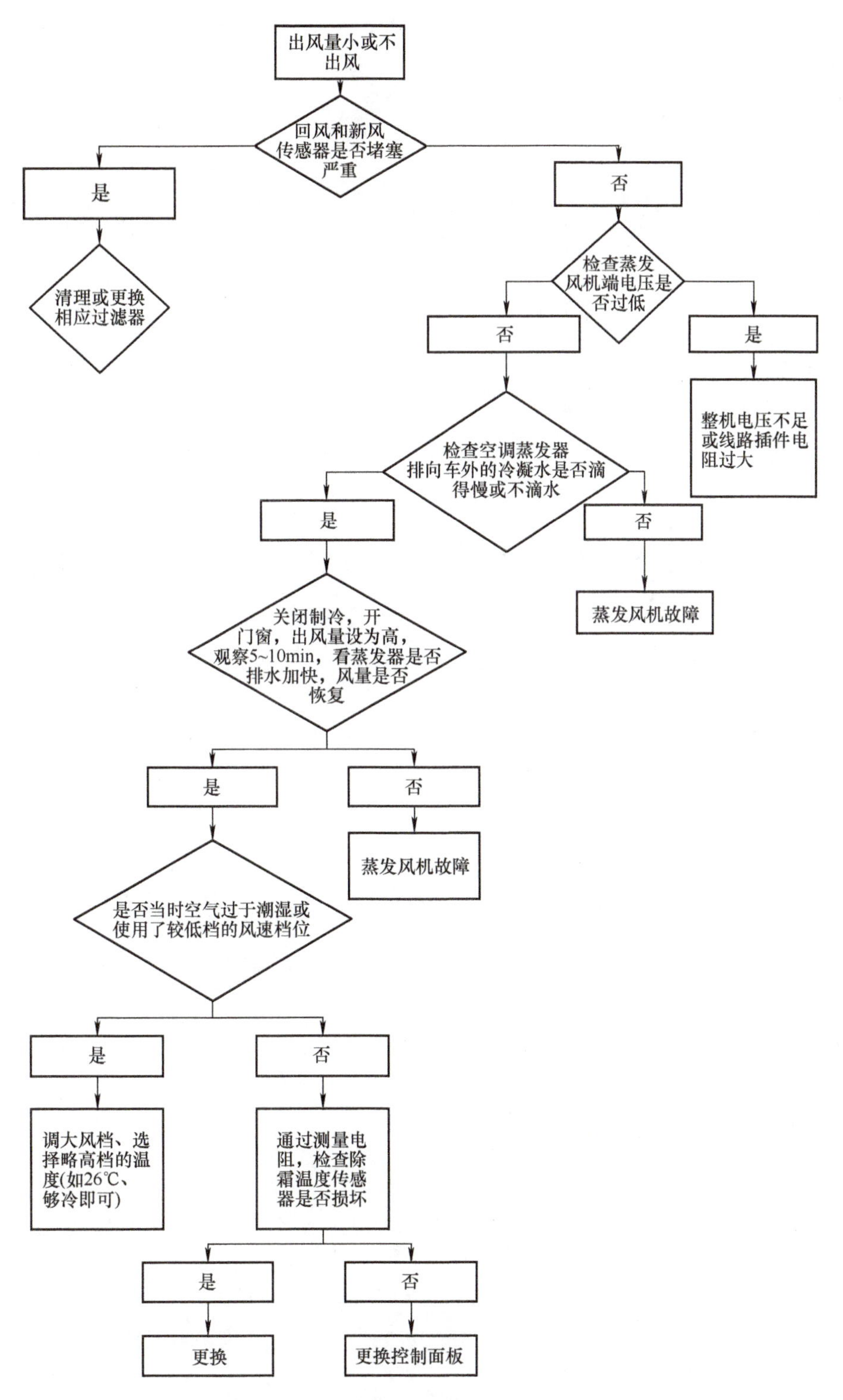

图7-18 鼓风机正常运转检修流程

装载机维修与服务（中英双语）

英文部分

Project 1

Introduction, delivery and inspection of the loader

Task 1.1 Introduction of the loader

☞ [Learning Objectives]

1) To be able to introduce the structure and working performance of the loader to the customer.

2) To be able to use PPT to introduce the loader to the target customers.

☞ [Work Task]

When a customer comes to buy a machine, how can the 856H loader be better introduced to them to demonstrate the professionalism? To recommend the loader to the customer, we need to know about the loader. The knowledge of the loader should be grasped when recommended to customers.

☞ [Relevant Knowledge]

1. Definition of the loader

The loader is a machine for earthwork construction that is widely used in highways, railways, construction, hydropower, ports, mines and other construction projects. It is mainly used for shoveling bulk materials such as soil, gravel, lime, coal, etc. It can also make light shoveling operations for ore, hard soil, etc. It can be used for bulldozing, lifting, loading and unloading by replacing different auxiliary working devices.

In the construction of roads, especially high-grade highways, the loader is used for filling and excavating road bases, aggregating and loading asphalt mixture and cement concrete yards, and other operations. In addition, it can also be used for pushing and transporting soil, scraping the ground and towing other machines. For the advantages of fast operation speed, high efficiency, good mobility and light operation, the loader has become one of the main types of machines for earthwork construction in engineering construction.

2. The structure of the loader and the names of each component

(1) The structure of the loader

The loader is mainly composed of the engine, the torque converter, the transmission, front and rear drive axles.

It can be divided into the following systems:

1) The power system: The original power of the loader is generally provided by the diesel engine with the features of reliable work, hard power characteristic curve, less fuel, etc., which meet the requirements of the harsh working conditions and variable load of the loader.

2) The transmission system mainly includes travel device, gearbox, etc.

3) The hydraulic system: The function of the system is to convert the mechanical energy of the engine into hydraulic energy with fuel oil as the medium, and then transmit it to the oil cylinder and oil motor to convert it into mechanical energy.

4) The control system: The control system is a system to control the engine, the hydraulic pump, the multi-way reversing valve and the actuating element. The hydraulic control drive mechanism is a device that converts small power electrical or mechanical energy into strong power hydraulic and mechanical energy in a hydraulic control system. It consists of hydraulic power amplifying elements, hydraulic actuating elements and loads, and is the core of the hydraulic system for static and dynamic analysis.

(2) The names of each component of the loader

The names of each component of the loader are shown in Fig. 1-1.

Fig. 1-1 The names of each component of the loader

1—Front work light 2—Cab 3—Front combination light-LH 4—Bucket cylinder 5—Rocker arm 6—Bucket 7—Counter balance weight 8—Rear wheel 9—Rear fender 10—Ladder 11—Front combination light-RH 12—Front frame 13—Front fender 14—Front wheel 15—Boom 16—Rear work light 17—Hydraulic oil tank 18—Engine cover 19—Rear combination light-RH 20—Rear frame 21—Rear combination light-LH

3. Classifications of the loader

1) Classified by the travelling mode: the rubber-tired loader, with fast driving speed and flexible mobility, can be driven on urban roads and is easy to use; the crawler loader, low grounding ratio pressure and high traction, but slow driving speed and inflexible transfer.

2) Classified by the form of frame structure: the integral frame loader (as shown in Fig. 1-2); the articulated loader (as shown in Fig. 1-3).

Fig. 1-2 The integral frame loader

Fig. 1-3 The articulated loader

3) Classified by the loading method: front discharge, rear discharge, rotary, and side discharge.

① Front discharge: shovel loading and unloading at its front end. It has the advantages of simple structure, reliable, safe and convenient operation, strong adaptability and wide application.

② Rear discharge: loading at its front end and discharging at its rear end. Its mechanical transport distance is short and the operation efficiency is high; however, the safety is poor and the application is less.

③ Rotary: the working device is installed on a rotary table which can be rotated 90° - 360°. Side unloading does not need to adjust the mechanical position with high operational efficiency, but its structure is complex with large mass and poor lateral stability. So it's suitable for operations in a narrow site.

④ Side discharge: loading at its front end and unloading at the side. During loading operation, there is no need to adjust the mechanical position, and it can unload the material directly to the transport vehicle parked on its side with high operational efficiency; however, the lateral stability is poor when unloading.

4) Classified by the transmission mode: the mechanical transmission, the hydraulic mechanical transmission, the hydraulic transmission, the electric transmission.

5) Classified by the power of the engine: the small loader (power <74 kW); the medium-sized loader (power 74~162 kW); the large loader (power 162~515 kW); the super-large loader (power >515 kW).

4. The basic parameters and performance of the loader

(1) Main performance parameters

1) The rated load: the maximum bucket load under the premise of ensuring the stable operation of the loader.

2) The rated power: the power output of the engine without fans, air filters, generators or other accessories.

3) Operating mass: the mass of the main engine when the loader is equipped with working devices and attachments, a 75 kg driver, and when oil and water are filled.

4) Standard capacity of bucket: also called rated bucket capacity, the sum of the bucket's flat loading capacity and the volume of the tip part of the pile (1:2 slope all around) under ordinary materials, as shown in Fig. 1-4.

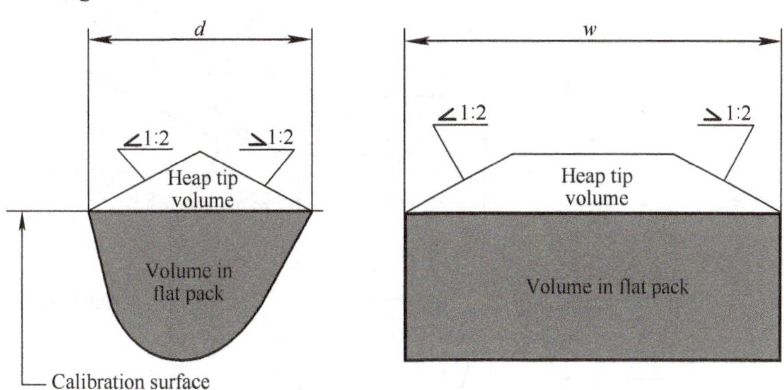

Fig. 1-4 Standard bucket capacity

5) Maximum prying force: the maximum upward plumbing force generated by the hydraulic cylinder of the working device at 100 mm after the bucket cutting edge when the bottom surface of the bucket cutting edge is 20 mm above the bottom surface reference. There are two kinds of prying forces: the bucket prying force and the luffing jib prying force, as shown in Fig. 1-5. Generally, the bucket prying force is greater than the luffing jib prying force, so the bucket prying force should be the standard prying force.

6) Tilting load (straight): the minimum load allowed at the bucket capacity centre when the luffing jib is extended flat to the farthest bucket capacity centre, so that the rear wheels of the loader are off the ground, as shown in Fig. 1-6.

7) The tilting load (full steering): when the steering angle is played to the maximum position and the luffing jib is extended to the farthest from the bucket centre, the minimum load allowed at the bucket centre while the rear wheels of the loader are off the ground, as shown in Fig. 1-7.

8) The traction force: the force used by the loader to haul or shovel soil, in addition to meeting the rolling resistance generated by the tires.

9) The sum of three times: the time used to lift the luffing jib from the lowest point to the highest point at maximum speed when the bucket is fully loaded with the rated load; the time used to unload the material at the highest point of the bucket; and the time used to float down the luffing jib from the highest point to the lowest point when the bucket is empty.

10) The maximum speed: speed on horizontal concrete roads, in the highest gear, at the maximum throttle.

(2) Main dimensional parameters of the loader (as shown in Fig. 1-8)

1) The minimum radius of turn (outside of bucket): the radius of the circle drawn on the outside of the bucket at the full steering position, as shown in Fig. 1-9.

2) A is the maximum height of bucket lower hinged pin: the distance from the centre point of the bucket lower hinged pin to the ground when the bucket is raised to the highest point on the hori-

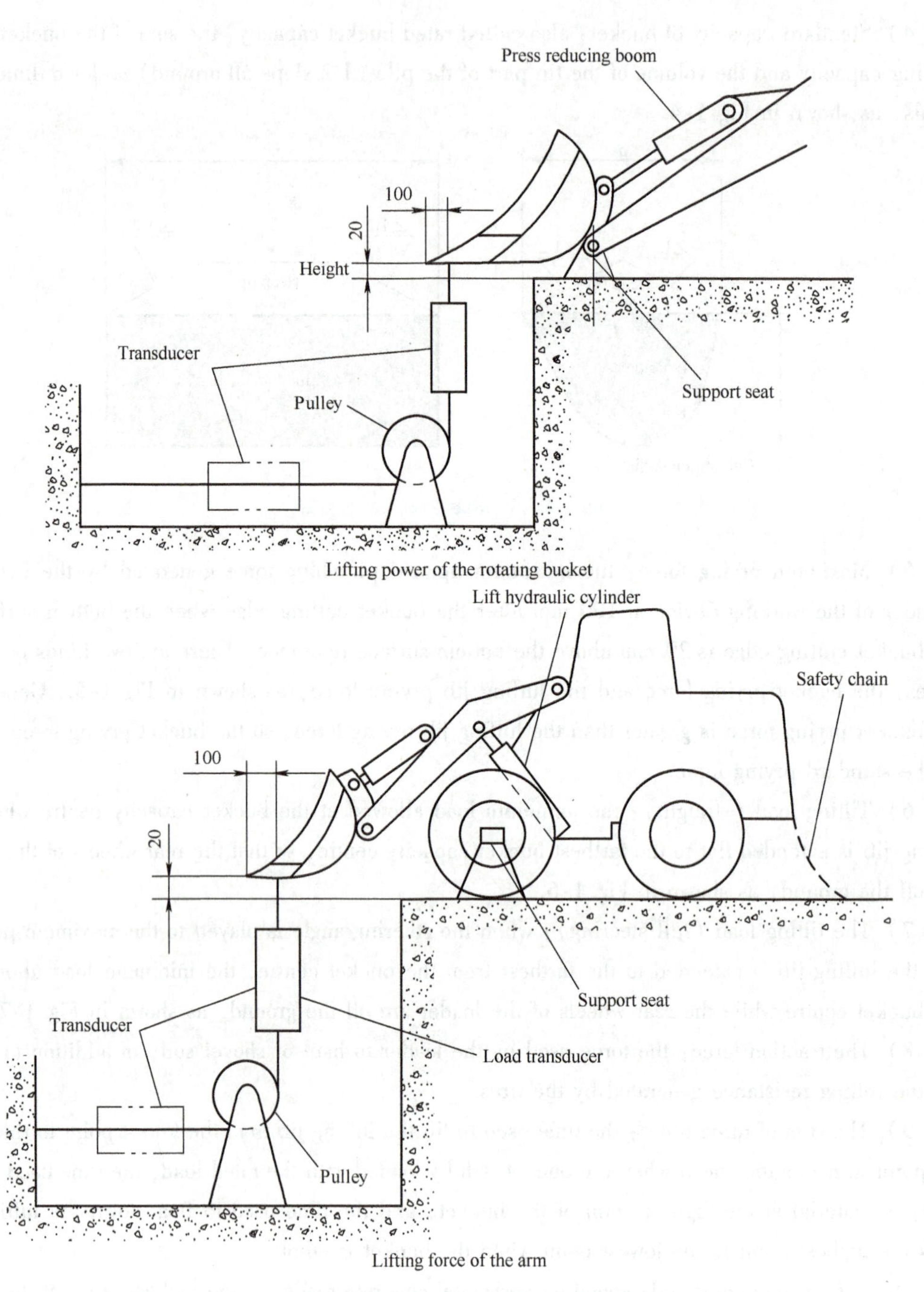

Fig. 1-5 The bucket prying force and the luffing jib prying force

zontal ground.

3) *B* is the unloading height: the distance from the tip of bucket teeth to the ground when the bucket is raised to the highest position and at the maximum discharge angle;

4) *C* is the unloading distance: the distance from the tip of bucket teeth to the outer edge of the front wheel when the bucket is raised to the highest position and at the maximum discharge an-

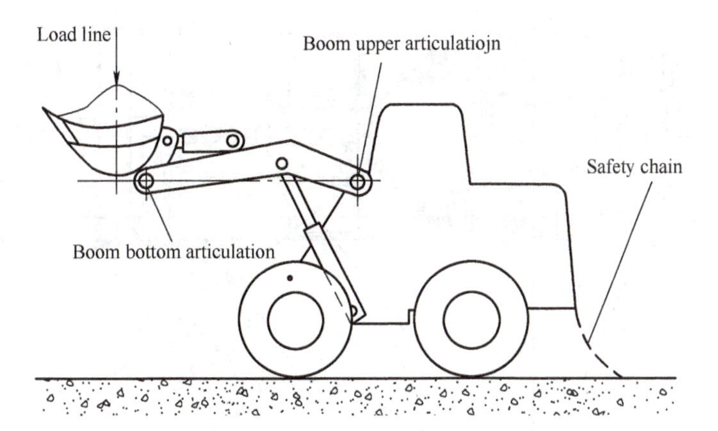

Fig. 1-6　The linear tilting load

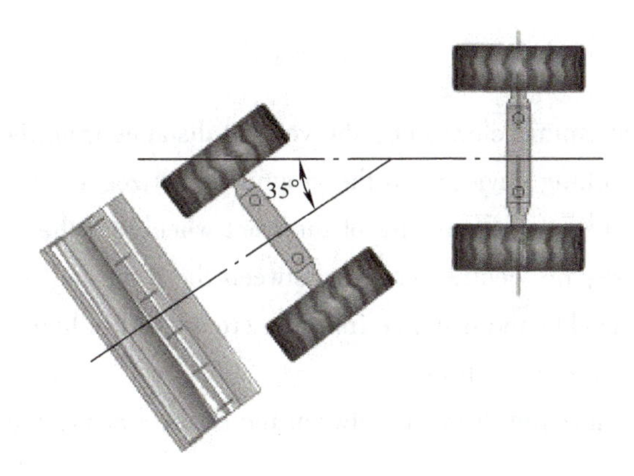

Fig. 1-7　The full steering tilting load

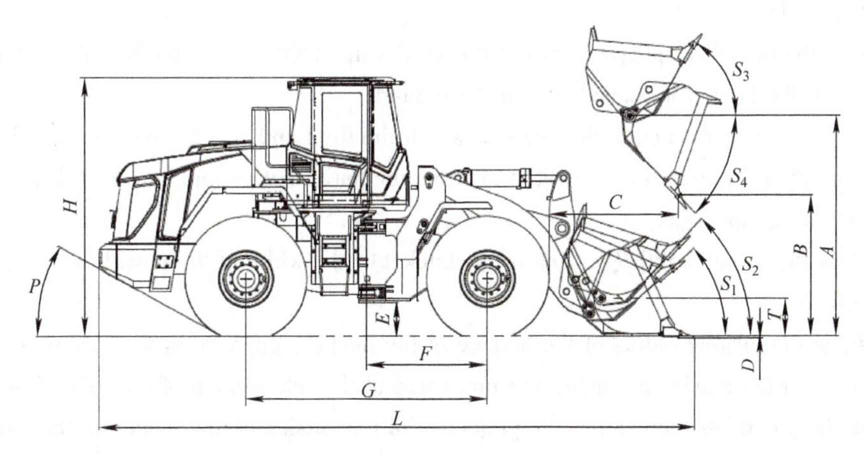

Fig. 1-8　Main dimensional parameters of the loader

gle;

5) D is the undercutting depth: the depth below the ground when the bucket is levelled and the bucket teeth's depth is below the ground;

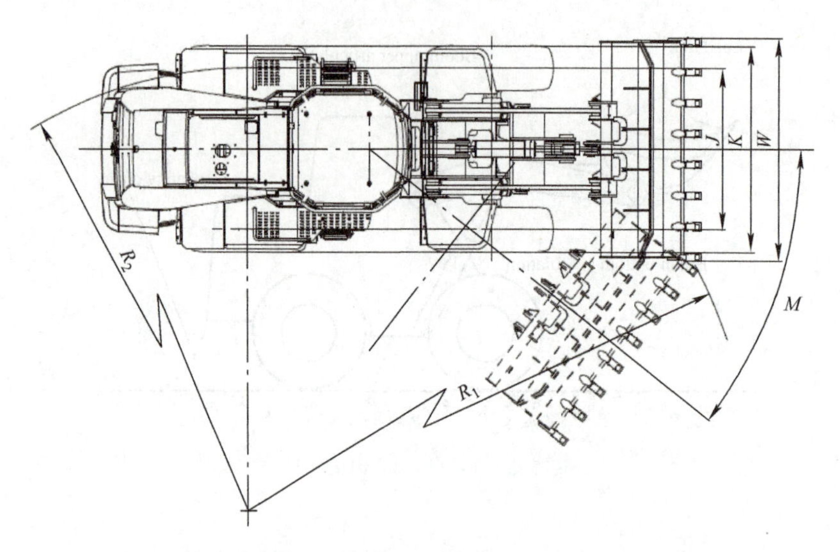

Fig. 1-9 The turning radius of the loader

6) *E* is the minimum ground clearance: the vertical distance from the lowest point of the chassis (excluding tires and working devices) to the ground on horizontal ground.

7) *F* is the distance between the centre of the front wheel and the hinged centre.

8) *G* is the wheelbase: the centre distance between the front and rear drive axles.

9) *H* is the overall height: the distance from the ground to the highest position of the loader at the standard inflation pressure of the tires.

10) *J* is the wheel tread: the distance between the centre lines of the left and right tires on the same drive axle.

11) *K* is the tire outer width: the distance between the outer edges of the left and right tires on the same drive axle.

12) *L* is the overall length: the distance from the tip of the bucket teeth to the outermost edge of the tail when the bucket is placed flat on the ground.

13) *M* is the steering angle: the angle at which the front and rear frames are turned opposite to the rear frame when the front frame is set straight on the horizontal ground and then the front frame is turned to the maximum angle.

14) *W* is the overall width (outside of the bucket): the width of the outside of the bucket, wider than the tire.

15) R_1 is the turning radius of the outside of the bucket: opposite to the rear frame, the front frame deflected to the maximum angle, the projection of the intersection of the axis of the front and rear axle on the ground as the centre, the projection of the outside of the bucket on the ground as the radius to draw a circle, the radius of this circle is the outside of the bucket turning radius.

16) R_2 is the turning radius of the tire centre: opposite to the rear frame, the front frame deflected to the maximum angle, the projection of the intersection of the axis of the front and rear axle on the ground as the centre, the projection of moving track of rear wheel on the ground as the radius to draw a circle, the radius of this circle is the turning radius of the outside of the tire centre.

17) P is the departure angle: to make a tangent line from the lowest point of the rear of the car to the rear of the tire outline (against the ground part), the departure angle is the angle between this tangent line and the horizontal plane. The departure angle is generally not less than 30°.

18) S_1 is the in-situ bucket retraction angle: the angle is formed by the bucket bottom and the ground when the luffing jib is at the lowest position and the bucket is at the maximum bucket retraction position.

19) S_2 is the bucket retraction angle at transportation position: the angle is formed by the bucket bottom and the ground when the luffing jib rises to height T and the bucket is at the maximum bucket retraction position.

20) S_3 is the bucket retraction angle at the highest position: the angle is formed by the bottom of the bucket and the centre of the lower hinge pin when the luffing jib is lifted to the highest position, the bucket is retracted to the maximum position.

21) S_4 is the unloading angle at the highest position: the angle is formed by the bottom of the bucket and the centre of the lower hinged pin when the luffing jib is lifted to the highest position and the bucket opens to the maximum unloading position.

22) T is the height of the bucket hinged pin at the transport position.

5. Safe operating regulations for the loader

1) Before getting on and off the machine, clean the oil or sludge on the handrail or ladder. Only get on or off the machine where there are stairs or handrails. When getting on and off the machine, face the machine, pull the handrail and step on the ladder with your feet to maintain three-point contact (two feet, one hand or two hands and one foot). It is forbidden to jump off the machine and get on and off the machine when the machine is moving. Never use any lever as a handrail when getting on and off the machine. When carrying tools or other items, do not climb up or down the machine. Use ropes to lift the required tools onto the operating platform.

2) When the loader is driving, the bucket shall be retracted and be 400-500 mm away from the ground. During the implementation, pay attention to whether there are roadblocks or high-voltage lines. Except for the specified drivers, no other personnel are allowed to ride, and the bucket is strictly prohibited from carrying people.

3) When the loader is driving on the road, avoid sudden reverse driving, and when the bucket is lifted with a load, sharp turning and sudden braking are not permitted.

4) When driving on the road, loaders must observe traffic rules and drive carefully, and it is forbidden to skid in neutral when going downhill.

☞ [Task Implementation]

Prepare a PPT introducing the CLG856H loader, which needs to contain the following contents:

1) General introduction of the 856H loader.

2) Performance parameters.

3) Structure.

4) Configuration.

Task 1. 2 Delivery and inspection of the loader

[Learning Objectives]

1) To be able to deliver the loader to the customer for inspection as required.

2) To be able to compile a loader delivery and inspection report.

[Work Task]

A new loader, which needs to be delivered to the customer, how to complete the delivery inspection.

[Relevant Knowledge]

1. Contents and standards for delivery and inspection

(1) The key parts to be inspected in the whole machine inspection and test run are shown in Table 1-1.

Table 1-1 Inspection point check – list for the loader

No.	Acceptance items	Result		No.	Acceptance items	Result	
		Normal	Abnorma			Normal	Abnormal
1	Engine oil			19	Operation of the air conditioning system		
2	Tension of the air conditioning belt						
3	Tension of the fan belt			20	The audio system		
4	Engine idle and medium – speed operation			21	Charging indication		
				22	Battery		
5	Engine oil pressure			23	Wheel – side brake		
6	Water temperature of the engine			24	Parking brake		
7	Bucket movement			25	Connection of counter balancing weight		
8	Luffing jib movement						
9	Turning action			26	Door lock		
10	Hinged device of front and rear frames			27	Hood locks		
11	Wheel			28	Whether the oil pipe connectors are leaking		
12	Travel forward at a low speed or a high speed						
				29	Radiator set		
13	Travel backward at a low speed or a high speed			30	Exterior paint		
				31	Is the control lever flexible and in place		
14	The electrical control system						
15	Lights			32	Tools, information and accessories accompanying the vehicle		
16	Wiper						
17	Horn			33	GPS alarm unit		
18	Instrument			34	Other (items not listed above)		

(2) The inspection contents of the whole machine

1) Engine fuel level inspection. The fuel level meter is located on the instrument assembly of the cab. The fuel level meter is divided into two areas: green and red. 1 indicates that the fuel level is full, and 0 indicates that the fuel level is 0. When the fuel level indication is lower than "0. 2", fuel should be added in time.

2) Engine oil level inspection. Too much or too little engine oil can cause engine damage.

Drive the machine to a flat site, turn off the engine and wait for 10 minutes; allow the engine oil in the crankcase fully flow back to the engine oil pan; then open the engine hood, the oil level dipstick is located on the right side of the engine; after that pull out the dipstick, wipe the dipstick with a clean cloth, and reinsert it into the engine oil level port to the end and pull it out again to check, the oil level should be between the "L" scale and the "H" scale of the dipstick as shown in Fig. 1-10; if the oil level is below the "L" scale, please replenish the oil; if the oil level is above the "H" scale, please loosen the drain plug at the bottom of the oil pan to drain some of the oil; at last return the dipstick and close the engine cover.

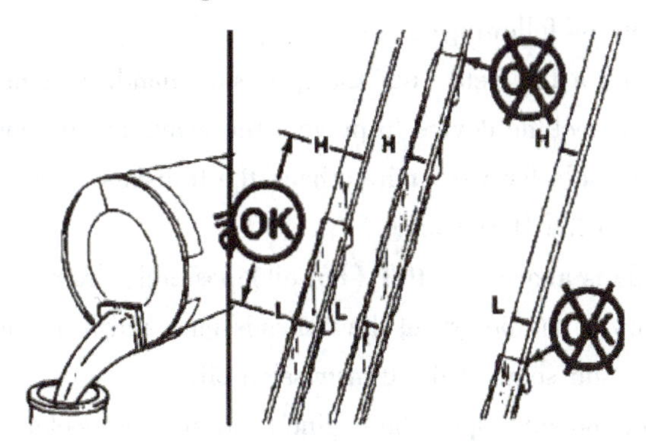

Fig. 1-10 Oil level scale value

3) Hydraulic oil inspection. Before the inspection, ensure that hydraulic cylinders, hydraulic pipelines, radiators, and other hydraulic components are filled with hydraulic oil; drive the loader to a flat site, and the front and rear frames are aligned without an included angle; the bucket is retracted to the limit and the engine lifts the luffing jib to the highest position at full speed; push its control lever to the "down" position when the machine is idling, so that the luffing jib is lowered to the lowest position at a constant speed, put the bucket level on the ground, then turn off the engine and remove the key. Push the operating lever back and forth, left and right to release the pressure; when there are no bubbles in the level gauge, check the hydraulic oil tank level gauge. At this time, the oil level should be within the green range of the level gauge, that is, between the MAX line and the MIN line, as shown in Fig. 1-11; when checking the oil level, if it is found that the oil level is higher than the green range of the liquid level gauge (observation of bubbles in the liquid level gauge), that is, the oil level is above the maximum oil level line, and the oil cannot be drained; after the bubbles are eliminated, check according to the above steps; if the oil level is lower than the green range, the hydraulic oil must be replenished immediately, and then check again according to

the above check method.

4) Brake fluid inspection. Park the loader on a flat site, turn off the engine, pull up the parking brake to prevent the machine from moving or turning, rotate and open the booster cover, and check the level height of the booster oil cup. If the level of the oil cup is lower than the screen surface inside the oil cup, refill the synthetic brake fluid (DOT4) until the level is flush with the screen surface inside the oil cup and the oil level is about 3/4 of the way up the oil cup.

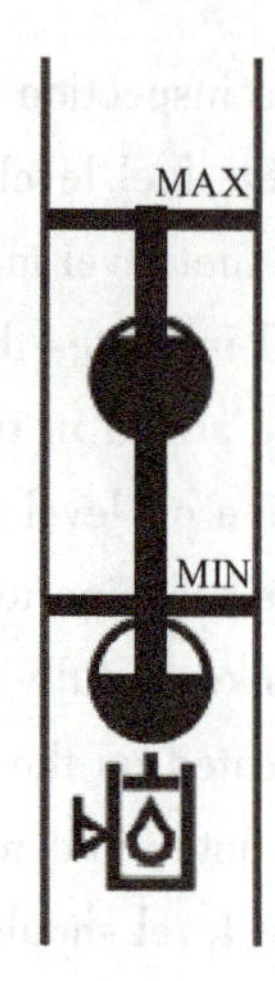

Fig. 1-11 Liquid level – meter scale

5) Transmission oil inspection. Before starting the engine with a low water temperature, the cold oil level should be checked; the purpose of checking the cold oil level is to make sure there is enough oil when starting. It is even more important for loaders that have been out of use for a considerable period of time. The steps for checking the transmission oil level are as following:

① Park the machine on a flat field, put the gear shift handles in neutral, apply the parking brake, and install the bogey locking device to prevent the machine from moving and turning.

② Start the engine and idle for 3 ~ 5 min, check the transmission level meter. The normal oil level of the transmission is in the HOT zone.

③ If the oil level scale is above the HOT (hot oil) position, drain some of the transmission oil by loosening the drain screw at the bottom of the transmission, and if the oil level scale is below the COLD (cold oil) position, add some of the transmission oil.

6) Check the radiator coolant. Open the engine hood and the coolant level observation window is located on the auxiliary water tank. Check the coolant level through the level observation window. When the coolant is lower than the centre line of the liquid level observation window, replenish the coolant.

7) Instrument inspections.

① Whether the oil pressure indicator light is off after the engine is started (when the light is on, it is a low pressure alarm, and there is also a brake air pressure alarm).

② Transmission pressure. The pointer of transmission oil pressure gauge is in the range of 1. 1 ~ 1. 5 MPa (green area);

③ Barometer. The pointer of the brake barometer is in the range of 0. 5 ~ 0. 784 MPa (the green area range, the alarm light is on when it is lower than 0. 4 MPa);

④ The chronograph goes normally and reads no more than 10 h.

8) Maneuverability inspection.

① The engine runs normally, idling speed, sound and smoke exhaust are normal.

② Steering performance. The left and right steering are flexible and in place at idle speed without heavy steering, lag, swing, abnormal noise, deviation, auto rotation, idling phenomenon, when checking the steering limit, the limit block should be able to be effectively connected.

③ Braking performance. Braking is flexible and reliable.

④ Gear control. The stroke of the shift control lever is appropriate, the shift of each gear is stable, the gear is accurate, and there is no off gear.

⑤ The operation of the hydraulic operating handle is light and stable without sticking or binding.

9) Electrical inspection. The wiper, the air conditioner and the reversing alarm, etc. work normally; front and rear headlights, turn signals, work lights, hazard warning lights and other lights are normal; there are no bump and scratch on electrical appearance; battery power shows green.

10) Appearance inspection. The appearance of the machine should be symmetrical, assembled proportionally, coordinated, without any bump and scratch; its sheet metal parts should be flat without bump and bend, knock, bump and scratch; its metal plating and oxidation treatment layer shall not flake and rust, the surface is clean; the machine's hoses, steel pipes and hydraulic parts connection are not dripping and seeping oil; there shall be no mistake or omission, the assembly shall be appropriate and correct, and the appearance shall be free of obvious defects.

11) Tire inspections.

① Tire side: cracks, scratches or other defects with a depth greater than 3 mm, a length greater than 5 mm and a width greater than 1 mm are not allowed.

② Tire crown: cracks, scratches or other defects with a depth greater than 10 mm, a length greater than 10 mm and a width greater than 1 mm are not allowed;

③ Tire shoulder: cracks, scratches or other defects with a depth greater than 5 mm, a length greater than 10 mm and a width greater than 1 mm are not allowed;

12) Check the materials accompanying the machine. The materials, spare parts and tools accompanying the machine are complete and available.

2. Precautions for operation and storage of the whole machine

(1) Notes on the operation of the whole machine The machine must be operated by specially trained or qualified personnel; the loading and unloading vehicles must be directed by someone and the operator is in good physical condition; the operator must operate the machine in strict accordance with the requirements of the "Operation and Maintenance Manual", and the throttle and handle should be operated slowly and without excessive force.

(2) Storage precautions The battery should be checked regularly and started once a month for no less than 20 min each time. If it exceeds 3 months or is in a cold area, the battery must be removed and stored separately and charged regularly; the exposed plating of the battery, metal, processing surface, etc. should be coated with lubricating grease to prevent rust and corrosion, and when operating again, the dirt on the exposed transition layer of the oil cylinder shall be removed; the cooling water in the tank should be drained, including the engine body and the drain valve of torque converter oil cooler, or replaced with antifreeze; the tires should be kept at 0.333 ~ 0.353 MPa for the front wheels and 0.275 ~ 0.294 MPa for the rear wheels.

☞ [Task Implementation]

Prepare a delivery and inspection report, as shown in Table 1-2.

Table 1-2 Delivery and inspection report

Date of acceptance			Equipment model	
No.	Inspection items	Acceptance contents		Results
Ⅰ.	Appearance acceptance	1. Lighting is normal		
		2. The instruments are normal, complete and effective		
		3. Tire screws are tightened without missing		
		4. Drive shaft screws are tightened without missing		
		5. There is no looseness of the horizontal and vertical tie rods of the steering machine		
		6. There are no oil, gas or water leaks in any part		
		7. There is no deformation in all parts of the loader		
Ⅱ.	Check the oil level and water level of each part	1. The water level of the tank is normal		
		2. The level of engine oil is normal		
		3. The diesel oil level is normal		
		4. The brake oil is normal		
		5. The hydraulic oil level is normal		
Ⅲ.	The Engine	1. The water temperature is normal		
		2. The engine runs normally without any abnormal noise		
		3. Each auxiliary mechanism works normally		
Ⅳ.	The hydraulic transmission	1. The hydraulic pump pressure is normal		
		2. The travelling system is normal		
		3. The lifting of luffing jib cylinder is normal without any gliding		
		4. The lifting of bucket cylinder is normal		
		5. The temperature of hydraulic oil is normal		
Ⅴ.	The operating system	1. The transmission is normal		
		2. The braking system is normal		
		3. The travelling system is normal		
Ⅵ.	Safety protection	1. The product quality certificate should be provided		
		2. Operators are licensed to work		
		3. Safety technical operating procedures are set in the cab		
Acceptance conclusions				
Signature of the inspector	Department of materials and equipment	Safety and quality department	Engineering department	Site management personnel

Project 2

Maintenance of the loader

Task 2.1 Preparation of regular maintenance and service plans

[Learning Objectives]

1) Be familiar with the contents and requirements of regular maintenance.

2) To be able to make regular maintenance plans.

[Work Task]

Prepare a key maintenance plan for the loader for 1000 h.

[Relevant Knowledge]

The proper maintenance of the loader, especially preventive maintenance, is the easiest and most economical maintenance. Because the Correct maintenance prolongs for the service life of the loader and reduces the use cost, it compensates for the time and cost required in the planned maintenance.

To carry out the correct maintenance of the loader, first of all, the daily work of the loader must be done well, and the necessary adjustments and repairs must be done in time according to the situation reflected by the loader in the use process. Secondly, different maintenance schedules should be made according to the relevant contents introduced below and the special working conditions and experiences of diverse users.

The equipment manufacturer generally requires that the maintenance work be carried out in accordance with the time cycle that expires first in the use of the work time table or calendar (day, week, month, etc.). In extremely harsh, dusty or wet working environments, more frequent lubrication maintenance is required than that specified in the regular maintenance. During maintenance, the maintenance items listed in the original requirements should be repeated. For example, when performing a maintenance program of 500 working hours or 3 months, the maintenance programs listed in 250 working hours or 1 month, 50 working hours or weekly and every 10 working hours or daily should be performed at the same time.

The maintenance of the loader is mainly divided into daily inspection, new loader running – in and regular maintenance.

1. Daily inspection

Before working on the loader every day, the following items should be checked.

1) Whether the instruments and meters in the cab are normal, and whether the light system and horn in the loader are working properly.

2) Whether the coolant, diesel and oil levels of the engine are normal and whether there are leaks.

3) Whether there are leaks in hydraulic pumps, hydraulic cylinders and hydraulic lines and connectors.

4) Whether the surface of the parts is covered with contaminants, and whether the parts are loose, damaged, or fallen off, etc.

5) Whether the fasteners are loose.

6) Whether there are cotton yarn, waste paper, plastic film, and other debris in the outlet and inlet of the air conditioning evaporator.

2. The new loader running – in

The running – in of the new loader plays an important role in prolonging the service life of the machine, eliminating hidden troubles, and avoiding the occurrence of major failures. After purchasing the machine, the user must operate and maintain the machine according to the regulations on running – in of new loaders, and then the machine can be used normally.

The general requirement of the new loader's running – in period is 100 h. During the run – in period, the following matters shall be paid attention to in the operation of the loader:

1) After each starts, first make the engine idle for 5 min. The loader shall start at a steady low speed, and then gradually increase the speed. It is better to run in all gears of the vehicle both forward and backward.

2) Except in case of emergency, sudden start, sudden acceleration, sudden steering and sudden braking shall be avoided, and the operation shall not be too violent or too hasty. Try to load loose materials, the loading weight shall not exceed 70% of the rated load, and the driving speed shall not exceed 70% of the rated maximum speed.

3) Attention should be paid to the lubrication of all parts, and lubricating oil and grease should be replaced or added according to the prescribed time cycle. Attention should be paid to the transmission, torque converter, front and rear axle, wheel hub, parking brake, intermediate support shaft, as well as hydraulic oil, engine coolant, engine oil temperature. In case of overheating, find out the cause and troubleshoot.

4) Regularly check the fastening of bolts and nuts of parts.

The following items shall be checked after 8 h, 50 h and 100 h of running – in of the new loader, as shown in Table 2-1.

1) Regularly check the fastening of bolts and nuts of parts. Especially the head bolts of diesel

engine cylinder, exhaust bolts, front and rear axle fixing bolts, rim nuts, drive shaft connecting bolts, diesel engine fixing bolts, transmission fixing bolts, front and rear frame articulating bolts, etc.

2) Check the fastening of the engine fan belt and air conditioner compressor belt, and the oil level of the transmission, drive axle lubricant and diesel engine as well as coolant level.

3) Check whether the hydraulic system and the braking system have leakage, and whether the connection of each control lever and throttle lever is stable.

4) Check the temperature and connection of each part of the electrical system, the power supply status of the generator, the working condition of the lighting and steering signal light, etc.

Table 2-1 Maintenance items for new loader running – in

Items	8 h	50 h	100 h	Remarks
Bolts, nuts	◎	◎	◎	Anti – loosening signs are in place
Fan belts	◎	◎	◎	Deflection ≤8mm
Generator belts	◎	◎	◎	Deflection ≤8mm
Air conditioner compressor belts	◎	◎	◎	Deflection ≤8mm
Gear oil	◎	◎	◎	In the scale range
Hydraulic oil	◎	◎	◎	In the scale range
Coolant	◎	◎	◎	In the scale range
Oil – water separators	◎	◎	◎	Water blow – off
Air conditioning refrigerant	◎	◎	◎	
Lighting	◎	◎	◎	
Each hinge pin of the working device	•	•	•	Grease filling
Main filter element and safety filter element of the air filter		•	◎	
The fuel pre – filter, the filter		•	◎	
The engine oil and the oil filter		•	◎	

Note: Those marked with "◎" in the table are for inspection, and those marked with "•" are for replacement.

3. Regular maintenance

Regular maintenance of the loader is divided into 10 h or 1 day, 50 h or 1 week, 100 h or 1 month, 500 h or 1 quarter, 1, 000 h or 6 months and 2, 000 h or 1 year, etc. The maintenance should generally be carried out in accordance with the time cycle that expires first in the working time table or calendar (day, week, month, etc.). In extremely harsh, dusty or humid working environments, more frequent lubrication maintenance is required than that specified in periodic maintenance. For the contents of regularmaintenance and repair are shown in Table 2-2.

Table 2-2 Regular maintenance and repair items

Items		10 h	50 h	100 h	250 h	500 h	1000 h	2000 h	Remarks
Engine	Fueloillevel/Filter element	◎	◎	◎	•	•	•	•	Drain every 50 h
	Oil/Filter	◎	◎	◎	•	•	•	•	
	Air filter/Filter element	◎	◎	◎	◎	◎	•	•	
	Coolant/Liquid level	◎	◎	◎	◎	◎	◎	•	
	Oil - water separator blow - off	◎	•	•	•	•	•	•	
	Crankshaft respirator tube	◎	◎	◎	◎	◎	◎	◎	
	Radiator/Liquid level	◎	◎	◎	◎	◎	◎	◎	
	Cooling fan/Belt	◎	◎	◎	◎	◎	◎	◎	Deflection≤8 mm
	Engine belts	◎	◎	◎	◎	◎	◎	◎	Deflection≤10 mm
	Inlet/exhaust pipe	◎	◎	◎	◎	◎	◎	◎	
Hydraulic system	Hydraulic oil/Filters	◎	◎	◎	◎	•	◎	•	
	Hydraulic pump	◎	◎	◎	◎	◎	◎	◎	
	Hydraulic hoses/Connectors	◎	◎	◎	◎	◎	◎	◎	
	Hydraulic cylinders	◎	◎	◎	◎	◎	◎	◎	
	Hydraulic valves/Connectors	◎	◎	◎	◎	◎	◎	◎	
Transmission system	Transmission oil	◎	◎	•	◎	◎	•	•	
	Gear oil for front and rear axle	◎	◎	•	◎	◎	•	•	
	Hub reduction gear oil	◎	◎	◎	◎	◎	•	•	
	Tire pressure	◎	◎	◎	◎	◎	◎	◎	Front tire pressure: 0. 333 ~0. 353 MPa Back tire pressure: 0. 275 ~0. 294 MPa
Electrical system	Air pressure	◎	◎	◎	◎	◎	◎	◎	
	Brake oil	◎	◎	◎	◎	•	•	•	
	Friction plate	◎	◎	◎	◎	◎	◎	◎	
	Gas storage tank /Blow - off	◎	◎	◎	•	•	•	•	
	Front and rearlights/Switches	◎	◎	◎	◎	◎	◎	◎	
	Fuses	◎	◎	◎	◎	◎	◎	◎	
	Horns	◎	◎	◎	◎	◎	◎	◎	
	Gauges/Indicators	◎	◎	◎	◎	◎	◎	◎	
	Wire harnesses/Switches/Plugs	◎	◎	◎	◎	◎	◎	◎	
	Battery /Electrolyte level		◎	◎	◎	◎	◎	◎	
	Rain scraper/Wiper	◎	◎	◎	◎	◎	◎	◎	
	Air conditioning/Refrigerant/Filter element		◎	◎	◎	◎	•	•	
Grease	Hinge points of front and rear axle	◎	•	•	•	•	•	•	
	Each hinge pin of the working device	◎	•	•	•	•	•	•	
Others	Connection bolts/Nuts	◎	◎	◎	◎	◎	◎	◎	
	Connecting rods/Tie rods	◎	◎	◎	◎	◎	◎	◎	

Note: ① item " ◎" is to check on time and replace as needed; ② item " • " is to replace or operate on time; ③the oil (liquid) position shall be inspected after 15 min of shut - down; ④ time cycle: 8 h - daily inspection; 50 h - weekly inspection; 100 h - semi - monthly inspection; 250h - monthly inspection; 500 h - quarterly inspection; 1000 h - semi - annual inspection; 2000 h - annual inspection.

4. Notes

The models of added coolant, engine oil, gear oil, hydraulic oil and diesel oil shall be consistent with the original ones, and the products designated by the loader manufacturer or the original manufacturer shall be selected as far as possible. The materials replaced with the season or application environment should be operated in accordance with the specifications of the manual or instruction manual.

☞ [Task Implementation]

Prepare a 1000 h regular maintenance plan for the CLG856H loader.

Before the maintenance plan is prepared, the specific time of the loader in the last maintenance, maintenance items and consumables used for maintenance should be inquired to understand the working condition of the loader after maintenance as well as the current working condition.

The contents of the maintenance plan mainly includes:

1) Check whether the connecting bolts and nuts of each component are loose or not.

2) Check the looseness of the engine fan belt and air conditioning compressor belt.

3) Check whether the hydraulic system and the braking system have leakage, and whether the connection of each control lever and throttle lever is stable.

4) Check the hinge points of the front axle and rear axle, the hinge points of the working device and fill with lubricant.

5) Replace engine oil and filter element, air filter and filter element, fuel filter element and air conditioner filter element of the diesel engine.

6) Replace the lubricating oil of the transmission, the front axle reducer, rear axle reducer and hub reduction gear, etc.

Task 2. 2 Maintenance of working devices

☞ [Learning Objectives]

1) To be able to refill grease according to the maintenance requirements.

2) To be able to replace bucket teeth according to the maintenance requirements.

☞ [Work Task]

Replace the bucket teeth of the loader's working device.

☞ [Relevant Knowledge]

The shovelling, loading and unloading of the loader are achieved by the movement of the working device. The working device is composed of parts such as a boom, a boom cylinder, a bucket, a bucket cylinder, a rocker arm, a tie rod, etc. The bucket is used for shovelling materials. The rear end of the boom is connected to the front chassis by a boom pin. The front end is equipped with a bucket, and the middle part is connected to the boom cylinder. When the boom cylinder is retracted, the boom rotates around its rear end pin to realize the lifting or lowering of the bucket. The middle part of the rocker arm is connected to the boom, and the two ends are connected to the tie rod and the bucket cylinder, respectively. When the bucket cylinder is retracted, the rocker arm rotates around its middle support point, and the lever is used to turn the bucket up or down.

1. Equivalent motion mechanism of the working device

The working device can generally be divided into forward – turning six – link rods, reverse – turning six – link rods, and forward – turning eight – link rods, as shown in Fig. 2-1.

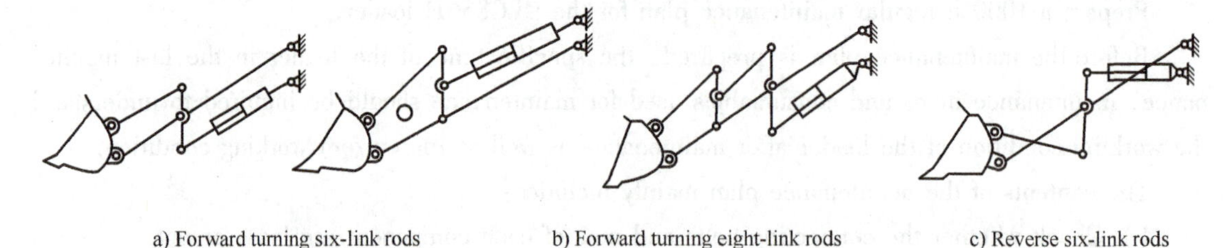

a) Forward turning six-link rods　　b) Forward turning eight-link rods　　c) Reverse six-link rods

Fig. 2-1 Schematic diagram of the structure of the working device

The working device of the loader is mainly based on reversing six – link rods. The reversing six – link rods is also known as the "Z" – shaped structure, mainly including six parts: the boom cylinder, the swivel cylinder, the rocker arm, the boom, the bucket, and the connecting rod. And it's also called a reversing 6 – bar linkage as the direction of the swivel cylinder movement is opposite to that of the bucket movement. It is a single rocker arm reversing six – link rods, as shown in Fig. 2-2.

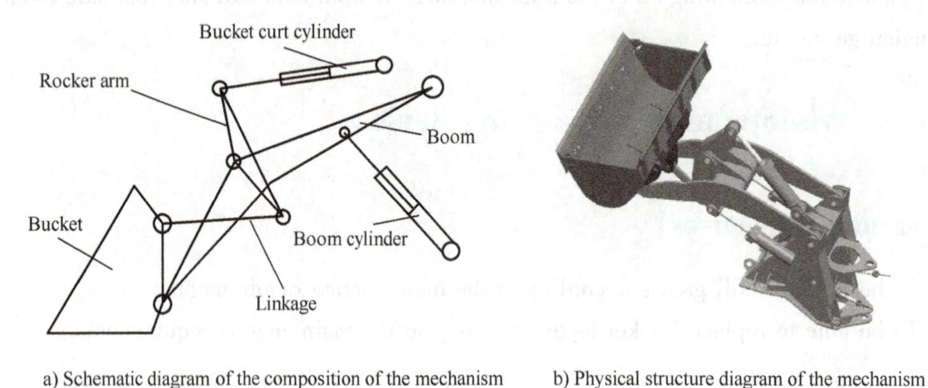

a) Schematic diagram of the composition of the mechanism　　b) Physical structure diagram of the mechanism

Fig. 2-2 The working device of the single rocker arm reversing six – link rods

In addition to the common single – rocker reversing six – link mechanism, a double – rocker reversing six – link mechanism is also used on some models, especially for large – tonnage loaders. The double – rocker arm means that there are also double bucket cylinders. The structure of the double rocker arm is different from the single one, but the principle is the same.

2. The fit clearance of pin shaft of the working device

Pins are used to connect the boom and the bucket of the loader, as well as the hydraulic cylinder of the boom, the hydraulic cylinder of the bucket and the working device. The pin is cylindrical in shape with clearance fit between the pin sleeve. It is mainly used as a connection between the two components to transmit transverse force or torque. During excavator operations, the pin and sleeve always rotate relatively frequently. If the lubrication is poor, the wear between the pin and the pin sleeve will be aggravated, resulting in larger fit clearance and affecting the working performance of working devices. The mating clearance of each pin shaft of working devices is shown in Table 2-3.

Table 2-3 Pin shaft fit state of the working device

No.	Pin shaft	Nominal size/mm	Assembly clearance/mm	Max allowable clearance after wearing/mm	Measures to be taken after the allowable value is exceeded
1	The hinge pin of the tie rod and rocker	$\phi90$	0.220 ~ 0.394	0.90	Replace pin shaft or bushing
2	The hinge pin of the tie rod and bucket	$\phi90$	0.220 ~ 0.394	0.90	Replace pin shaft or bushing
3	The hinge pin of boom and bucket	$\phi63$	0.190 ~ 0.338	0.80	Replace pin shaft or bushing
4	The hinge pin of boom and rocker	$\phi110$	0.240 ~ 0.414	1.00	Replace pin shaft or bushing
5	The hinge pin of swivel cylinder and rocker arm	$\phi90$	0.220 ~ 0.394	0.90	Replace pin shaft or bushing
6	The hinge pin of boom and carriage	$\phi90$	0.220 ~ 0.394	0.90	Replace pin shaft or bushing
7	The hinge pin of boom cylinder and boom	$\phi75$	0.220 ~ 0.348	0.85	Replace pin shaft or bushing

☞ [Task Implementation]

The loader bucket is a flat bucket type with teeth, and the main cutting edge and side cutting are wear - resistant plates, which can extend the service life of the bucket. The tooth body is welded to the main cutting edge, and the connecting pin of the tooth sleeve is fixed. The connection form between the tooth sleeve and the tooth body is shown in the cutaway view of *A—A* in Fig. 2-3. When the bucket teeth are worn excessively and the inner holes or deformation of the tooth sleeves are exposed, it will seriously affect the normal operation of the bucket and should be replaced.

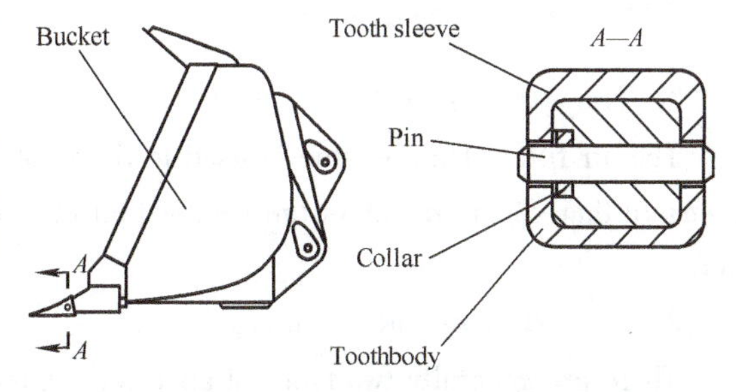

Fig. 2-3 Tooth sleeve installation diagram

The method and steps for replacing bucket teeth are as follows:

1) Park the bucket. Park the loader safely on a flat ground, place a wooden block or hard cushion of suitable height on the ground, make the bottom surface of the bucket fall flat on the cushion or wooden block, and turn off the engine.

2) Separate the tooth sleeves. Prepare a copper rod punch with a suitable outer diameter, hold one end of the punch against the bucket tooth pin, and gently beat the other end of the punch with a hammer. After pushing the bucket tooth pin out along the axis, remove the tooth sleeve and the clamp ring.

3) Clean up the bucket teeth. Use a rag or brush to clean up the dirt and rust on the surface of the tooth body, bucket tooth pin and clamp ring. If necessary, use diesel to clean it, and then dry it.

4) Install a new tooth sleeve. First, install the clamp ring in the groove on the side of the tooth body, and then install the new tooth sleeve on the tooth base. From the side of the clamp ring, penetrate the bucket tooth pin into the clamp ring, tooth body and tooth sleeve, and gently beat the bucket tooth pin with a hammer to make the bucket tooth pin flush with the tooth sleeve.

Task 2.3 Maintenance of the engine

Subtask 2.3.1 Inspection and replacement of air filters

☞ [Learning Objectives]

1) Select the air filter correctly according to the maintenance manual.

2) Replace the air filter in accordance with the maintenance specifications.

☞ [Work Task]

Replace the filter element of the air filter according to the specifications.

☞ [Relevant Knowledge]

1. The function of the air filters

The air filter of the engine is used to filter solid particles such as dust and impurities in the air to ensure that clean air enters the combustion chamber of the engine and reacts chemically with the fuel.

2. Classifications of air filters

There are generally two types of air filters: paper - based and oil - based filters. Because paper filters have the advantages of high filtration efficiency, lightweight, low cost, and easy maintenance, they have been widely used. The filtration efficiency of paper - based filter elements is as high as 99.5% or more, and the filtration efficiency of oil - based filters is 95% - 96% under normal circumstances.

When the engine is running, the air intake is intermittent, which causes the air in the air filter housing to vibrate. If the air pressure fluctuates too much, it will sometimes affect the air intake of the engine. In addition, the intake noise will also be increased at this time. In order to suppress the intake noise, the volume of the air filter housing can be increased, and some partitions are arranged in it to reduce resonance.

The filter elements of air filters are divided into two types: dry filter elements and wet filter elements. The dry filter element is made of filter paper or non – woven fabric. In order to increase the air passing area, most of the filter elements are processed with many small wrinkles. When the filter element is slightly polluted, compressed air can be used to blow it clean. When the filter element is severely polluted, it should be replaced with a new one in time. For dry – type filter elements, once they are immersed in oil or moisture, the filtration resistance will increase sharply. Therefore, avoid contact with moisture or oil when cleaning, otherwise must be replaced with a new one.

The wet filter element is made of sponge – like polyurethane material. When installing, drop some motor oil and knead well by hand to absorb extraneous objects in the air. If the filter element is polluted, it can be cleaned with cleaning oil. If the filter element is heavily polluted, it should be replaced with a new one.

If the filter element is severely blocked, the inlet resistance will increase and the engine power will decrease. At the same time, due to the increase in air resistance, it will also increase the amount of fuel sucked in, resulting in excessive mixing ratio, which will deteriorate the operation of the engine, increase fuel consumption, and easily produce carbon deposits.

The engine needs to inhale a large amount of air during operation. If the air is not filtered, the dust suspended in the air is sucked into the cylinder, which will accelerate the wear of the piston group and the cylinder. Larger particles entering between the piston and the cylinder will cause serious "scuffing of cylinder".

☞ [Task Implementation]

If the air filter is severely clogged, it will cause a negative pressure in the front pipeline of the diesel engine intake pipe and the alarm indicator light to light up. At this time, the main filter element of the air filter must be cleaned.

According to the diesel engine maintenance specifications, the main filter element of the air filter should be inspected regularly after it has been in use for 500 h. If the surface of the filter element is stuck with dust or damaged, it should be cleaned or replaced.

1. Methods and steps for the main filter element replacement

1) Turn off the engine, open the side door of the engine compartment, clean the dust on the outer surface of the air filter, then loosen the clasp, open the end cover, and gently pull out the main filter element from the filter holder, as shown in Fig. 2-4.

2) Use a rag to clean the dust on the inner surface of the air filter housing, and then use compressed air ($\leqslant$300 kPa) to blow along the circumference of the housing to make the dust emanate from the inside of the housing to the outside, as shown in Fig. 2-5.

3) Clean the main filter element, use compressed air to flush outward along the crease from the

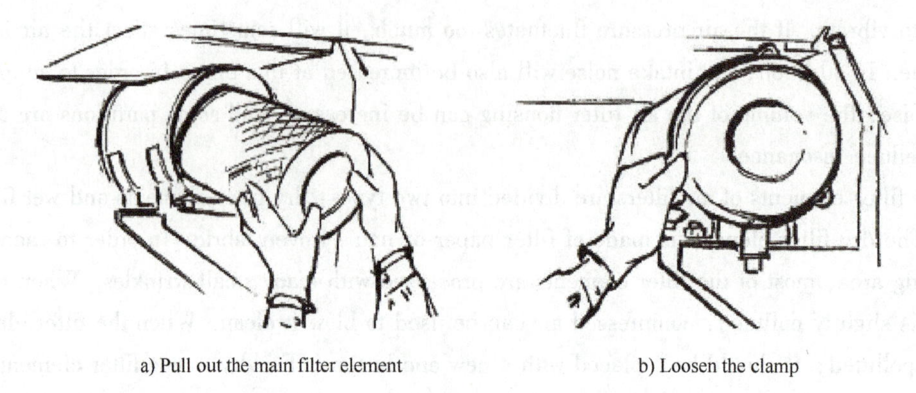

a) Pull out the main filter element　　b) Loosen the clamp

Fig. 2-4 Schematic diagram of removing the main filter element

inside of the main filter element, and then flush repeatedly along the crease from the outside and inside of the main filter element, respectively, to blow away the dust adhered to the surface of the main filter element, as shown in Fig. 2-6.

Fig. 2-5 Clean the innerwall of the air filter housing

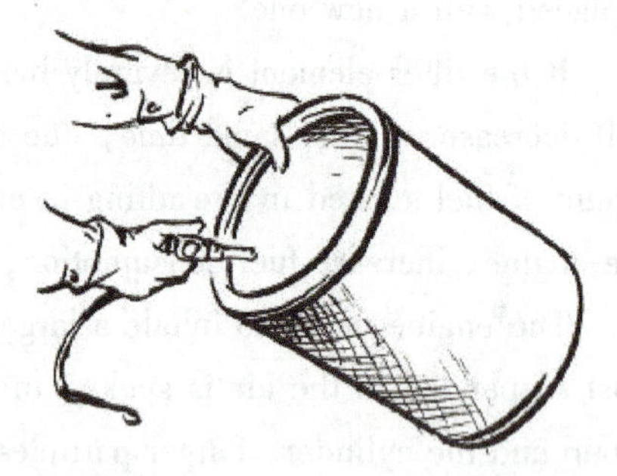

Fig. 2-6 Clean the main filter element

4) After the main filter element is cleaned, the lighting can be used to check whether dust remains on the surface of the filter element. Put a light or a flash light into the filter element (as shown in Fig. 2-7), so that the light is dispersed through the small hole of the filter element. If the filter element is clean, the light emitted from its grid is uniform. If small holes or particles are found on the surface of the filter element, as well as washers or seals are damaged, the main filter element should be replaced.

Fig. 2-7 Inspection by light

5) Install the main clean filter element into the air filter housing to ensure uniform sealing contact at the end of the main filter element. Tools should not be used to avoid damaging the main filter element.

6) Clean and install the end cap of the air filter, ensuring that the gasket of the inner cover of the air filter is in uniform contact with the air filter housing. When installing the end cover, make sure the dust discharge port of the end cap face downward. When the end cover is completely

closed, snap the holdfast sub to make the end cover firmly on the filter housing.

2. Notes for main filter element replacement

The main filter element should be replaced after it has been cleaned for 6 times. Even if it has not been cleaned for 6 times, it should be replaced once a year. When replacing the main filter element, replace the safety element inside the main filter element at the same time.

If the air filter obstruction warning light is still on or black smoke is still emitted after cleaning the main filter element, the safety filter element should be replaced with a new one.

Subtask 2. 3. 2 Check belt tension of the compressor

☞ [Learning Objectives]

1) To be able to judge the belt tension according to the technical specifications.

2) To be able to adjust the belt tension force according to the specifications.

☞ [Work Task]

Adjust the compressor belt tension according to the specifications.

☞ [Relevant Knowledge]

Belt drive is a form of mechanical transmission. The drive belt can drive the rotating motion and power of the diesel engine crankshaft to the rotating shaft of the air conditioning compressor through the belt. Since the belt drive transmits power through friction, the drive belt should be adjusted to a moderate tension to obtain the right friction. The tension of the drive belt is adjusted by the tensioning pulley. If there is no tensioning pulley, the tension is adjusted directly by adjusting the wheel treads. The tensioning device of diesel air conditioning compressor belt is shown in Fig. 2-8.

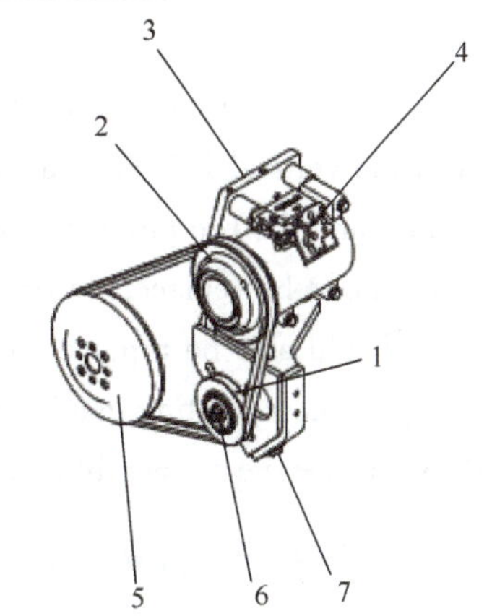

Fig. 2-8 The tensioning device of the compressor belt

1—Belt tensioning pulley 2—Pulley
3—Compressor mounting bracket 4—Compressor
5—Engine pulley 6—Locking nut 7—Adjusting bolt

If the surface of the air conditioning compressor belt and the side of the pulley groove are bright, and there is a squeaking noise when starting the air conditioning, it means that the belt has slipped seriously and the belt has worn out. Replace the belt or pulley. If the belt is too loose, it should be adjusted, otherwise it will easily cause poor cooling of the air conditioning system.

In fact, after the air conditioning is in operation for a period of time, the compressor belt will crack and become disconnected, or when the internal core wire reaches its fatigue life, it is prone to fracture, so it should be replaced in time.

In the absence of professional tools, the tension of the belt can be judged by hand pressing.

Turn off the engine and use the thumb to press down slightly on the belt and the belt edge perpendicular to the cutting midpoint of the two belt pulleys (air conditioning compressor pulley and engine pulley). The pressing force is about 58.5 N. At this time, the deflection A generated by the belt should be 5 ~ 8 mm. Otherwise, it is necessary to adjust the belt tension.

[Task Implementation]

1. Tension adjustment steps for air conditioning compressor belts

1) Use a wrench to loosen the lock nut 6 and adjust the position of the adjusting bolt 7 according to the belt tightness to make the belt tension change.

2) Use the thumb to exert a slight force of about 58.5 N and press the belt downward against the vertical belt edge at the midpoint of the tangent edge of the two pulleys, as shown in Fig. 2-9.

3) If the deflection $A > 8$ mm, tighten the adjusting bolt 7 and lock the locking nut 6 after the belt has been tensioned to a suitable position.

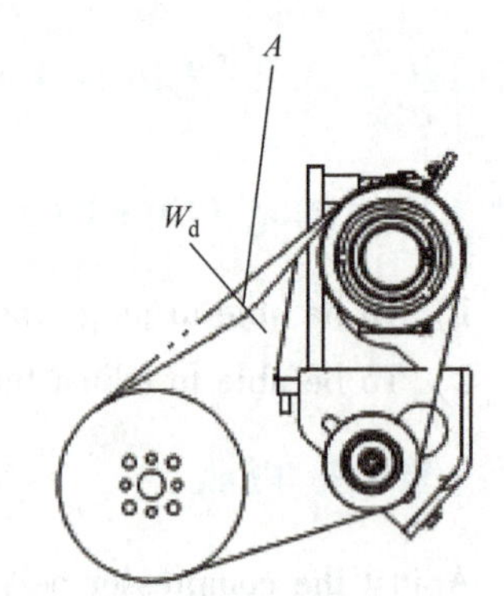

Fig. 2-9 Belt tensioning force adjustment schematic diagram

W_d—Adjustment force A—Deflection

2. Notes

When the belt cannot be tensioned, i. e., when a load 58.5 N is added to the vertical belt edge at the midpoint of the tangent point of the belt and the two pulleys, and the belt deflection $A > 8$ mm, the belt should be replaced in time.

The belt should also be replaced when the belt has a burr or a crack.

According to the operating conditions of the whole machine and the wear of the belt, it is recommended to replace the air conditioning belt every 1000 h.

Subtask 2.3.3 Oil replacement

[Learning Objectives]

1) To be able to select the correct oil.

2) To be able to change the oil according to the specifications.

[Work Task]

Change the engine oil according to the specifications.

[Relevant Knowledge]

Engine oil is mainly composed of base oil and additives. The base oil is the main component of the engine oil and determines the basic properties of the engine oil. Additives can make up for and improve the deficiencies in the performance of base oil and give it some new performances, which is an important part of the engine oil.

1. The function of engine oil

(1) Lubrication When the engine works, the parts do relative movement between them. Through the lubrication of the oil, a layer of adsorption oil film is formed between the moving parts, which can reduce the wear between the parts inside the engine. So the mechanical efficiency can be improved and the working life of the engine can be extended.

(2) Cooling and heat dissipation During the lubrication process, the engine oil can take away the excess heat generated by the friction between the engine parts, which plays an indirect role in heat dissipation, and prevent the parts from burning out due to excessive temperature.

(3) Cleaning During the repeated relative movement of various engine components, some metal debris and carbon particles will be produced due to friction, which can be taken away by the engine oil to prevent particle wear and the aggravation of aging and deformation of parts.

(4) Anti - corrosion and anti - rust The oil is evenly distributed on the surface of the parts during operation, and the additives with anti - rust effect contained in the oil will form an adsorption layer on the metal surface, which can make it isolated from the outside air, water and gas, avoiding oxidation and rusting of the parts and reducing corrosive wear.

(5) Seal The oil has a certain physical viscosity. When adhered to the surface of parts, it can isolate different working environments in the combustion chamber and have a good sealing effect. In addition, when diesel fuel is compressed in the engine combustion chamber, it will produce high - pressure gas, which in turn pushes the piston to do its work. The engine oil can seal the gap between the piston and the valve, reduce pressure loss and improve mechanical efficiency.

(6) Vibration damping When the diesel engine starts, accelerates or undergoes a large change in load, the piston pin, the bearing bush of the crankshaft, large and small ends of the connecting rod and other parts are subjected to violent changes brought by vibration. The oil film formed by engine oil of appropriate viscosity covering these parts can absorb part of the energy of the impact load, reduce the negative impact of the vibration of the parts on the normal operation of the engine, and buffer the load.

2. Selection of engine oil

There are two main bases for choosing the quality level of diesel engine oil, one is based on the sum of mechanical load and thermal load of the engine, which is expressed by the reinforcement factor, the other is based on the harshness of engine working conditions.

There are 11 levels of diesel engine oil, namely CA, CB, CC, CD, CE, CF, CF -4, CG -4, CH -4, CI -4, CJ -4, etc. "C" represents diesel engine oil. " -4" means that the oil is suitable for 4 - stroke diesel engines. "S" stands for gasoline engine oil. If the letters "S" and "C" exist at the same time, it means that the oil is a universal type for gasoline and diesel engines.

The performance of the engine oil will be better than that of the previous one with each letter increment of the engine oil quality level. As for what engine oil is good for diesel engines, it actually needs to be decided according to the specific worksituation.

3. Analysis of the causes of engine oil loss

When the engine is working, the cylinder wall, piston ring and piston are all filled with oil.

When the fit clearance of the three is abnormal, the oil between the cylinder wall and the piston will rush into the combustion chamber, which is the main cause of oil channelling.

After the engine works for a certain time, the radial size of the piston ring decreases, the elasticity is weakened, the opening gap becomes larger, and the sealing and oil scraping effects become worse. The clearance between the piston ring and the ring groove is increased, which strengthens the oil pumping effect of the piston ring, and the oil is scraped into the combustion chamber when the piston goes up.

After oil channelling, it is easy to produce carbon and accelerate cylinder wear, forming a vicious circle. The piston ring burns the engine oil after it gets stuck. When the piston ring gets stuck in the ring groove, the piston ring loses elasticity and the seal deteriorates. Not only does the piston bring a large amount of engine oil into the combustion chamber when it goes up, but also when the piston is doing work, a large amount of high-pressure gas rushes into the crankshaft, which increases the pressure in the crankshaft. The splashing oil enters the air inlet through the crankshaft ventilation device, causing the engine oil and air to enter the combustion chamber together for combustion.

The overhaul quality of the engine has a greater impact on the cylinder oil channelling. Oil channelling will occur when there are cylindrical degree errors, when the surface roughness of the cylinder are too large, when the piston ring is not matched with the piston according to the technical requirements, when the piston deviates from the cylinder during assembly, or when the centreline of the cylinder and the centre line of rotation of the crankshaft exceed the specified requirements.

At the same time, it will accelerate the cylinder channelling to varying degrees if the piston carbon removal is not clean, if the piston ring is installed reversely, if the ring mouth is not reasonably distributed, or if the ring mouth clearance is too large. Oil channelling occurs soon after the engine runs, which is partly caused by the unqualified quality of the piston ring.

☞ [Task Implementation]

After using the oil for a period of time, the amount and characteristics of the oil will change, so it is necessary to check the oil level and replace the oil regularly. When the oil is replaced, the oil filter should be replaced at the same time.

1. Preparation

Park the loader reliably on a flat site and turn the engine off. Wait for about 10 min and allow the engine oil in the crankshaft fully flow back to the engine oil pan.

2. Check the oil level

Open the engine hood and find the oil level dipstick. The location of the engine oil filler and the oil level dipstick are shown in Fig. 2-10.

Pull out the oil level dipstick, wipe it with a clean cloth, re-insert it into the engine oil level port to the end, and then pull it out to check that the oil level should be between the "L" scale and the "H" scale of the dipstick, as shown in Fig. 2-11. If the oil level is under the "L" scale, it means the amount of oil is too little; if the oil level is above the "H" scale, it means the amount of oil is too high.

Fig. 2-10 Schematic diagram of oil level dipstick

1—Oil level dipstick 2—Oil filler

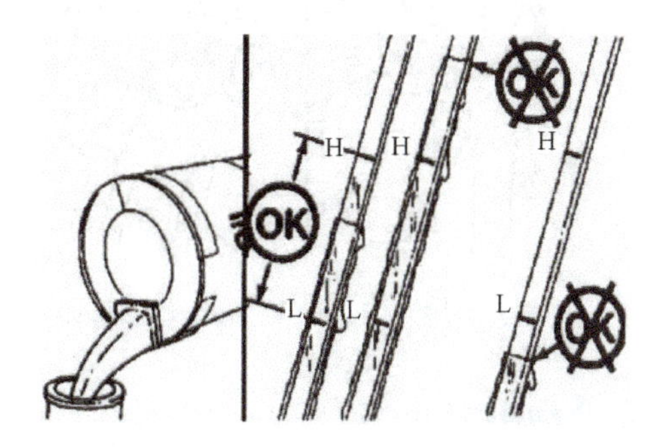

Fig. 2-11 Schematic diagram of oil level check

H—The upper limit of oil level L—The lower limit of oil level

3. Discharge oil

Remove the oil emission plug at the bottom of oil pan of the engine and discharge the oil into the container prepared in advance, as shown in Figure 2-12. The oil should be discharged cleanly, and if necessary, clean oil should be poured into the oil chamber for cleaning. The capacity of the container must be larger than the amount of the oil contained to avoid contaminating the site by the overflowing oil .

After the oil is completely drained, screw on the oil drain bolt, as shown in Fig. 2-13.

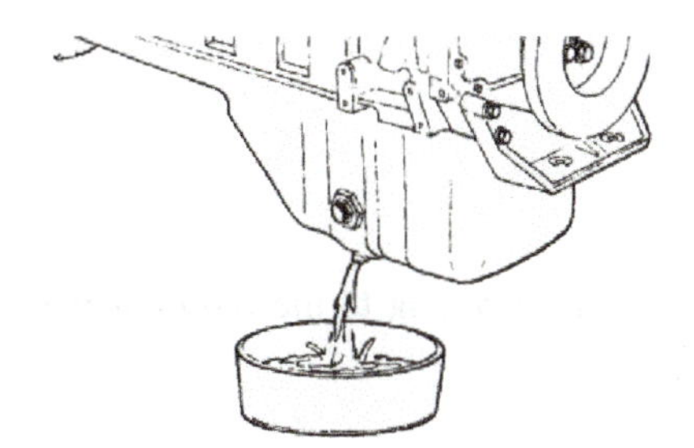

Fig. 2-12 Drain the oil

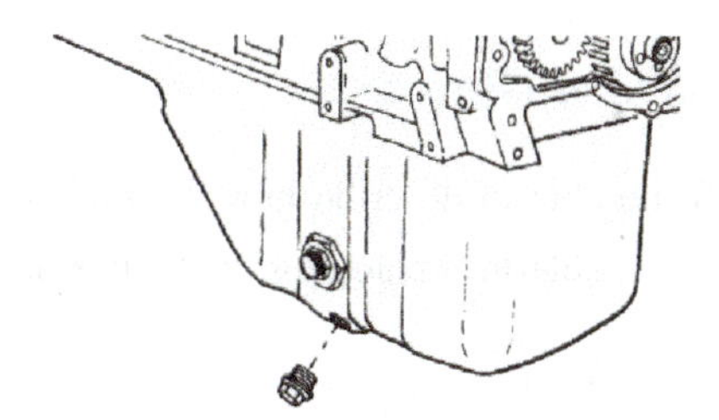

Fig. 2-13 Screw on the oil drain bolt

4. Replace the oil filter

Clean the outer surface of the oil filter, remove the oil filter with a wrench, and clean the contact surface of the gasket on the mounting base, as shown in Fig. 2-14. If the old O – ring seal is stuck to the mounting base, it should be removed by a tool.

Install the new O – ring, fill the oil filter with clean oil, and apply a clean coat of the oil to the gasket surface.

Install the oil filter onto the mounting base (as shown in Figure 2-15), tighten it with your hands until the oil filter gasket surface touches the mounting base, and then use the belt wrench to tighten the oil filter to the requirements.

Note that the oil filter should be filled with clean oil before installation, otherwise the engine would be damaged due to the lack of engine oil.

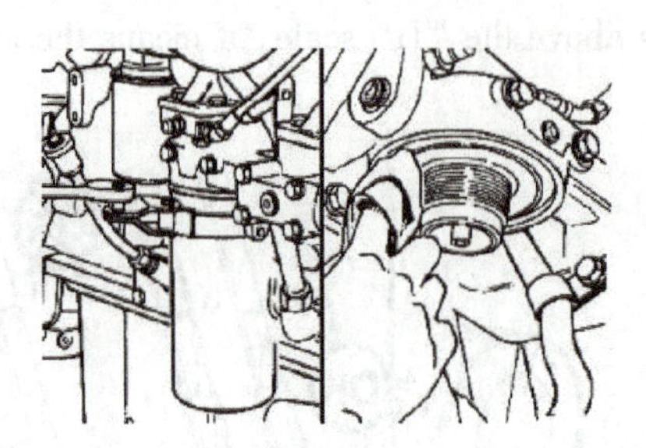

Fig. 2-14 Clean the mounting base

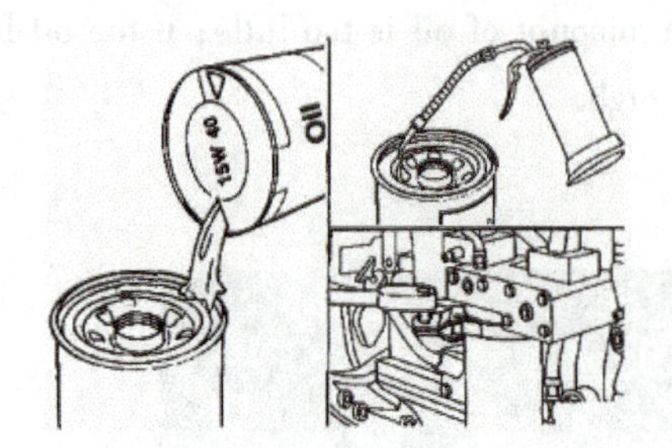

Fig. 2-15 Install the oil filter

5. Replace the oil

Use a rag to clean the oil filler, unscrew the plug of the oil filler, and fill clean oil from the oil filler until the oil level at the scale "H" of the dipstick. Run the engine at the idle speed and check whether the oil filter and oil drain plug have leaked. Turn off the engine, wait about 10 min, let the oil fully flow back to the oil pan, and check the oil level of the engine again. If the oil is insufficient, please replenish the oil until the oil level dipstick is at scale "H".

6. Tidy up the site

After replacing the oil filter and the oil, screw the plug of the oil filler, clean promptly the oil scattered during filling the oil, pack up the tools and clean the site.

Subtask 2. 3. 4 Replace fuel filter

☞ [Learning Objectives]

1) To be able to describe how the fuel filter works.

2) To be able to replace the fuel filter and pre – filter according to the requirements.

☞ [Work Task]

The fuel filter and pre – filter of a loader should be replaced due for replacement and maintenance.

☞ [Related Knowledge]

1. Role of the fuel filter

The role of the fuel filter is to filter out harmful particles and water in the fuel gas system of the engine to protect the fuel pump nozzle, cylinder liner, piston ring, etc. , and reduce wear and avoid blockage.

2. Types of the fuel filters

According to different working media, fuel filters can be divided into diesel filters, gasoline filters, natural gas filters and other types. The engine installed in the loader is generally a diesel engine, so the diesel filter is used.

The structure of the diesel filter is roughly the same as the oil filter. And there are two types of diesel filters: replaceable and spin – on. However, its requirements for working pressure and oil temperature resistance are much lower than the oil filter, while its requirements of filtration efficiency are much higher. Filter paper is mostly used in the filter element of diesel filter, while hair felt or high – molecular material is also used.

Diesel filter can be further divided into pre – filter and diesel fine filter. Pre – filter also called diesel water separator, is mainly used to separate the water from the diesel fuel. Water is extremely harmful to the diesel fuel supply system, might rust, wear, block or even deteriorate the combustion process of diesel fuel. When burning sulphur in diesel fuel would react with water to generate sulphuric acid, which might corrode engine parts.

3. How the fuel filter works

There are hand oil pump (filter seat), filter parts, water level sensor, temperature sensor, heater, seal, pipe connector and other parts in the diesel fuel filter, filter elements are core parts of filtration. Figure 2-16 is a diesel fuel filter.

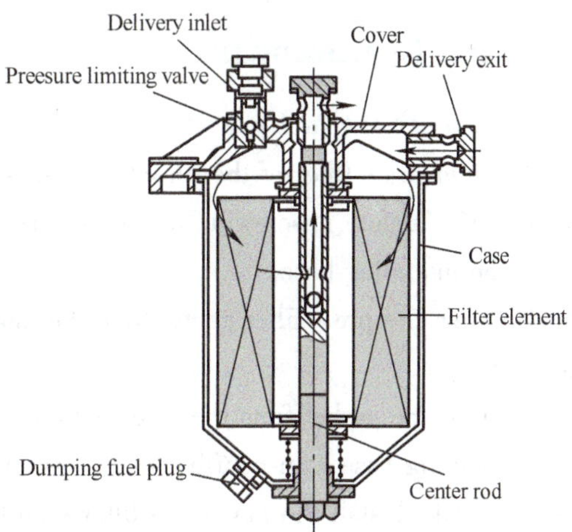

Fig. 2-16 Diesel fuel filter

The filter element is of a porous structure. When the fluid passes through the filter element, the impurity particles with particle size in the fluid are intercepted by the porous body on the surface; while the particles smaller than the porous body pass through the porous body with the fluid. The pore size of the filter material and the diameter of impurity particles are not wholly round, but equivalent pore size or equivalent particle size. So the particle size it intercepted is much smaller than the pore size of the filter material.

Impurity particles in the fluid make irregular Brownian motion all the time. When they move to the pore wall of the porous body, they are stuck in the small area of the pore wall, and filtered out by the porous body.

When the oil flow with tiny water droplets enters the filter material, due to the "affinity" of the filter material to the oil, the oil flow passes smoothly through the micro – pores of the filter material; because of the "alienation" of the filter material to the water, the tiny water droplets could not pass through the micro – pores of the filter material. Due to the infiltration effect of tiny water droplet on the fluff of the filter material, the tiny water droplets are retained on the fluff of the filter material. The intercepted tiny water droplets gradually increase, and grow into big water droplets. Under the action of oil pressure, the water droplets are pressed to the wedge – shaped passage of the filter element, forming bigger water droplets. Under the action of oil pressure, after the water droplets pass through the filter material, they sink downward under the action of their own mass and fall in the water collection cup.

4. The maintenance of the fuel filter

The fuel filter needs regular maintenance. If the filter is not replaced regularly, it might be clogged with contaminants and restrict fuel flow, leads to a significant decrease in engine performance because the engine couldn't draw enough fuel to run properly.

Diesel fuel contains water, and water is denser than diesel fuel, so the fuel filter with a bowl shape is often installed as the first filter in the diesel engine, allowing water to settle at the bottom of the fuel filter and then draining the water by opening a valve at the bottom of the fuel filter to ensure that the diesel fuel in the combustion chamber contains as little water as possible.

Oil and water separation sensors are installed in some fuel filters. When the separated water level in the fuel filter reaches a certain height, sensors send a signal to the engine control unit, prompting the driver to drain the water manually.

☞ [Task Implementation]

1. Replace the fuel pre – filter

The fuel pre – filter of the diesel engine is installed on the side of the fuel pump, as shown in Figure 2-17. Before unscrewing the pre – filter, clean the periphery of the pre – filter and the surface of the mounting base.

Unscrew the pre – filter gently from the mounting base with a wrench and remove it, as show in Fig. 2-18.

Remove the gasket from the threaded connector of the mounting base and clean the sealing face of the mounting base with nonfiber cloth, as shown in Fig. 2-19. Check the gasket and replace it if it is damaged. Install the new pre – filter onto the pre – filter seat.

Fig. 2-17 Fuel pre – filter

Fig. 2-18 Remove the pre – filter

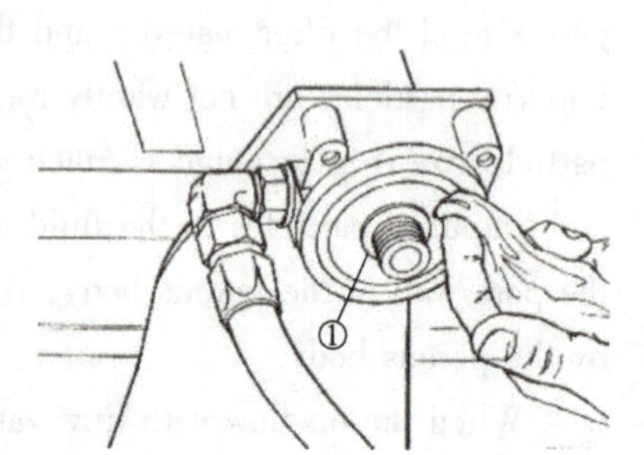

Fig. 2-19 Clean the sealing face

2. Replace fuel filter

Fuel filter and pre – filter are installed on the same side of the engine, as shown in Fig. 2-20. Before removing the fuel filter, clean the area around the filter and the mounting base, and then remove the fuel filter from the mounting base with a wrench.

Fig. 2-20 Fuel filter

Clean the fuel filter mounting base and remove all old gaskets.

Install a new gasket onto the threaded joint of the fuel filter's mounting base, apply a coat of oil to the sealing face of the fuel filter, fill the filter with clean fuel (Fig. 2-21), and then install it onto the fuel filter's mounting base, screw the fuel filter onto the mounting base with your hands, and tighten it 1/2 to 3/4 turns after the filter's gasket is on the mounting base. Don't over-tighten it by mechanical means to avoid damaging the filter, as shown in Fig. 2-22.

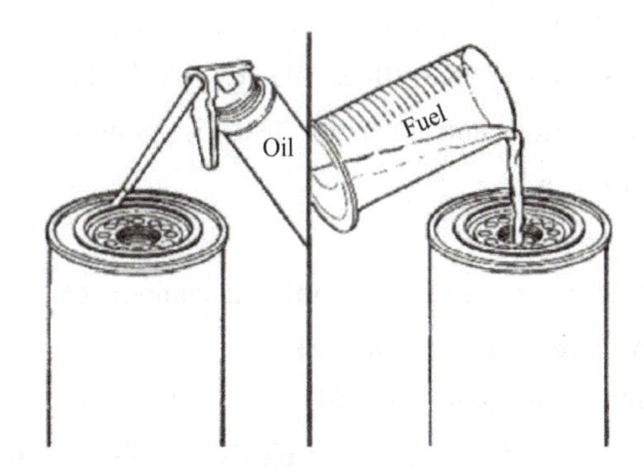

Fig. 2-21 Apply oil and fill fuel

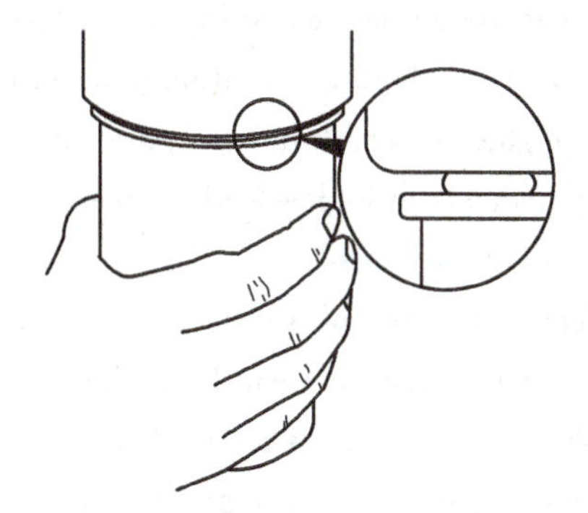

Fig. 2-22 Install fuel filter

Subtask 2.3.5 Replace the coolant

☞ [Learning Objectives]

1) To be able to select the coolant.

2) To be able to replace the coolant according to the specification.

☞ [Work Task]

A loader shows high water temperature. After the check, the coolant is found being reduced and to the maintenance and replacement time. The coolant should be replaced.

☞ [Related Knowledge]

1. Role of coolant

Coolant can protect the engine's cooling system from rust and corrosion, effectively inhibit the formation of scale, prevent overheating of the water tank, reduce the evaporation of coolant, provide lubrication for the pump thermostat and other components.

(1) Anti-freeze in winter The freezing point of selected coolant is generally required to be about 10 ~ 15 ℃ lower than the minimum temperature in the use area to prevent expansion and cracking of the water tank and the engine cylinder caused by the freezing of the coolant after parking the car in winter.

(2) Anti-corrosion In the cooling system, radiator, water pump, cylinder block and cylinder head, sub-pipe and other components are made of steel, cast iron, brass, copper, aluminium, welding tin and other metals. Due to different electrode potential of different metals, they are prone

to electrochemical corrosion in the electrolytes. A certain amount of anti - corrosion additives are added in the coolants to prevent corrosion in the cooling system.

(3) Anti - scale In the cycle the coolants would produce a certain amount of scale. The increase of scale might lead to the blockage of the circulating pipes and affect the heat dissipation function of cooling system. Therefore, when selecting and adding coolants, appropriate ratio of coolants should be chosen according to the specific situation.

(4) Anti - boiling Boiling point of coolants is usually more than 105 ℃. Coolants can withstand a higher temperature than pure water without boiling, and meet the heat dissipation and cooling requirements of high - load engines to a certain extent.

2. Types of coolants

Coolant consists of water, antifreeze and additives. According to different components of antifreeze, common coolants can be divided into glycol type and glycerol type.

Glycol type coolant is formulated by using glycol as antifreeze and adding a small amount of anti - foam, anti - corrosion and other comprehensive additives. Since glycol is easily soluble in water, it could be formulated into coolant with various freezing points, and its lowest freezing point could be - 68 ℃, this coolant has characteristics of high boiling point, low foam tendency, good viscosity temperature, anti - corrosion and anti - scale, and is a relatively ideal coolant. At present, at home and abroad almost all the coolants used in engines and sold in the market use glycol coolant.

Glycerol coolant has characteristics of high boiling point, small volatility, not easy to catch fire, non - toxic and little corrosion, but poor effect of reducing freezing point, high cost, expensive, so it is only used in a few Nordic countries .

Coolant circulates in the engine cooling system to take away the excess heat energy generated during engine operation, so engine could operate at normal working temperature. When coolant is not enough, engine water temperature would be too high to damage engine parts. When the coolant is insufficient, it should be added in time.

3. Selection of coolant

Coolant models are generally classified according to the freezing point, mainly including - 25 ℃, - 30 ℃, - 35 ℃, - 40 ℃, - 45 ℃ and - 50 ℃. When choosing coolant, the freezing point should be a bout10 - 15 ℃ lower than the minimum temperature of the vehicle operating environment.

☞ [Task Implementation]

According to regular maintenance requirements, coolant should be replaced after being used for 2000 h.

1. Inspection of coolant

Under normal circumstances, the level of engine coolant should be between the upper scale mark and the lower scale mark of the reservoir. If the liquid level is very high, there might be something wrong with the cooling system, such as a large amount of gas or other liquids mixed in the coolant. If the coolant drops below the lower scale mark, replenish the coolant.

When it is necessary to check the coolant, wait until the temperature of the engine coolant drops below 50 ℃, and then slowly unscrew the water filler cap of the water radiator to release the internal pressure of the cooling system, so as to prevent high temperature steam or high temperature coolant from spraying out of the water filler.

Check whether the coolant level is within 10 mm below the water filler, and replenish the coolant if necessary. Check the seal of the water filler cap of the coolant radiator, and replace it if it is damaged.

2. Replace coolant

When replacing the coolant, make sure to drain all the coolant inside the engine, add clean water, start the engine, idle for 10 min, drain the water after flameout, and then add new coolant.

When refilling coolant, it should be avoided when the engine is at a high temperature.

1) Wait until the coolant temperature drops below 50 ℃, then slowly unscrew the water filler cap of the water radiator to release the pressure.

2) Open the drain valve at the bottom of the water radiator (Fig. 2-23), drain the engine coolant and hold it in a container. After the engine coolant is drained, close the drain valve at the bottom of the water radiator.

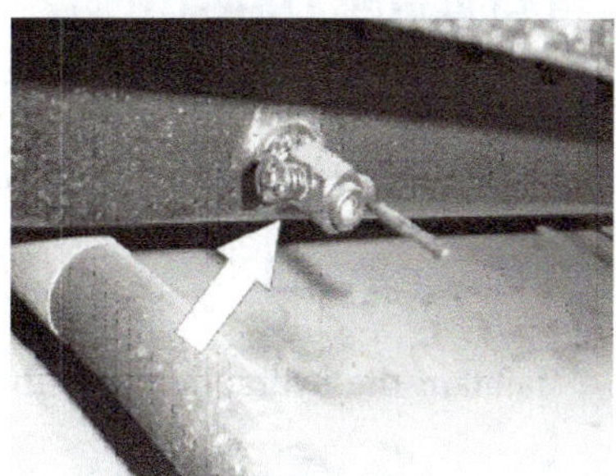

Fig. 2-23 Coolant drain valve

3) Check all water pipes and pipe clamps of the cooling system for damage and replace them if necessary. Check the water radiator for leakage, damage and dirt accumulation, and clean and repair it as needed.

4) Add the cleaning liquid mixed with water and sodium carbonate to the engine cooling system. The mixing ratio is 0. 5 kg sodium carbonate per 23 L of water. The fluid level should reach the level of normal use of the engine and remain stable for 10min.

5) Keep the water filler cap of the water radiator open and start the engine. When the coolant temperature rises above 80 ℃, run the engine for another 5 min, then shut down the engine and drain the cleaning fluid.

6) Add clean water to the engine cooling system to the normal use level and keep it unchanged for 10 min Keep the water filler cap of the water radiator open and start the engine. When the coolant temperature rises above 80 ℃, run the engine for another 5 min, then shut down the engine and drain the water in the cooling system. If the drained water is still dirty, the system must be cleaned again until the drained water becomes clean. After the cleaning water is drained, close all drain valves.

7) Fill coolant. Open the water filler cap of the water radiator and slowly add the coolant until the liquid level reaches within 10 mm below the water filler of the water radiator and remains stable for 10 min.

8) Keep the water filler cap of the water radiator open, start the engine, first run at low idle speed for 5 min, then run at high idle speed for 5 min, and make the coolant temperature above 85 ℃.

9) Check the coolant level again. If necessary, replenish the coolant until the level reaches within 10 mm below the water filler of the water radiator.

10) Replace the water radiator coolant, at the same time, the coolant in the sub - tank should be replaced.

Task 2.4 Maintenance of hydraulic system

Subtask 2.4.1 Maintenance of hydraulic oil

☞ [Learning Objectives]

1) To be able to make regular maintenance plan for the hydraulic system.

2) To be able to inspect and replace hydraulic oil as required.

☞ [Work Task]

Maintain the hydraulic oil of the loader according to the specification requirements.

☞ [Related Knowledge]

1. Role of hydraulic oil

Hydraulic oil is the hydraulic medium used in the hydraulic system using liquid pressure. It plays the role of energy transmission, anti - wear, system lubrication, anti - corrosion, rust prevention and cooling in the hydraulic system. For hydraulic oil, first of all, it should meet the requirements of hydraulic devices for liquid viscosity under working temperature and starting temperature. Since the viscosity change of hydraulie oil is directly related to hydraulic action, transmission efficiency and transmission accuracy, it is also required that the viscosity temperature performance and shear stability of the hydraulic oil should meet the various requirements of different applications.

2. Classification of hydraulic oil

In GB/T 7631.2 -2003, HH, HL, HM, HR, HG, HV and HS hydraulic oils belong to mineral oil hydraulic oils. There are many kinds of such oils, and their usage accounts for more than 85% of the total hydraulic oil. Most of the hydraulic oils commonly used in the hydraulic systems of automobiles and construction machinery also belong to this category.

3. Selection of hydraulic oil

The variety of hydraulic oil should be selected according to the working environment and working conditions. When selecting the hydraulic oil used by the hydraulic equipment, it should be comprehensively considered and judged from the aspects of working pressure, temperature, working environment, hydraulic system and component structure, material and economy. Environmental factors include ground, underground, indoor, outdoor, coastal, cold areas, high temperatures, and open fires. Operating conditions include pump type, pressure, temperature, material, sealing material and operating time. Oil properties mainly refer to the physical and chemical properties of hydraulic oil. Economy mainly refers to the use time, oil change period and price of hydraulic oil.

4. Maintenance of hydraulic oil

The hydraulic oil of the loader hydraulic system should have the following properties:

1) Suitable viscosity and good viscosity temperature performance to ensure accurate and sensitive power transmission under the condition of changing working temperature, and ensure the normal lubrication of hydraulic components.

2) Good rust resistance and anti – oxidation safety, not easy to oxidize and deteriorate under high temperature and high pressure, and long service life.

3) Good anti foam property, the foam produced by oil under continuous working conditions of mechanical stirring is easy to disappear so as to make the power transmission stable and avoid the accelerated oxidation of hydraulic oil.

The hydraulic oil of loader hydraulic system generally adopts anti – wear L – HM series hydraulic oil, which is divided into four grades: 22, 32, 46 and 68 according to kinematic viscosity. Anti – wear hydraulic oil has good anti – rust and anti – oxidation properties, especially in anti – wear.

☞ [Task Implementation]

1. Check hydraulic oil quantity

Before checking the hydraulic oil quantity of loader, under normal circumstances, hydraulic cylinder, hydraulic pipeline, radiator and other hydraulic parts should be filled with hydraulic oil.

Drive the machine to a flat site, align the front and rear frame without included angle, retract the bucket to the limit position, lift the boom to the highest position at full speed, and push the boom control lever to the "down": position at idle speed to make the boom drop to the lowest position at a uniform speed. Put the bucket horizontally on the ground, then turn off the engine and remove the key, and push the control lever back and forth, left and right to release the pressure.

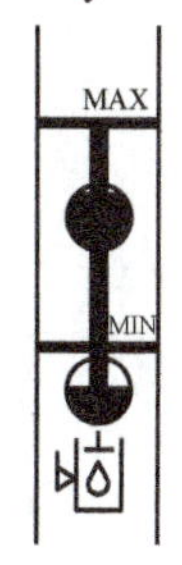

Fig. 2-24 Level scale of oill evel meter

Keep the hydraulic oil temperature within 50 ± 5 ℃, and checkthe hydraulic oil tank level meter when there is no bubble on the level meter. At this time, the oil level should be within the green range in the level meter, that is, between the MAX line and the MIN line, as shown in Fig. 2-24.

When checking the oil level, if it is found that the oil level is higher than the green range of the liquid level meter (when there are bubbles on the liquid level meter), even the oil level is above the maximum oil level line, and the oil cannot be drained. Check whether the oil level is within the green range of the liquid level meter after the bubbles are eliminated. If the oil level is lower than the green range, the hydraulic oil must be replenished promptly, and the oil level must be checked again after replenishment.

2. Replenish hydraulic oil

The hydraulic oil tank of the loader is generally installed on the side or behind the cab. As shown in Fig. 2-25, the hydraulic oil tank is installed behind the cab. Park the loader safely on a flat ground, open the mounting bolts of the upper channel plate of the hydraulic oil tank and the mounting bolts of the air filter cover plate on the rear cover, and remove the channel plate and cover plate, as shown in Fig. 2-26.

Fig. 2-25 Installation position of hydraulic oil tank

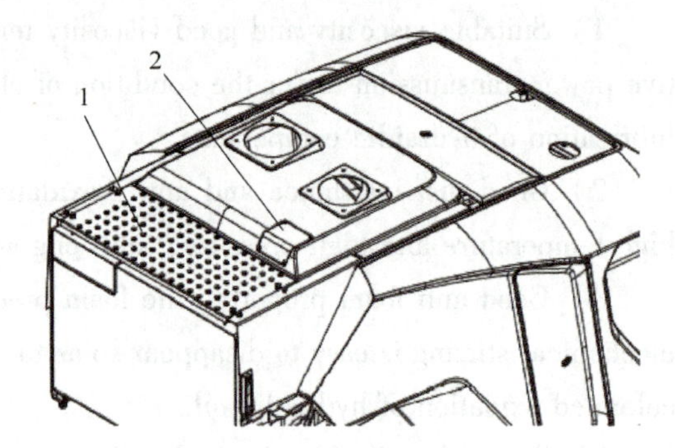

Fig. 2-26 Hydraulic oil tank cover

1—Channel plate 2—Cover plate

Clean the contaminants on the surface of air filter and oil filler flange. Slowly loosen the air filter cover of the hydraulic oil tank (as shown in Fig. 2-27) to release the pressure, remove the mounting bolt of the oil filler flange, remove the flange (as shown in Fig. 2-28), add hydraulic oil from the oil filler port, and check the hydraulic oil level .

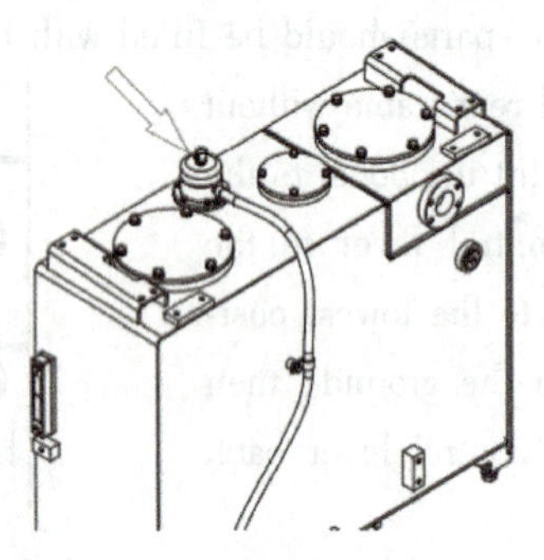

Fig. 2-27 Air filter

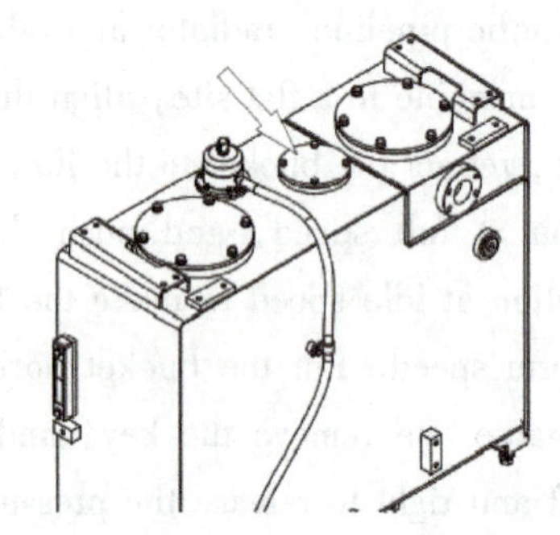

Fig. 2-28 Oil filler flange

After the hydraulic oil level meets the requirements, install the oil filler flange of the hydraulic oil tank, tighten the air filter cover, and install the upper channel plate of the hydraulic oil tank and the air filter cover plate.

3. Replace hydraulic oil

Park the loader on a flat and open site, pull up the parking brake and install the bogie locking device, as shown in Fig. 2-29. Start the engine and run it at idle speed for 10 min, during which repeatedly lift or lower the boom, tilt the bucket forward or backward, etc.. Heat up the hydraulic oil and keep it within the range of (50 ±5)℃.

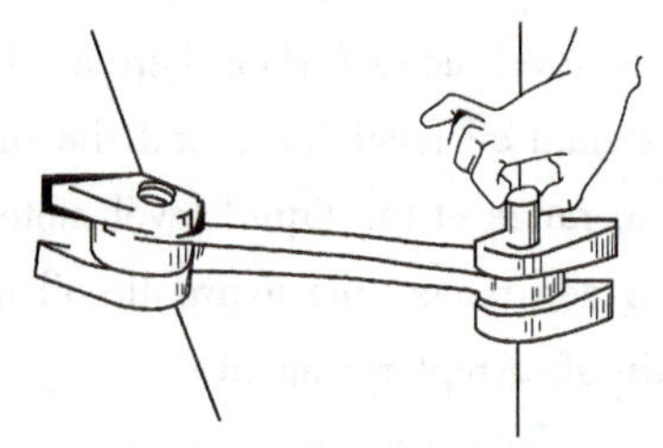

Fig. 2-29 Lock bogie

Raise the boom to the highest position, tilt the bucket back to the maximum position, and shut down the engine. Operate the working device at the unloading position to turn the bucket turn forward under the action of self weight and drain the oil in the bucket cylinder; after the bucket is turned to the position, operate the working device in the lower position, and the boom will descend under the action of self weight to drain the oil in the boom cylinder.

Disassemble the fixing clamp bolt of the hydraulic oil tank drain pipe located on the inner side of the left web of the rear frame, lead out the drain pipe from the hinge, unscrew the open oil plug, drain the hydraulic oil, and use a container to hold it. At the same time, open the air filter cover to speed up the oil discharge process. The web and fixed pipe clamp are shown in Fig. 2-30.

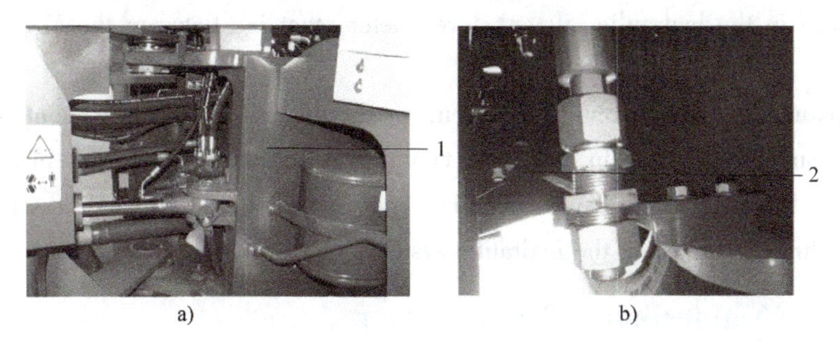

Fig. 2-30 The web and fixed pipe clamp

1—Web 2—Fixed pipe clamp

Open the oil drain valve of the hydraulic oil radiator (as shown in Fig. 2-31) to drain the residual hydraulic oil in the hydraulic oil radiator. Take out the oil return filter element end cap and other components of the hydraulic oil tank, take out the oil return filter element and replace it with a new one, as shown in Fig. 2-32.

Fig. 2-31 Oil drain valve

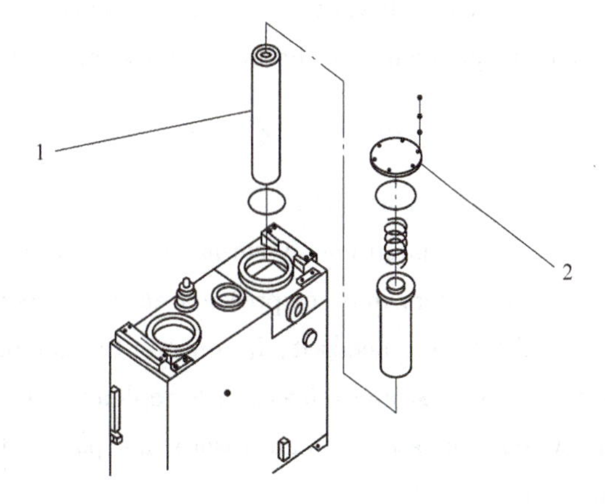

Fig. 2-32 Accessories of oil return filter element

1—Oil return filter element 2—End cover

Open the oil filler flange and take out the internal oil filter element. Clean the oil filler flange and refuelling filter with non – flammable solvent, dry them or blow them thoroughly with compressed air.

Install the oil drain plug, oil return filter element, refuelling filter and drain valve of the hydraulic oil radiator of the hydraulic oil tank. Add clean hydraulic oil from the oil filler port of the hydraulic oil tank to make the oil level reach the upper scale of the oil level meter in the hydraulic oil tank, and screw on the oil filler flange cover.

Remove the frame locking device and start the engine. Operate the operating lever of the working device, lift the boom, tilt the bucket forward and backward 2 ~ 3 times, and turn left and right to the maximum angle, to fill the oil cylinder and oil pipe with hydraulic oil. Then run the engine at idle speed for 5 min to remove air from the system.

Turn off the engine, open the oil filler cap of the hydraulic oil tank and add clean hydraulic oil to the green range in the hydraulic oil tank level meter, that is, between the MAX line and MIN line.

When disassembling any hydraulic oil circuit or disassembling any components containing liquid, the liquid must be collected in a suitable container. Dispose of all liquids in accordance with local regulations. During the oil change operation, pay attention to the cleanliness of the hydraulic oil and prevent dirt from entering the hydraulic system.

Subtask 2. 4. 2 Maintenance of accumulator

☞ [Learning Objectives]

1) To be able to replace the accumulator as required.

2) To be able to inflate the accumulator as required.

☞ [Work Task]

A loader can't lower the boom after flameout. After inspection, there is no pressure in the pilot oil supply accumulator, so the accumulator needs to be inflated.

☞ [Related Knowledge]

1. Role of accumulator

Accumulator is an energy storage device in the hydraulic and pneumatic system. It converts the energy in the system into compression energy or potential energy at the appropriate time and stores it. When the system needs it, it converts the compression energy or potential energy into hydraulic or pneumatic energy and releases it to replenish the supply system. When the instantaneous pressure of the system increases, it can absorb this part of the energy to ensure the normal pressure of the whole system.

2. Functions of accumulator

When the loader moves at a low speed, the flow required by the load is less than that of the hydraulic pump, and the excess flow of the hydraulic pump is stored in the accumulator. When the flow required by the load is greater than that of the hydraulic pump, the liquid is discharged from the accumulator to make up for the insufficient flow of the hydraulic pump.

When it is shut down but still needs to maintain a certain pressure, the hydraulic pump can be stopped and the accumulator compensates for the leakage of the system to maintain the pressure of the system. Accumulator can also be used to absorb the pressure pulsation of the hydraulic pump or the hydraulic shock pressure generated in the system. The pressure in the accumulator can be generated by compressed gas, a heavy hammer or a spring. Accordingly, the accumulator is divided into three types: gas type, heavy hammer type and spring – loaded type. Direct contact between gas and liquid in a gas accumulator is called contact type. It has a simple structure and a large capacity, but it is easy to mix gas in the liquid, which is often used on hydraulic press. The non – contact between gas and liquid is called isolation type. It is often isolated by skin bag and diaphragm. The volume change of skin bag is large and the volume change of the diaphragm is small. It is often used to absorb pressure pulsation. The heavy hammer type accumulator has a larger capacity and is often used in rolling mills and other systems for energy storage.

☞ [Task Implementation]

When the loader works for 50 h, 100 h, 250 h, 500 h and 1000 h, check the nitrogen pre – charge pressure of the accumulator, and then check it every 2000 h. The accumulator used by the pilot oil supply valve is generally located on the left side of the loader, below the cab and inside the left web, as shown in Fig. 2-33.

Before adding pressure to the accumulator, park the loader on a flat and open ground, place the bucket on the ground, place the variable speed joystick in the neutral position, and turn off the engine.

Press the parking brake handle to release the parking brake and release the brake. Continuously pull the operating handle of the working device in any direction to discharge the high – pressure oil in the accumulator. Remove the charging valve protective cap from the upper end of the accumulator. Tighten the switch 1 of the accumulator charging tool (right – hand rotation), close the exhaust valve 2, tighten the bonnet of the check valve 3, and fix the charging tool on the charging valve at the upper end of the accumulator with the slotted nut 4, as shown in Fig. 2-34.

Fig. 2-33 Installation position of accumulator

Fig. 2-34 charging tool

1—Switch 2—Exhaust valve 3—Check valve 4—Nut

Slowly open the switch 1 (left – hand rotation) of the charging tool. After the pressure reading becomes stable, the reading on the pressure meter is the nitrogen pre – charge pressure of the accumulator, and its value is 1. 0 ±0. 05 MPa. If the pressure is low, supplement nitrogen . If the pressure is high, it can be relieved to the required pressure value through exhaust valve 2.

Turn off the switch 1 (right – hand rotation) of the charging tool, then remove the charging tool of the accumulator from the accumulator and install the protective cap of the charging valve.

Task 2. 5 Maintenance of electrical system

Sub – task 2. 5. 1 Inspection and maintenance of battery

☞ [Learning Objectives]

1) To be able to replace battery correctly.

2) To be able to select a suitable accumulator to charge battery.

☞ [Work Task]

Maintain battery according to the specification requirements.

☞ [Relevant Knowledge]

Loaders generally use two batteries with a nominal voltage of 12 V in series. The negative pole of one battery is grounded through the main power switch, and the positive pole is connected to the negative pole of the other battery; the positive pole of the other battery is connected to the starter motor.

1. Function of loader battery

1) Provide a large current to the starter motor to start the engine. Therefore, the lead – acid battery for starting is generally adopted for the battery of loader, so as to provide a large current for starting the engine in a short time.

2) To supply power to all electrical loads on the vehicle when the generator does not generate electricity.

3) Absorb transient voltage in the system to protect electronic components. The sudden changes of the speed and load of the generator and the switching of inductive loads (such as electric horn, electromagnetic coil, etc.) will cause transient voltages in the system (the peak value is more than 100 V and the duration is milliseconds). Although the voltage regulator is set in the power supply system, due to the lag of the voltage regulator to the transient voltage regulation, the transient voltage in the power supply system cannot be suppressed, which has a strong impact on the electronic components in the circuit and even damages the electronic components. The characteristics of low impedance and large capacitance of battery make it have a strong absorption effect on transient overvoltage, so as to protect electronic components. Therefore, it is necessary to ensure the reliable connection of the charging circuit between the generator and the battery. During the normal operation of the engine, the connection between the generator and the battery must not be disconnected in any

way (such as closing the electric lock during the operation of the engine).

2. Composition of loader battery

A device that can repeatedly convert chemical energy to electrical energy is called a battery. Loaders generally use lead – acid battery. Lead – acid battery is generally composed of shell, positive plate, negative plate, separator, battery slot, electrolyte and terminal block.

☞ [Task implementation]

1. Inspection of battery

The battery is located in the battery box on the left side of the tail of the whole machine. Unscrew the 4 cover bolts to see the battery, as shown in Fig. 2-35.

Fig. 2-35 Schematic diagram of battery installation position

Check whether the battery pressing plate nut, battery terminal and cable connector are loose or not. If the wire head is loose, tighten the battery pressing plate nut, battery terminal and cable connector, as shown in Fig. 2-36.

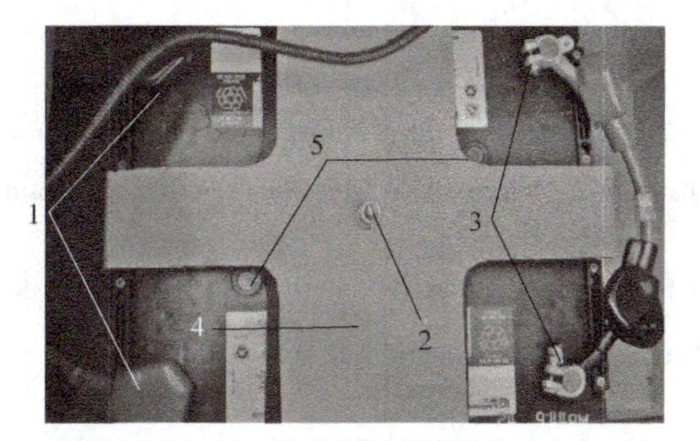

Fig. 2-36 Schematic diagram of battery cable connection

1—Battery terminal 2—Pressing plate nut 3—Battery terminal 4—Pressing plate 5—Electric eye

Check the battery status indicator (electric eye). If the electric eye is green, it indicates that the battery is sufficient and the car can be started normally. If the electric eye is black, it indicates that the battery is low and needs to be recharged. If the electric eye displays white, the battery is scrapped and needs to be replaced.

After checking the battery, close the battery box cover.

2. Battery replacement

Turn off the engine, remove the key and turn off the negative switch of the battery.

Clean battery terminal and battery surface with a clean cloth. Be careful not to touch the positive and negative poles of the battery at the same time. In addition, never use gasoline or any other organic solvents or cleaners.

Remove the battery connecting cable. First remove the cable connector connecting the negative pole of the battery to one end of the negative switch, then remove the connector connecting the negative pole of the battery to the negative end, and remove the negative cable; then remove the jumper wire between the two batteries, and finally remove the battery positive cable. Check whether the battery terminal is corroded and whether the battery cable is worn or damaged, if necessary, replace the battery cable.

Take out the old battery, place the new battery on the mounting seat, first connect the jumper wires of the two batteries, then connect the positive cable of the battery, then connect the negative cable, and finally connect the negative cable of the battery to the negative switch of the battery, as shown in Fig. 2-37. Note that all connecting bolts and nuts must be tightened.

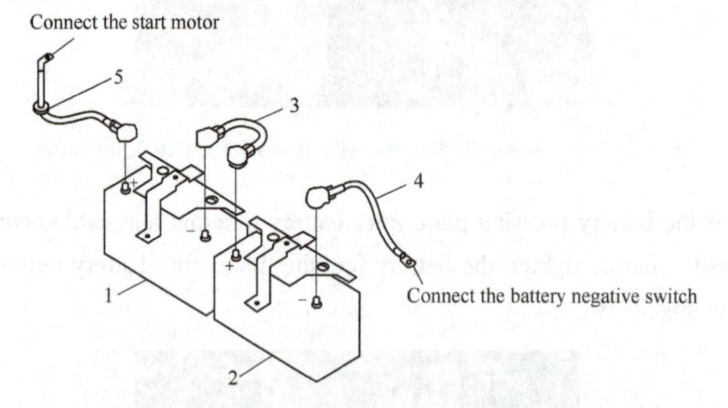

Fig. 2-37 Schematic diagram of battery cable connection

1, 2—Battery 3—Jumper cable 4—Negative cable 5—Positive cable

Insert the key and turn the battery negative switch on, turn the key switch to "ON", observe the instrument status, and then start the engine for a test run.

Subtask 2. 5. 2 Inspection and maintenance of generator

☞ [Learning Objectives]

1) To be able to judge the power on state of the generator.
2) To be able to maintain the generator according to the specification requirements.

☞ [Work Task]

Check the output voltage of the generator according to the specification.

☞ [Relevant Knowledge]

The engines of construction machinery are equipped with a generator. Driven by the engine, the generator converts mechanical energy into electrical energy. Generator and battery are connected in parallel to form the power supply system of the whole machine. Before the diesel engine is started, the battery supplies power to the electrical appliances of the whole machine. After the diesel engine is started, the generator mainly supplies power to the electrical appliances, and the generator charges the battery. When the energy provided by the generator cannot meet the needs of the electrical appliances of the whole machine, the generator and the battery supply power together. Outline of generator is shown in Fig. 2-38.

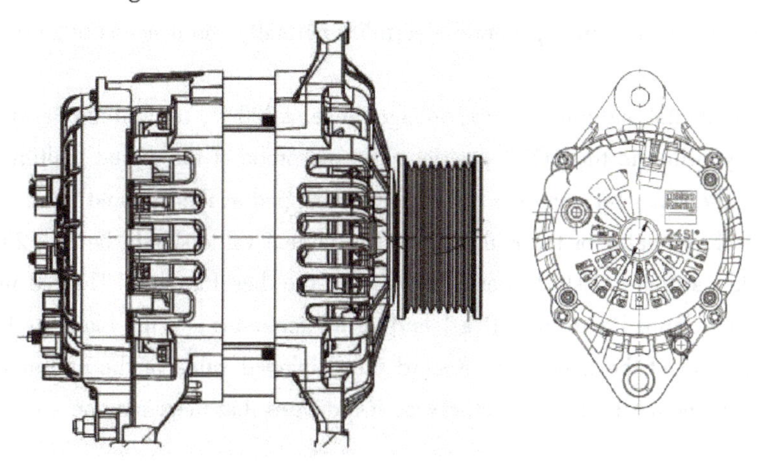

Fig. 2-38 Outline diagram of generator

The generator has three terminals, namely "B+" terminal, Ground terminal and "D+" terminal. "B+" terminal is the positive pole of the generator, which provides power for battery charging and electrical appliances on the vehicle. Ground terminal is the negative terminal of generator, which is connected with the negative terminal of vehicle battery. "D+" terminal is the excitation terminal, which is connected with charging indicator light, starter protection relay, etc, and the output current should not exceed 1 A.

The working voltage of the generator is 24 V, the rated current is 35 A, and built-in electronic voltage regulator. During the operation of the diesel engine, always check whether the wiring terminal of the generator is tight and reliable. Whether the positive or negative wiring is loose, the electrical system of the whole vehicle will not work normally, or even cause serious faults. Be sure to turn off the start switch before checking whether the wiring of these two terminals is tight and reliable.

Methods and treatment measures for judging whether the generator generates power normally:

1) Turn on the start switch (electric lock), observe the reading of the voltmeter carefully, start the diesel engine, and then observe the reading of the voltmeter. The latter reading should be higher than the previous one.

2) You can also check the DC voltage 200 V gear of the multimeter, open the electric lock,

measure the voltage at the generator terminal (the red probe is connected to the positive pole of the generator and black probe is connected to the ground), and record the reading of the multimeter.

3) Start the diesel engine, increase the speed of the diesel engine to the rated speed, measure the voltage at the generator terminal again, and record the reading of the multimeter. Compared with the two, the latter reading should be higher than the previous reading.

4) If the generator does not generate electricity, check whether the transmission belt of the generator is too loose and whether the generator is grounded firmly. Check with a wrench whether the wiring terminal of the generator is correct, tight and reliable.

☞ [Task Implementation]

Generator of one loader can't generate electricity normally, so it needs to be checked. The inspection steps are as follows:

1) Adjust the digital multimeter to the voltage range of 200 V, turn on the negative switch of the loader, put the key switch in the "ON" position, the red probe of the digital multimeter is lapped at the "B+" end of the generator, and the black probe is lapped at the "ground" end of the generator, and record the displayed value of the multimeter. The normal value should be 24 ~25 V.

2) Set the key switch to "start" position and start the diesel engine. The red probe of the digital multimeter is overlapped with the "B+" end of the generator and the black probe is overlapped with the "ground" end of the generator. Record the displayed value of the multimeter. The measured value should be higher than 25 V, otherwise it indicates that there may be a fault in the generator.

Task 2.6 Maintenance of air conditioning system

☞ [Learning Objectives]

1) To be able to maintain air conditioning system as required.

2) To be able to replace return air filter element and air filter element as required.

☞ [Work Task]

The air conditioner of one loader is not cool. After inspection, the air filter is dirty, resulting in blockage. The air filter needs to be replaced.

☞ [Relevant Knowledge]

The air filter of air conditioner is a device that collects dust from gas – solid two – phase flow and purifies the gas through the action of porous filter material. It purifies the air and sends it indoors to ensure the process requirements of clean rooms and the air cleanliness in general air – conditioned rooms.

A cab air filter is composed of an external air filter and an internal air filter. The blockage degree of the external air filter is related to the working environment and the opening time of the external air circulation. Keeping the cab clean can prolong the service life and maintenance cycle of the internal air filter.

1. Performance index of filter

(1) Filtering accuracy Refers to the maximum diameter of impurity particles allowed to pass through the air filter. The key to affecting the filtering accuracy is the filter element. Different filter elements can be selected according to the needs of the following components to achieve the corresponding filtering accuracy.

(2) Flow characteristics It refers to the relationship curve between the air flow through the filter and the pressure drop at both ends of the filter under a certain inlet pressure. In actual usage, it is best to select within the range of pressure loss less than 0.03 MPa. In the air filter, the main factors affecting the flow characteristics are the body and the filter element.

(3) Water separation efficiency It refers to the ratio of the separated moisture to the moisture contained in the air at the air inlet. Generally, the water separation efficiency of air filter shall not be less than 80%. The main factor affecting the water separation efficiency is the deflector. .

(4) Filter resistance Under the rated air volume, the resistance of the new filter is called the initial resistance. Under the rated air volume, the resistance when the dust capacity of the filter reaches large enough and the filter material needs to be cleaned or replaced is called the final resistance.

(5) Dust capacity of filter Under the rated air volume, when the resistance of the filter reaches the final resistance, the total mass of dust particles contained in the filter is called the dust capacity of filter.

2. Selection principle of filter

Select the appropriate air filter according to the specific situation.

1) Determine the efficiency of the last stage air filter according to the cleaning and purification standards, and reasonably select the combination stage and efficiency of each stage of the air filter. If general purification is required indoors, primary efficiency filter can be used. If medium purification is required indoors, two – stage filters of primary efficiency and medium efficiency should be adopted. If the room requires ultra clean purification, three – stage purification filtration of primary efficiency, medium efficiency and high efficiency should be adopted, and the efficiency of filters at all levels should be reasonably and properly matched. If the efficiency difference between two adjacent stages of filters is too large, the former filter cannot protect the latter one.

2) Correctly measure the dust content and dust particle characteristics of outdoor air. Because the filter is to filter and purify the outdoor air and send it indoors, the dust content of the outdoor air is a very important data. Especially in multi – stage purification and filtration treatment, the selection of pre – filter should be decided after comprehensive consideration of factors such as operating environment, cost of spare parts, operation energy consumption, maintenance and supply.

3) Correctly determine the characteristics of the filter. The characteristics of the filter are mainly the filtration efficiency, resistance, penetration rate, dust capacity, filtration wind speed and processing air volume. If conditions permit, filters with high efficiency, low resistance, large dust

capacity, moderate filtering wind speed, large processing air volume, convenient manufacturing and installation and low price should be selected as far as possible. This is the economic analysis need to comprehensively considering one – time investment, secondary investment and energy efficiency ratio in the selection of air filters.

4) Analyse the properties of dusty gas. The properties of dusty gas related to the selection of air filters are mainly temperature, humidity, acid and alkali and the amount of organic solvent. Because some filters are allowed to be used at high temperature, while some filters can only work at room temperature and constant humidity, and the amount of acid – base and organic solvent of dusty gas has an impact on the performance and efficiency of air filters.

☞ [Task Implementation]

1. Clean the internal air filter

Remove the cover plate bolt at the left rear inside the cab, then remove the cover plate and remove the internal air filter from the air conditioning evaporator, as shown in Fig. 2-39.

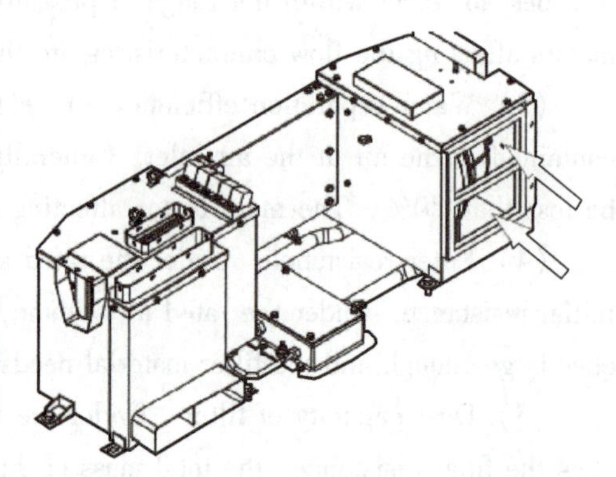

Fig. 2-39 Location of internal air filter

Clean the filter with high – pressure air. If there is oil on the filter or it is too dirty, flush it with neutral medium. After washing, it should be thoroughly dried before reuse.

Reinstall the cleaned internal air filter and fix the access door corer. In principle the internal air filter should be replaced every 2000 h.

2. Clean the external air filter

The cleaning of the filter can appropriately extend or shorten the maintenance cycle according to the amount of dust in the working environment of the machine. If the filter element inside the filter is blocked, the fresh air volume in the cab may be reduced.

The fresh air device of the air conditioner is located outside the left side of the cab. Remove the fixing bolts of the filter cover of the left door accessory of the cab, and then remove the filter cover and the external air filter, as shown in Fig. 2-40.

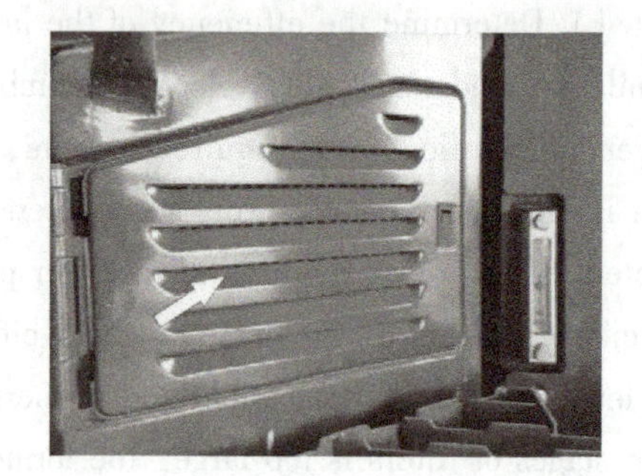

Fig. 2-40 Location diagram of external air filter

Clean the filter with high – pressure air. If there is oil on the filter or it is too dirty, flush it with neutral medium. After washing, it should be thoroughly dried before reuse.

If there is a risk of serious injury to the human body caused by pollutants, goggles, dust cover or other protective equipment must be used.

Put the cleaned external air filter into the groove of the filter housing, then install it back together with the filter housing, and tighten the bolts.

Project 3

Analysis and measurement of loader hydraulic system

According to the transmission requirements of the working device and each mechanism of the loader, the combination that organically connects various hydraulic components with pipelines is called the hydraulic system of the loader. Its function is to take the oil as the working medium, use the hydraulic pump to convert the mechanical energy of the engine into hydraulic energy and transmit it, and then convert the hydraulic energy into mechanical energy through the hydraulic cylinder and hydraulic motor to realize various actions of the loader.

A complete hydraulic system consists of five parts, namely the power element, executive element, control element, auxiliary element and hydraulic oil. The power element is to convert the mechanical energy of the prime mover into the pressure energy of the liquid, such as the hydraulic pump in the hydraulic system, which provides hydraulic oil to the entire hydraulic system. There are many forms of hydraulic pumps, generally including gear pumps, vane pumps, piston pumps. The commonly used hydraulic pump in modern loaders is gear pump. The actuator converts the pressure energy of hydraulic oil into mechanical energy, such as various oil cylinders and hydraulic motors. The control element is used to control the flow direction, pressure and flow of hydraulic oil in the hydraulic system, such as various hydraulic valves. Auxiliary components include oil tanks, oil filters, oil pipes and connectors, sealing rings, etc. Hydraulic oil is the working medium of transmitting energy in hydraulic system, including various mineral oils, emulsions and synthetic hydraulic oils.

The loader hydraulic system is divided into working hydraulic system and steering hydraulic system. According to the form of working pump and steering pump, it can be divided into quantitative hydraulic system (both working pump and steering pump are quantitative pumps), constant variable hydraulic system (working pump is quantitative pump and steering pump is variable pump) and full variable hydraulic system (both pumps are variable pumps). The following mainly describes the quantitative system.

1. Working principle of loader hydraulic system (CLG856H)

As can be seen from Fig. 3-1, the transmission path of the loader hydraulic system is as follows: the engine drives the hydraulic pump (working pump and steering pump) through the power take-off port of the gearbox. The oil output from the steering pump provides pilot control oil to the flow amplification valve through the pilot oil supply valve and steering gear, and supplies oil to the steering cylinder through the flow amplification valve. The hydraulic oil output by the working pump enters the distribution valve. The operator controls the distribution valve by operating the pilot valve. The hydraulic oil enters the boom cylinder and bucket cylinder through the distribution valve to realize the action of the working device.

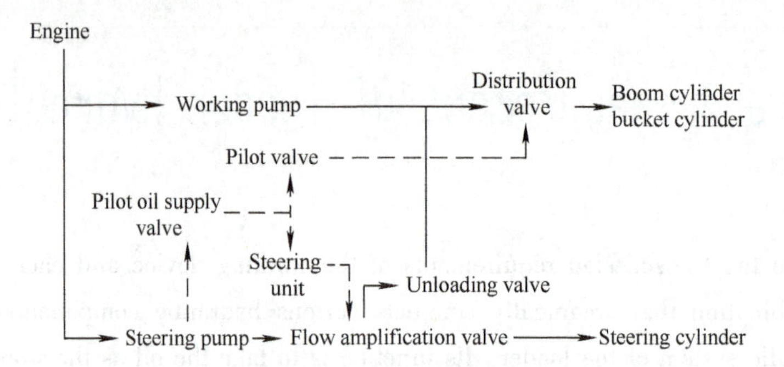

Fig. 3-1 Working principle of loader hydraulic system

2. Composition of loader hydraulic system

1) From the perspective of components, the loader hydraulic system is divided into the following five parts:

① Energy device (power element): working pump, steering pump.

② Actuator (actuator): boom cylinder, bucket cylinder and steering cylinder.

③ Control and regulation device (control element): pilot oil supply valve, pilot valve, one-way valve, distribution valve, flow amplification valve and unloading valve.

④ Auxiliary device (auxiliary components): hydraulic oil tank, radiator, filter, connector, hose, steel pipe, etc.

⑤ Transmission medium: hydraulic oil.

2) From the perspective of system (assembly drawing), the loader hydraulic system includes: working device hydraulic system, steering system, braking system (wet axle), hydraulic pipeline assembly, auxiliary devices, etc.

Task 3.1 Working principle analysis of hydraulic system

Subtask 3.1.1 Working principle analysis of pilot hydraulic system

☞ [Learning Objectives]

1) To be able to describe the control principle of working hydraulic system.

2) To be able to analyse the control circuit of pilot hydraulic system.

3) To be able to analyse and eliminate common faults through the principle of pilot circuit.

☞ [Work Task]

One loader has the problem of slow boom lifting. After inspection, there is no problem with the main valve pressure and pump. After inspection, the pilot pressure is abnormal. It is necessary to test and adjust the pilot pressure.

☞ [Relevant Knowledge]

The working hydraulic system is used to control the action of boom, bucket and other additional working devices in the loader. For example, control the working device of the loader to complete the actions of bucket retraction and unloading, boom lifting, lowering, floating and lowering. Specific working principle: the hydraulic oil is sucked out from the oil tank by the working pump and enters the distribution valve through the pipeline. There are boom valve core and bucket valve core in the distribution valve. The two valve cores are controlled by pilot pressure. When the pilot handle is operated, the pilot hydraulic oil enters one side of boom valve core or bucket valve core to push the valve core to move, and the hydraulic oil from the working pump enters the boom cylinder or bucket cylinder through the opening of the valve core to realize the action of the working device.

According to the different models, the working hydraulic system can be divided into quantitative hydraulic system, constant variable hydraulic system and full variable hydraulic system.

The following mainly describes the quantitative system, as shown in Fig. 3-2.

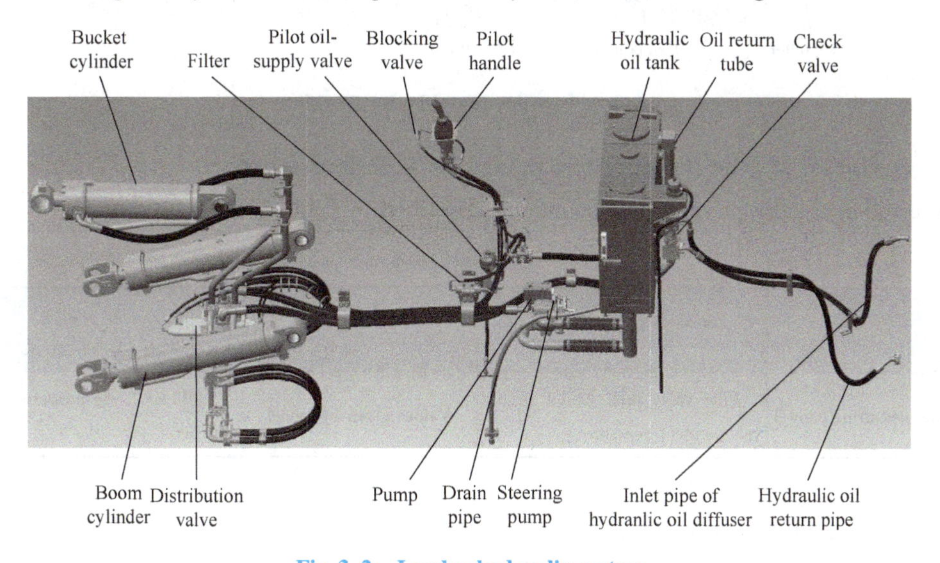

Fig. 3-2 Loader hydraulic system

1. Working principle of the pilot hydraulic circuit

The pilot operated valve is installed in the cab on the right side of the driver's seat. The pilot control valve is a single handle pilot valve, which is composed of two valve groups: boom control linkage and rotary bucket control linkage. By operating the boom control lever and the bucket control lever of the pilot control valve, the action of the boom slide valve or the bucket slide valve in the distribution valve can be operated, so as to control the working device of the vehicle. The operating positions of the boom handle include four positions: lifting, middle position, lowering and floating.

The operating positions of the bucket handle include three positions: bucket retraction, middle position and unloading. In the pilot operated valve, electromagnets are set in the three positions of boom lifting, boom lowering and bucket retraction. Through the connection with the boom and bucket automatic reset device on the front frame and rocker arm, the automatic limit of boom height and the automatic levelling of bucket can be realized.

2. Pilot pressure test

The pilot pressure test points are shown in Fig. 3-3.

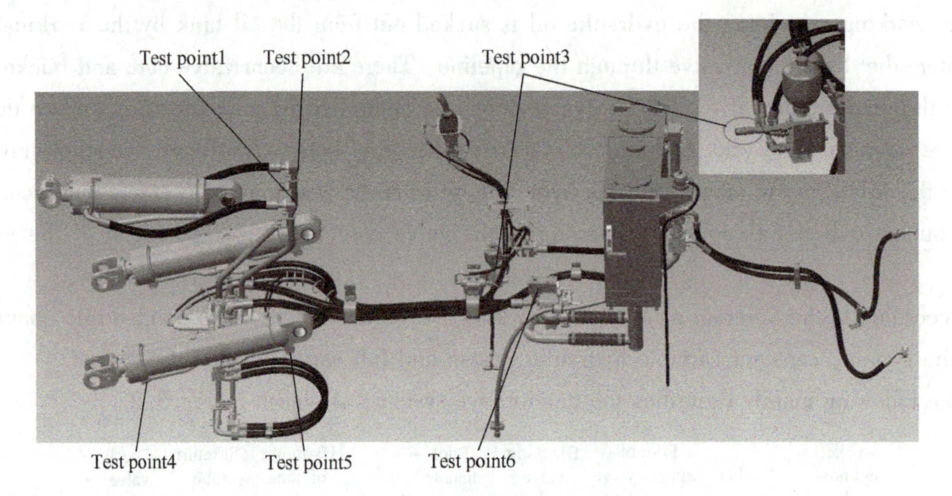

Fig. 3-3 Pilot pressure test points

3. Description of pilot pressure test points are described in Table.

Description of pilot pressure test points are described in Table 3-1.

Table 3-1 Description of test points

Pressure measuring point	Description of measuring points	Theoretical pressure value/MPa	Interface size
Pressure measuring point 3	Pilot valve inlet (pilot supply valve outlet) pressure	Values given by model	M18×1.5-6g pressure testing connector

[Task Implementation]

Pilot pressure test:

Place the whole machine horizontally, and install the pressure gauge at test point 3. After start-up, operate the pilot valve. At this time, the pressure value of the pressure gauge should be the design pressure value given by the model.

If the pressure is incorrect, adjust the pressure of the pressure reducing valve of the pilot oil supply valve (refer to the adjustment steps of the pressure reducing valve of the pilot oil supply valve for specific adjustment) to make it reach the normal value.

Subtask 3.1.2 Working principle analysis of rotary bucket hydraulic circuit

☞ [Learning Objectives]

1) To be able to describe the control principle of bucket hydraulic system.

2) To be able to analyse the hydraulic control circuit of bucket.

3) To be able to analyse and eliminate the common faults of the bucket circuit through the principle.

☞ [Work Task]

One loader has the problem of weak bucket. After testing, the system pressure is insufficient for the design value, so it is necessary to test and adjust the system pressure.

☞ [Relevant Knowledge]

When the loader does not act, the boom and bucket reversing valve in the distribution valve are in the middle position, the hydraulic oil output by the working pump returns to the oil tank through the distribution valve, the front and rear cavities of the boom and bucket are closed, and the boom and bucket remain in the original position.

The operator operates the pilot valve to lift, lower or float the boom, and rotate the bucket forward or backward. At this time, after the pilot oil from the pump passes through the pilot oil supply valve, the pilot valve enters the control end of the boom or bucket valve core of the distribution valve to push the valve core to move, so that the oil from the main oil circuit enters the boom cylinder or bucket cylinder through the distribution valve to realize the corresponding action.

1. Function and structure of distribution valve

The function of distribution valve is to control the movement direction of bucket cylinder and boom cylinder and the action of boom and bucket by changing the flow direction of oil, so as to meet the operation requirements of different working conditions of loader.

Structural form: integral double slide valve type.

Oil circuit form: series parallel priority bucket.

Main components: bucket reversing valve, boom reversing valve, and safety valve, as shown in Fig. 3-4.

2. Principle of distribution valve

As shown in Fig. 3-5, the bucket reversing valve is a three - position valve, which can control the three actions of bucket: forward tilt, backward tilt and hold. The boom reversing valve is a four - position valve, which can control the four actions of boom: rising, holding, falling and floating. The safety valve controls the system pressure. When the system pressure exceeds the set value, the oil overflows back into the oil tank to protect the system from damage. Port P on the distribution valve is the oil inlet and port T is the oil return port. A1 and B1 are respectively connected with the small and large chambers of the bucket cylinder, and A2 and B2 are respectively connected with the small

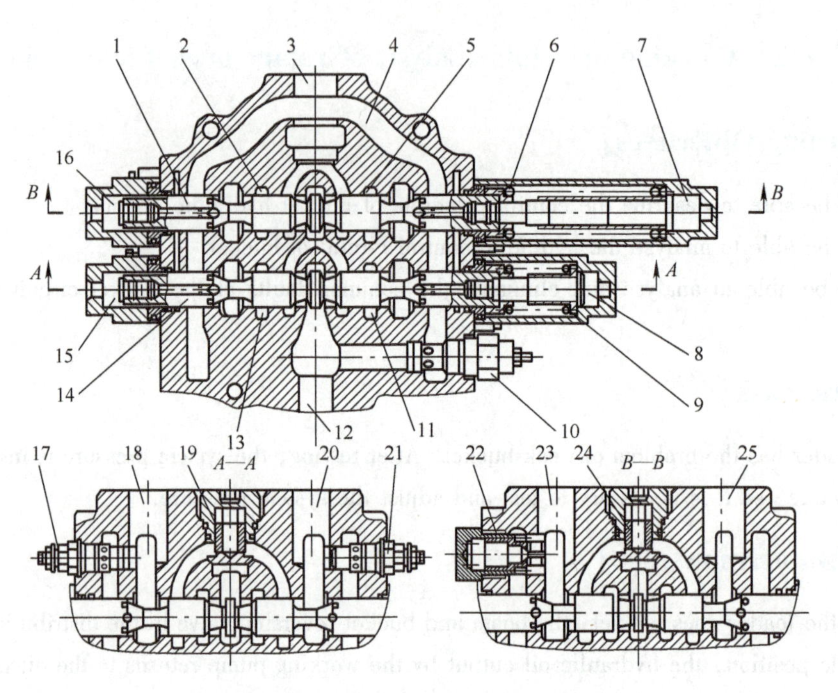

Fig. 3-4 Section of distribution valve

1—Boom valve stem 2—Boom linkage left oil inlet passage 3—Oil return port 4—Oil return passage
5—Boom linkage right oil inlet passage 6, 9—Spring 7—Boom valve stem lift chamber
8—Bucket valve stem retraction chamber 10—Main relief valve 11—Right oil inlet passage of bucket linkage
12—Oil inlet 13—Left oil inlet passage of bucket linkage 14—Bucket valve stem
15—Bucket valve stem unloading chamber 16—Boom valve stem lowering chamber
17—Small chamber overload valve of bucket cylinder 18—Working oil port of small chamber of bucket cylinder
19, 24—Check valve 20—Working oil port of large chamber of bucket cylinder
21—Large chamber overload valve of bucket cylinder
22—Oil refill check valve 23—Working oil prot of small chamber of boom cylinder
25—Working oil port of large chamber of boom cylinder

and large chambers of the boom cylinder.

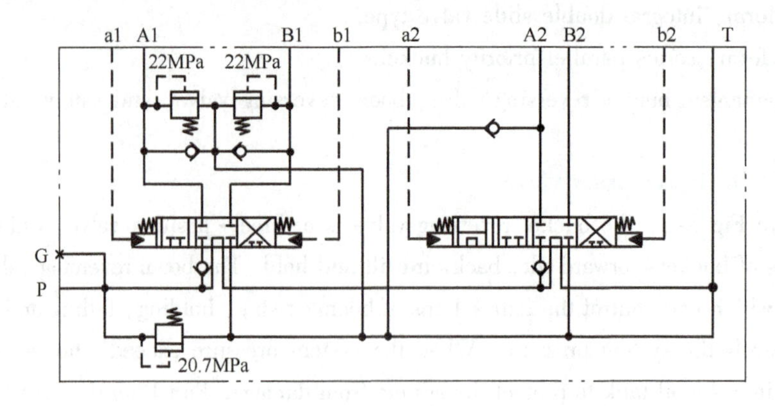

Fig. 3-5 Schematic diagram of distribution valve

3. Working principle analysis of double acting safety valve

The oil circuit connected with the oil passages of the large and small chambers of the bucket oil cylinder is equipped with a large and small chamber double action safety valve, which plays the role of overload protection and oil supplement for the large and small chambers of the bucket oil cylinder, so as to solve the problem of interference of the working device, stabilize the system and protect the relevant components of the system.

4. Analysis of hydraulic cylinder circuit of bucket

(1) Neutral position of bucket linkage When there is no pilot pressure oil at both ends of the bucket valve stem of the distribution valve, the bucket valve stem is in the middle position under the action of spring 9. The oil from the working pump enters the oil passage through the oil inlet 12 and supplies oil to the boom linkage. At this time, the two working oil ports 18 and 20 connected to the distribution valve at both ends of the large and small chamber of the bucket oil cylinder are closed by the bucket valve stem, and the bucket oil cylinder remains stationary. If the boom valve stem is also in the middle position at this time, the oil from the working pump flows back to the oil tank through the two heart - shaped oil passages of the bucket linkage and the boom linkage through the oil return port 3.

(2) Bucket linkage unloading position When the operating handle of the bucket is operated to the unloading position, the pilot hydraulic oil enters the unloading chamber 15 of the bucket valve stem, and the oil in the bucket retraction chamber 8 of the valve stem is connected to return oil through the pilot valve. Under the action of oil pressure, the valve stem overcomes the force of spring 9 and moves to the right to open the working oil port 18 connecting the small chamber of the bucket oil cylinder and the opening of the left oil inlet passage 13 of the bucket linkage. After jacking the check valve 19, the hydraulic oil of the working pump enters the working oil port 18 of the small chamber of the bucket oil cylinder through the left inlet passage 13 of the bucket linkage and reaches the small chamber of the bucket oil cylinder. And the oil in the large chamber of the bucket oil cylinder passes through the working oil port 20 and returns to the oil tank through the oil return passage 4 and the oil return port 3. The piston stem of the bucket oil cylinder retracts, and the bucket realizes the unloading action.

(3) Bucket retraction position of bucket linkage When the operating handle of the bucket is operated to the bucket retraction position, the pilot hydraulic oil enters the bucket retraction chamber 8 of the bucket valve stem, and the oil in the unloading chamber 15 of the bucket valve stem is connected to return oil through the pilot valve. Under the action of oil pressure, the valve stem overcomes the force of the spring 9 and moves to the left to open the working oil port 20 connecting the large chamber of the bucket oil cylinder and the opening of the right oil inlet passage 11 of the bucket linkage. After jacking the check valve 19, the hydraulic oil of the working pump enters the working oil port 20 of the large chamber of the bucket oil cylinder through the right oil inlet passage 11 of the bucket linkage and reaches the large chamber of the bucket oil cylinder. And the oil in the small chamber of the bucket oil cylinder passes through the working oil port 18 and returns to the oil tank through the oil return passage 4 and the oil return port 3. The piston stem of the bucket oil cylinder

extends, and the bucket realizes the bucket retraction action.

When the bucket valve stem moves to the right and reaches the maximum bucket retraction position, the hydraulic oil of the working pump cannot enter the boom linkage and the boom cannot work.

5. Pressure test of working system

The pressure test points of working hydraulic system are shown in Fig. 3-6.

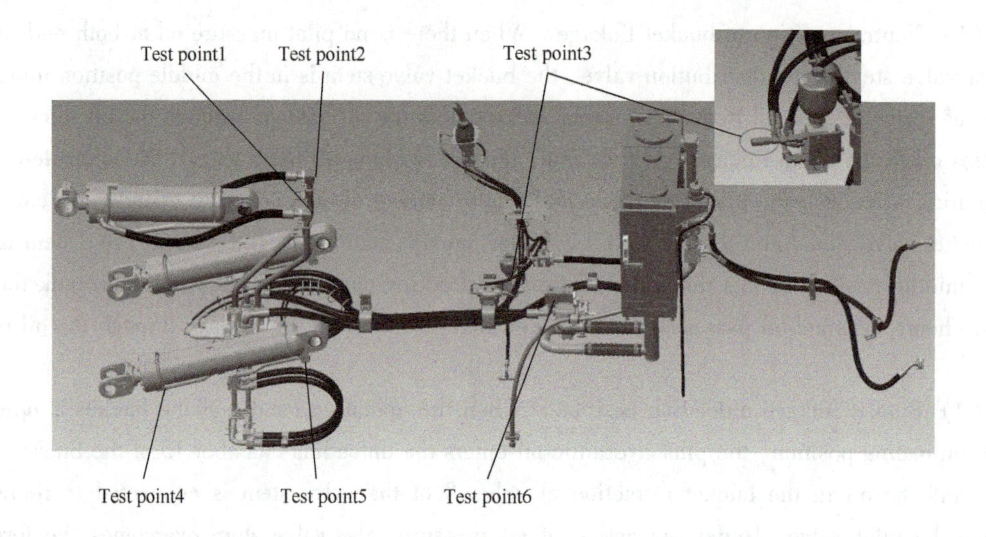

Fig. 3-6 Pressure test points of working hydraulic system

6. The description of pressure test points of working hydraulic system

The description of pressure test points of working hydraulic system is shown in Table3-2.

Table 3-2 Description of measuring points

Pressure measuring point	Description of measuring point	Theoretical pressure value/MPa	Interface size
Pressure measuring point 1	Pressure in the small chamber of the bucket oil cylinder	Values given by model	M14 × 1. 5 – 6H
Pressure measuring point 2	Large chamber pressure of bucket oil cylinder	Values given by model	M14 × 1. 5 – 6H
Pressure measuring point 4	Boom oil cylinder small chamber pressure	Values given by model	M14 × 1. 5 – 6H
Pressure measuring point 5	Boom oil cylinder large chamber pressure	Values given by model	M14 × 1. 5 – 6H
Pressure measuring point 6	Working pump outlet pressure	Values given by model	M14 × 1. 5 – 6H

☞ [Task Implementation]

Working hydraulic system pressure test:

Adjustment of working pressure: Place the whole machine horizontally, install the corresponding pressure measuring connector and pressure gauge at the position of test point 2 according to the test interface size. After start - up, operate the pilot valve to hold the bucket. At this time, the pressure value of the pressure gauge should be the design value of the corresponding model.

If the pressure is incorrect, adjust the working pressure of the main safety valve of the distribution valve (refer to the adjustment steps of the main safety valve of the distribution valve for specific adjustment) to make it reach the corresponding pressure value.

Subtask 3.1.3 Analysis of the working principle of the hydraulic circuit of boom

☞ [Learning Objectives]

1) To be able to describe the control principle of the hydraulic circuit of boom.

2) To be able to analyse the hydraulic control circuit of boom.

3) To be able to analyse the common faults of the hydraulic circuit of boom through the principle and eliminate them.

☞ [Work Task]

If the boom of a loader has weak lift and the system pressure is tested to be less than the design value, the system pressure needs to be tested and adjusted.

☞ [Relevant Knowledge]

1. The working principle analysis of make - up valve

The overload valve 21 and 17 in the large and small chambers of the bucket connecting the bucket cylinder contains a make - up check valve, and there is a make - up check valve 22 in the small chamber of boom cylinder coupling boom. Its role: when the speed of the bucket cylinder or boom cylinder piston rod is greater than the speed of the work pump output flow can provide, the pressure in one of the chambers of the cylinder is less than that in the tank, and the check valve opens. The oil from the tank is replenished to the chamber of the cylinder with lower pressure to ensure that the cylinder is sufficient oil and avoid cavitation in the cylinder.

2. The analysis of the working principle of the main safety valve

On the oil inlet of the integral distribution valve, the main safety valve is integrated to control the pressure of the whole main working hydraulic system. The main safety valve is a pilot - operated cartridge valve, and its pressure setting is the maximum system pressure of the working hydraulic system of the vehicle. When the pressure of the working hydraulic system rises and reaches the pressure set by the main safety valve, the main safety valve is opened and the hydraulic oil of the working pump is overflowed back to the tank through the oil return outlet 3. The work pump output pressure will be limited below the regulated value.

The set pressure of the main safety valve can be increased or decreased by increasing or decreasing the initial spring pressure on the pilot spool.

3. The analysis of boom hydraulic cylinder oil circuit

(1) Boom connector neutral position In the case that the bucket connector does not work, and when there is no pilot hydraulic oil at both ends of boom stem 16 and 7 of the distribution valve, boom stem is in the neutral position under the action of the spring 6. The oil from the work pump is supplied to boom connector through the inlet 12 via the bucket connector. At this time, the two working oil ports 23 and 25 of the distribution valve which are connected to the two ends of the large and small chambers of the movable arm cylinder are closed by the boom valve lever, and the boom cylinder remains unmoved. The oil from the working pump passes through the two heart-shaped oil passages of the bucket and boom, and flows back to the oil tank through the oil return port 3.

(2) Lifting position of boom connector In the case that the bucket connector does not work, and when the control handle of the boom move to the lifting position, the pilot hydraulic oil enters the lifting chamber 7 of boom stem. The oil in the lowering chamber 16 of boom stem is connected to the return oil through the pilot valve. Under the action of oil pressure, boom valve stem overcomes the force of spring 6, moves to the left, opens the opening of the working oil port 25 connected to the large chamber of boom cylinder and the opening of the boom connector right oil inlet passage 5. After the check valve 24 is opened, the hydraulic oil of the work pump through the oil passage 5 of the boom connector, enters the work port 25 of boom cylinder large chamber and to boom cylinder large chamber. The oil in the small chamber of boom cylinder passes through the working oil port 23 and returns to the oil tank via the return oil passage 4 and the return port 3. Boom finishes the lifting with the piston rod of boom cylinder extending.

(3) Lowering position of boom connector In the case that the bucket connector does not work, and when the control handle of the boom move to the descending position, the pilot hydraulic oil enters the descending chamber 16 of boom stem. The oil in the lifting chamber 7 of boom stem is connected to the return oil through the pilot valve. Under the action of oil pressure, boom valve stem overcomes the force of spring 6, moves to the right, opens the opening of the working port 23 connected to the small chamber of boom cylinder and the opening of the left inlet passage 2 of boom connector. After the check valve 24 is opened, the hydraulic oil of the work pump enters the working oil port 23 of the small chamber of boom cylinder through the left inlet passage 2 of boom connector, and reaches the small chamber of boom cylinder. While the oil in the large chamber of the boom cylinder passes through the working oil port 25 and returns to the oil tank via the oil return passage 4 and oil return port 3. Boom finishes the lowering action with the piston rod of boom cylinder retracting.

(4) Boom connector floating position When the control handle of the boom continue to move forward from the descending position, the pilot oil pressure in the down chamber 16 of boom valve stem continues to rise, pushing boom valve stem to continue to move to the right. At this time, the oil from the operational pump is connected to the return oil through the throttle groove on the shoul-

der in the middle of the valve stem, and the valve port between the working port of the large chamber of boom cylinder and the oil return passage is further opened; the right oil passage 5 of the lift is connected to the return oil, and because the right and left oil passage of boom connector is connected, so at this time, the large and small chamber of boom cylinder and the work port oil are connected to the tank. Under the self - weight of the working device, the boom realizes floating decline.

4. Working hydraulic system pressure test

The working hydraulic system pressure test points are shown in Fig. 3-7.

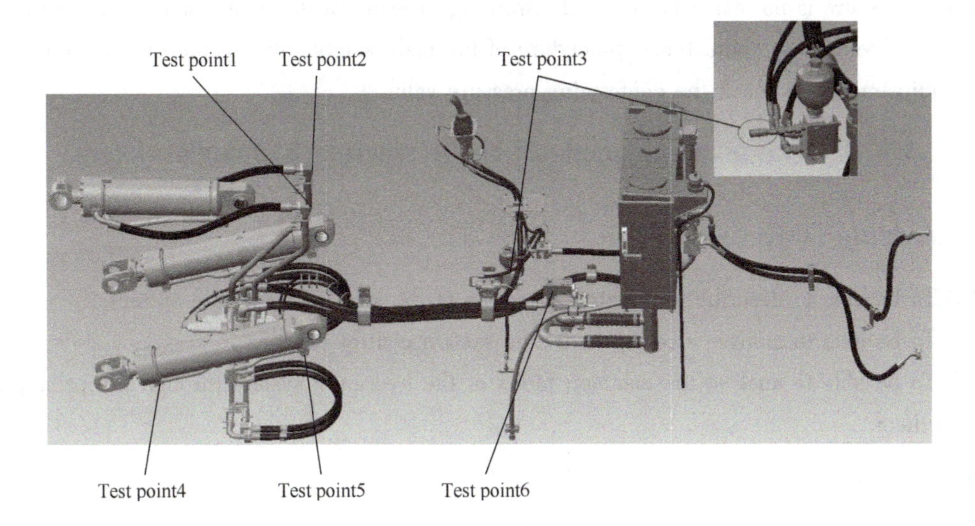

Fig. 3-7 Working hydraulic system test point

5. Description of working hydraulic system pressure test points

The description of the pressure testing points for the working hydraulic system is shown in Table 3-3.

Table 3-3 Test point description

Pressure testing point	Test point description	Theoretical pressure value/MPa	Interface size
Pressure testing point 1	Pressure in the small chamber of the bucket cylinder	Values according to model	M14 × 1.5 - 6H
Pressure testing point 2	Pressure in the large chamber of the bucket cylinder	Values according to model	M14 × 1.5 - 6H
Pressure testing point 4	Pressure in the small chamber of the boom cylinder	Values according to model	M14 × 1.5 - 6H
Pressure testing point 5	Pressure in the large chamber of the boom cylinder	Values according to model	M14 × 1.5 - 6H
Pressure testing point 6	Outlet pressure of working pump	Values according to model	M14 × 1.5 - 6H

☞ [Task Implementation]

The pressure test of working hydraulic system:

The whole machine is placed horizontally, and the corresponding pressure measuring connector and pressure gauge are installed at test point 5 according to the test interface size; after starting, the pilot valve is manipulated to hold the bucket, and the pressure value of the pressure gauge should be the design value of the corresponding model.

If the pressure is not correct, adjust the working pressure of the main safety valve of the distribution valve (refer to the adjustment procedure of the main safety valve of the distribution valve for specific adjustment) to reach the conforming pressure value.

Subtask 3.1.4 The principle analysis of the steering hydraulic system

☞ [Learning Objectives]

1) To be able to describe the control principle of the steering hydraulic system.

2) To be able to analyse steering hydraulic system control loops.

3) To be able to analyse the common faults of the hydraulic system through the principle and eliminate them.

☞ [Work Task]

If a loader has heavy steering, and it is checked that the pressure of the steering system is insufficient for the design value, the pressure of the steering system needs to be tested and adjusted.

☞ [Relevant Knowledge]

1. Steering hydraulic system of loader

Introduction of loader steering hydraulic system based on CLG856H. The steering system adopts the flow amplification system which is mainly used to control the steering of the whole machine when driving. The system is mainly divided into two parts: the steering control oil circuit and the main working oil circuit. The main working oil circuit is controlled by the steering control oil circuit to make small flow, low - pressure control large flow and high pressure. The components of the whole steering hydraulic system mainly include steering pump, steering gear, flow amplification valve, unloading valve, steering cylinder and related pipelines, as shown in Fig. 3-8.

This system can make the steering priority and dual pump flow confluence. Steering priority means that the oil from the steering pump is preferentially supplied to the steering system, and the excess oil is supplied to the working hydraulic system. And it's done by the flow amplification valve integrated with the priority valve. The priority valve is actually a floating valve core in the flow amplification valve and the floating valve core adjusts the flow of oil supplied to the steering system by sensing the load pressure of the steering system. The flow of oil supplied to the steering system is adapted to the steering load and speed, and thus the priority valve reserves the load sensing function

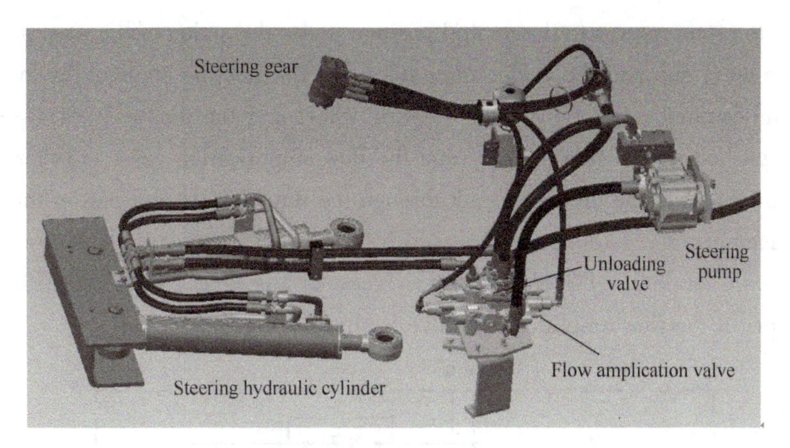

Fig. 3-8 Loader steering hydraulic system

of the ordinary flow amplifying valve.

Three main advantages of system confluence:

1) Reduction of the power loss of the working hydraulic system during non – operating conditions.

2) Troubleshooting of diesel engine stalling out when working hydraulic system and steering hydraulic system of loader work simultaneously.

3) Reduction of the cost and improvement of the reliability.

2. The principle of steering hydraulic system

The steering hydraulic system adopts the flow amplification system, and the oil circuit of the system is composed of the control oil circuit and the main oil circuit. The flow amplification means that there is a certain ratio between the flow change of the control oil circuit and that entering the steering cylinder in the main oil circuit through the full hydraulic steering gear and the flow amplification valve, so that low pressure and small flow control high pressure and large flow. Driver operation is smooth and easy, and the system power is fully utilized with good reliability.

The steering gear is a closed – core non – responsive type, and the neutral position is disconnected when the steering wheel is not turned. At this time, the main valve stem of the flow amplifier valve is kept in the neutral position under the action of the reset spring, and the oil circuit between the steering pump and the steering cylinder is disconnected, and the main oil circuit is unloaded back to the tank by the flow control valve in the flow amplifier valve. When the steering wheel is turned, the oil discharged by the steering gear is proportional to the rotational speed of the steering wheel. After the pilot oil enters the flow amplification valve, acting on the end of the main stem of the flow amplification valve controls the displacement of the main stem. By controlling the size of the stem opening, the flow entering the steering cylinder is controlled. Since the flow amplification valve adopts pressure compensation, the flow entering the steering cylinder is basically independent of the load, but only related to the size of the opening on the stem. After the steering is stopped, the pilot pressure oil entering one end of the main stem of the flow amplifier valve is connected to the oil tank through a small throttle hole at the other end, and the oil pressure at the two ends of the stem tends to be balanced, the stem returns to the neutral position under the action of the reset spring, thus cutting off the main oil circuit and the loader stops steering. The steering angle of the loader can be ensured through the continuous rotation of the steering wheel and the feedback. The feedback of the system is accomplished by the steering gear and the flow amplification valve together. The flow amplifier valve returns part of the oil to the tank through the throttle hole and part of the oil to the tank through the radiator.

3. Analysis of the left and right steering circuit of the loader flow amplification valve

The principle of the flow amplification valve is show in Fig. 3-9.

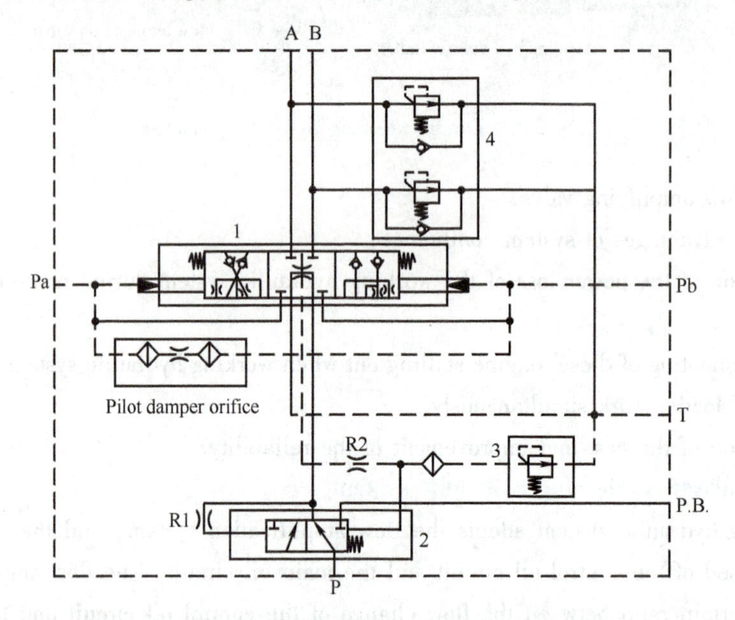

Fig. 3-9 Flow amplification valve schematic diagram

1—Steering spool 2—Priority spool 3—Main safety valve 4—Overload filler valve

1) The flow amplification valve is mainly composed of steering spool, priority spool, main safety valve and overload filler valve.

2) The steering spool can change the direction of the oil circuit and amplify flow; the priority spool makes steering priority and realize the confluence of dual pump; the main safety valve limits the highest pressure of the system to protect the system components; the overload filler valve is used to overcome the sudden increase of the local pressure in the large/small chamber of the steering cylinder caused by driving obstacles during the whole machine running to keep the driving stability.

3) When steering right, the pilot oil from Pa port (connected to steering R) pushes the steering spool of flow amplifier valve to move right, and the hydraulic oil from steering pump enters the large chamber of left steering cylinder and small chamber of right steering cylinder through the spool and A port of valve body to steer right. When steering left, the pilot oil from Pa port (connected to steering R) pushes the steering spool of flow amplifier valve to move left, and the main pressure oil from steering pump enters the small cavity of left steering cylinder and large cavity of right steering cylinder through the spool and A port of valve body to achieve steering left.

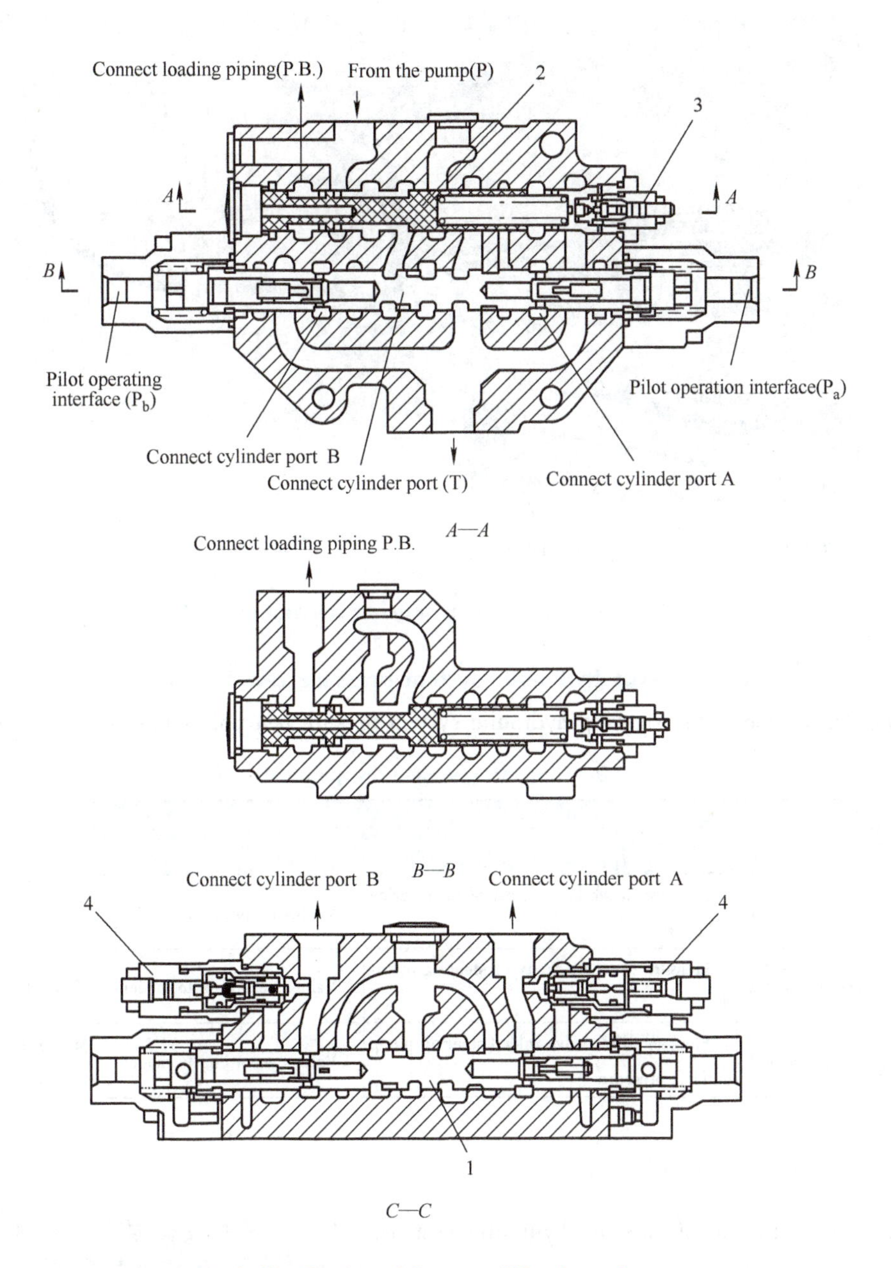

Fig. 3-10 Section of flow amplification valve

1—Steering spool 2—Priority spool 3—Main safety valve 4—Overload filler valve

4) In the process of separate steering, the priority spool works in the right position, and all the hydraulic oil from the steering pump enters the flow amplification valve; when working, the priority spool works in the left position, and most of the oil enters the working hydraulic system from the P_b port to realize the confluence of dual pump, as shown in Fig. 3-10.

4. Measuring point position of steering hydraulic system

The measuring point position of the steering hydraulic system is shown in Fig. 3-11.

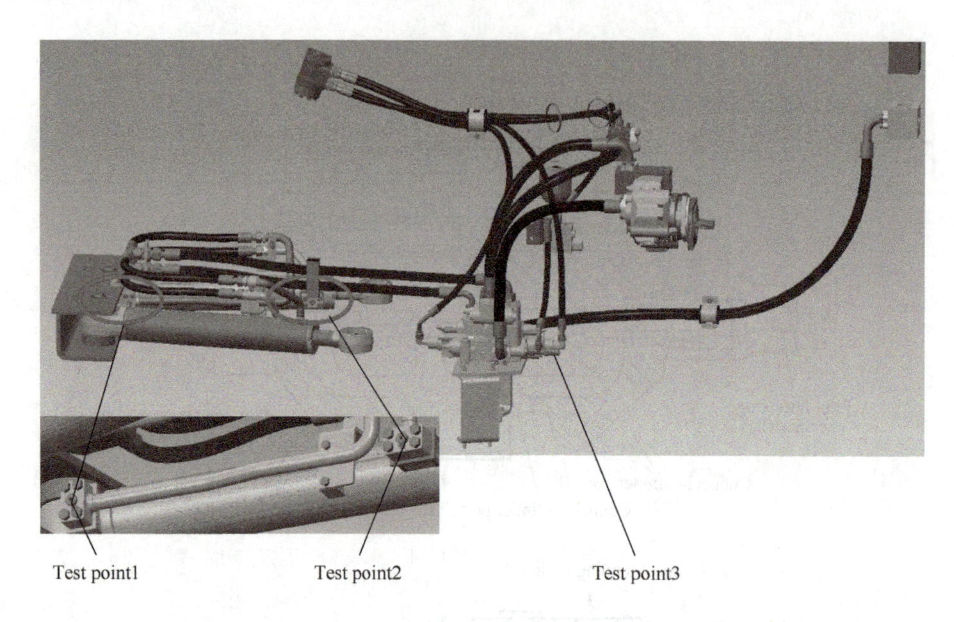

Fig. 3-11　Measuring point position of steering hydraulic system

5. Description of steering hydraulic system pressure test points

The description of the steering hydraulic system presure test points is shown in Table 3-4.

Table 3-4　Steering system test points description

Pressure testing point	Test point description	Theoretical pressure value (MPa)	Interface size
Pressure testing point 1	Pressure in the large cavity of the steering cylinder	Values according to model	M14 × 1.5 – 6H
Pressure testing point 2	Pressure of small cavity of steering cylinder	Values according to model	M14 × 1.5 – 6H
Pressure testing point 3	Flow amplifier valve inlet (steering pump outlet) pressure	Values according to model	M18 × 1.5 – 6g pressure measuring connector

[Task Implementation]

The pressure test of steering hydraulic system, as shown in Fig. 3-12.

The steps for adjusting the pressure of the safety valve are as follows:

1) Install the front and rear frame bumpers so that the frame cannot rotate to each other. Or steering to the limit position.

2) Install a pressure gauge (range is 25 MPa) at test point 3 on the flow amplification valve.

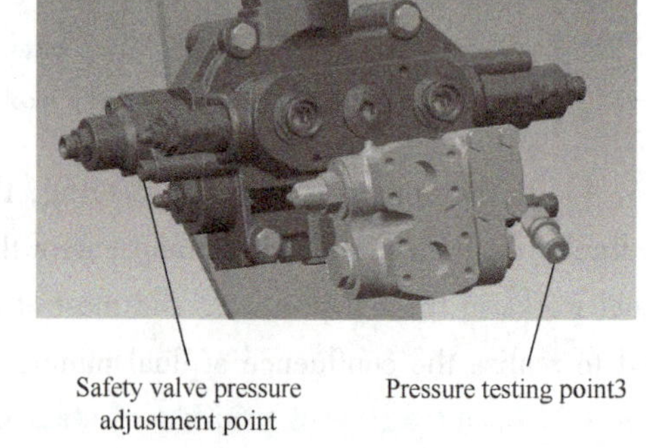

Fig. 3-12　Pressure test points of steering system

3) Start the whole machine and make it idle at

high speed. Turn the steering wheel quickly until the safety valve is opened, and the pressure gauge should indicate the design value given by the model.

4) If the pressure is not correct, it can be adjusted at the pressure adjustment point of the safety valve with a hexagonal wrench (tighten to increase pressure, loosen to decrease pressure).

5) After proper adjustment, remove the pressure gauge and bumper.

Subtask 3.1.5 The principle analysis of braking system

☞ [Learning Objectives]

1) To be able to describe the control principle of braking system.

2) To be able to analyse braking system control loops.

3) To be able to analyse the common faults of the hydraulic system through the principle and eliminate them.

☞ [Work Task]

For a dry - braking loader, the user reports that the brakes are soft, so the brake system pressure needs to be tested and checked.

☞ [Relevant Knowledge]

1. The basic components of the braking system and working principle

The braking system of the loader is used for speed reduction or stopping when driving, as well as stopping for a longer period on flat ground or ramps. It is divided into two parts, one is the driving brake, the other is the parking brake. Driving brake is used for frequent general driving speed control and stopping, also called foot brake. Parking brake is used for braking after parking, or emergency braking when the driving brake fails.

Loader braking system can be divided into dry braking system and wet braking system. Dry brake is also called air top oil brake, most loaders currently use this system; wet brake also called full hydraulic brake, is used in premium machines and large machines. Air top oil four - wheel disc brake has the characteristics of stable, safety and reliability, simple structure, easy maintenance, good recovery from water, etc. Full hydraulic wet brake has the advantages of stable, short response time, easy operation, safety and reliability, braking performance is not affected by the operating environment and so on. This section mainly introduces the air top oil dry brake system.

2. The composition of CLG856H loader dry braking system

The CLG856H loader driving brake adopts the air - top oil four - wheel calliper disc brake, and the parking brake adopts the shoe brake, whose braking position is at the front of the output shaft of the gearbox. The driving method of parking brake is hand - pulled flexible shaft control which is different from the pneumatic control of CLG856 and ZL50C, so there is no emergency brake function.

The driving brake system consists of air compressor, combination valve, air storage tank, air brake valve, booster, calliper disc brake and auxiliary pipelines. The parking brake system consists of hand brake and flexible shaft. The arrangement of brake system components is shown in Fig. 3-13.

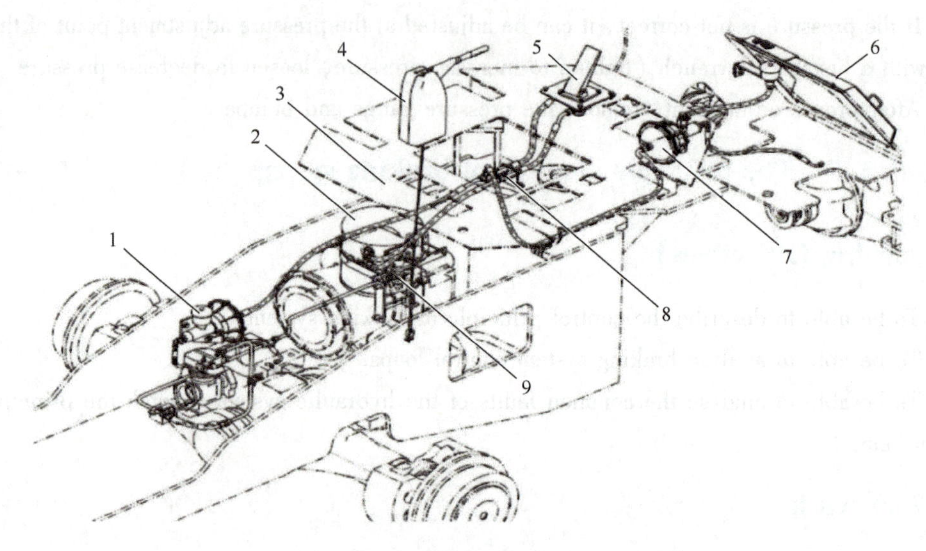

Fig. 3-13 Component arrangement position

1—Rear booster 2—Air storage tank 3—Flexible shaft 4—Hand brake
5—Brake valve 6—Cover plate 7—Front booster 8—Joint 9—Combination valve

3. The working principle of CLG856H loader dry braking system

1) The working principle of the driving brake system (as shown in Fig. 3-14). The air compressor is driven by the engine to convert air into compressed air, and the compressed air is stored in the air storage tank after passing through the combination valve. When the pressure in the air storage tank reaches the maximum working pressure of the braking system (usually about 0.78 MPa), the combined valve opens the unloading port, and the compressed air output by the air compressor is directly discharged to the atmosphere. When the pressure in the air storage tank is lower than the minimum working pressure of the braking system (generally about 0.71 MPa), the combined valve will open the outlet to the air storage tank, close the unloading port and supplement the compressed air in the air storage tank until the pressure in the air storage tank reaches the maximum working pressure. When braking, after the operator steps on the brake valve, the compressed air enters the cylinder of the booster through the brake valve and pushes the booster piston, which sends the brake fluid from the booster cylinder into the clamping sub - pump of the wheel - side brake 5 and squeezes the friction pads to brake.

2) The working principle of braking system (as shown in Fig. 3-15) The operator pulls up the handle of the parking brake, pulls the flexible axle upward, and the flexible axle pulls the brake shoe assembly of the brake to brake; at the same time, the button is pressed down, and the compressed air entering the gearbox gear - shift control valve through the nylon tube is cut off, and the power of the gearbox is cut off. When the handle is pulled up, the hand brake also outputs an electrical signal to the P lamp.

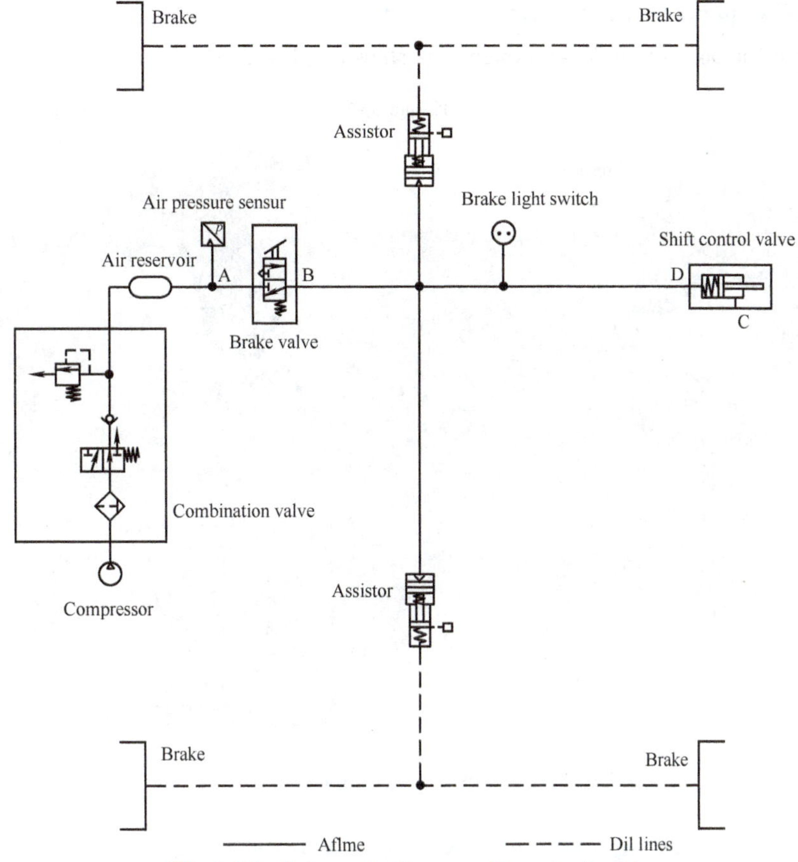

Fig. 3-14 Schematic diagram of service braking

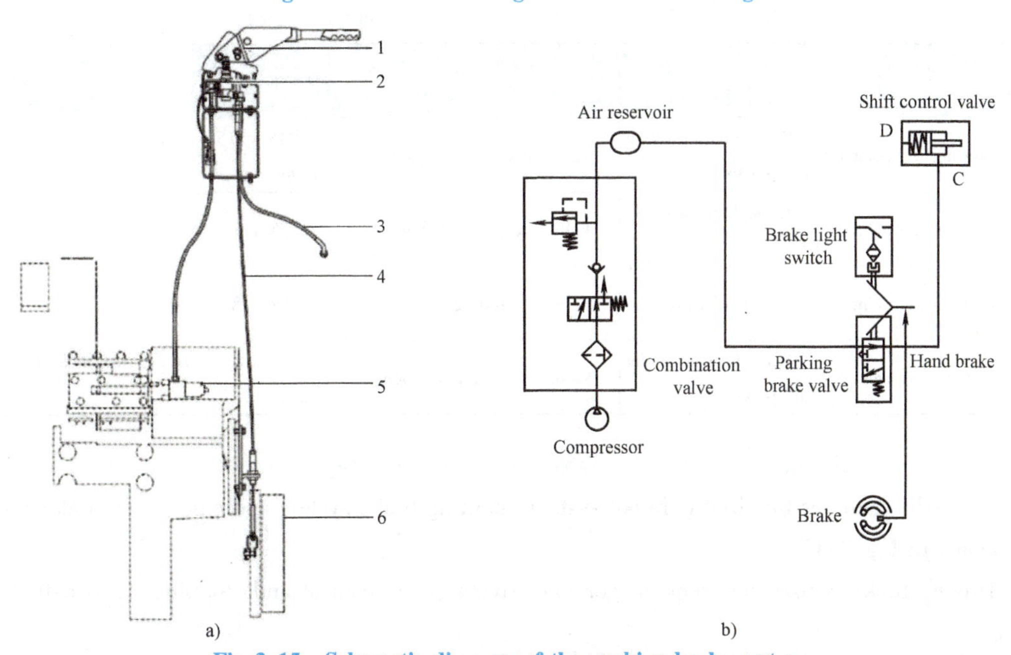

Fig. 3-15 Schematic diagram of the parking brake system

1—Hand brake 2—Push button valve 3—Nylon tube 4—Flexible shaft
5—Gear - shift control valve 6—Brake assembly

4. Measuring point position of brake system

Measuring point position of brake system, as shown in Fig. 3-16.

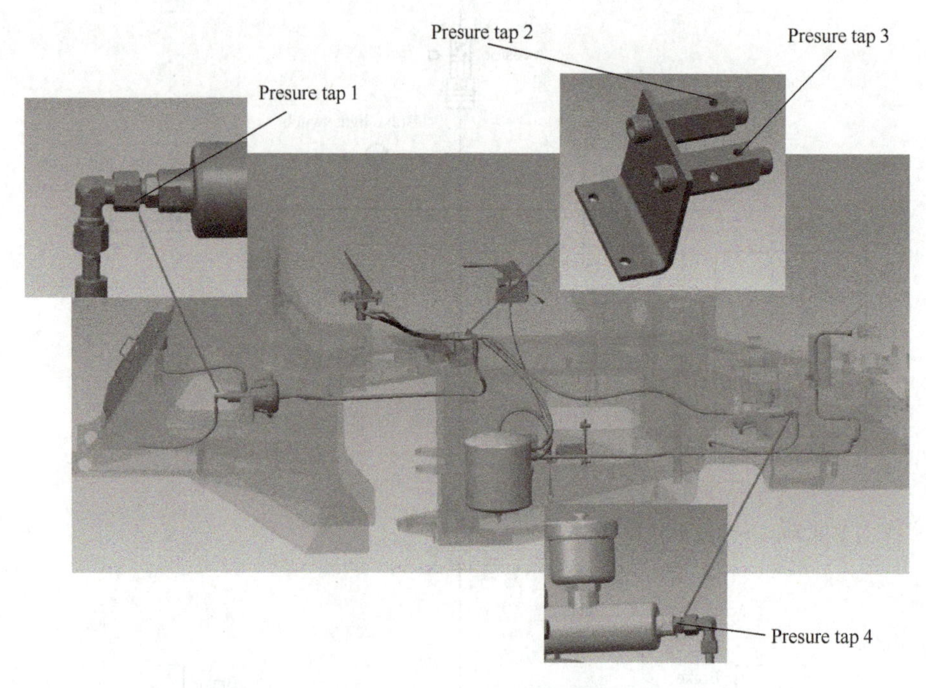

Fig. 3-16 Measuring point position of brake system

5. Description of steering hydraulic system pressure test points

Table 3-5 Test points description of steering system

Pressure testing point	Test point description	Theoretical pressure value/MPa	Interface size
Pressure testing point 1	Front booster outlet fluid pressure	Values according to model	M18 × 1. 5 – 6g pressure measuring connector
Pressure testing point 2	Driving brake system pressure	Values according to model	NPT 1/8
Pressure testing point 3	Booster inlet pressure	Values according to model	NPT 1/8
Pressure testing point 4	Rear booster outlet fluid pressure	Values according to model	M18 × 1. 5 – 6g pressure measuring connector

6. The working principle of CLG856H loader full hydraulic braking system

It usually includes the driving brake system, parking brake system and emergency brake system (as shown in Fig. 3-17)

Driving brake is used for frequent general driving speed control and stopping, also called foot brake.

Emergency and parking brake is used for braking after parking, or emergency braking when the driving brake fails. In addition, when the system pressure is lower than the safety pressure, the system automatically stops the loader to ensure safety.

Hydraulic components include brake pump, brake valve, liquid filling valve, brake valve block, accumulator, wet brake (inside the transmission axle), parking brake cylinder, parking brake (box belt), pressure switch and pipeline, etc.

The characteristics of the wet wheel – side brake: fully enclosed to avoid the influence of external uncertainties, such as sand intrusion and friction discs stained with oil, etc. Oil – immersed cooling, the friction disc is immersed in oil to avoid the reduction of the braking capacity due to heat recession of the friction material high temperature rise of the friction disc caused. Reduction of the requirements for the brake oil.

The characteristics of the fully hydraulic brake system: cancel the air circuit and simplify the system. It avoids the corrosion of pipelines and brake components caused by water in the air circuit system, thus improving the reliability of the system. It improves response speed.

The characteristics of the double circuit braking: the rear axle brake pipeline is independent, which is safer and more reliable.

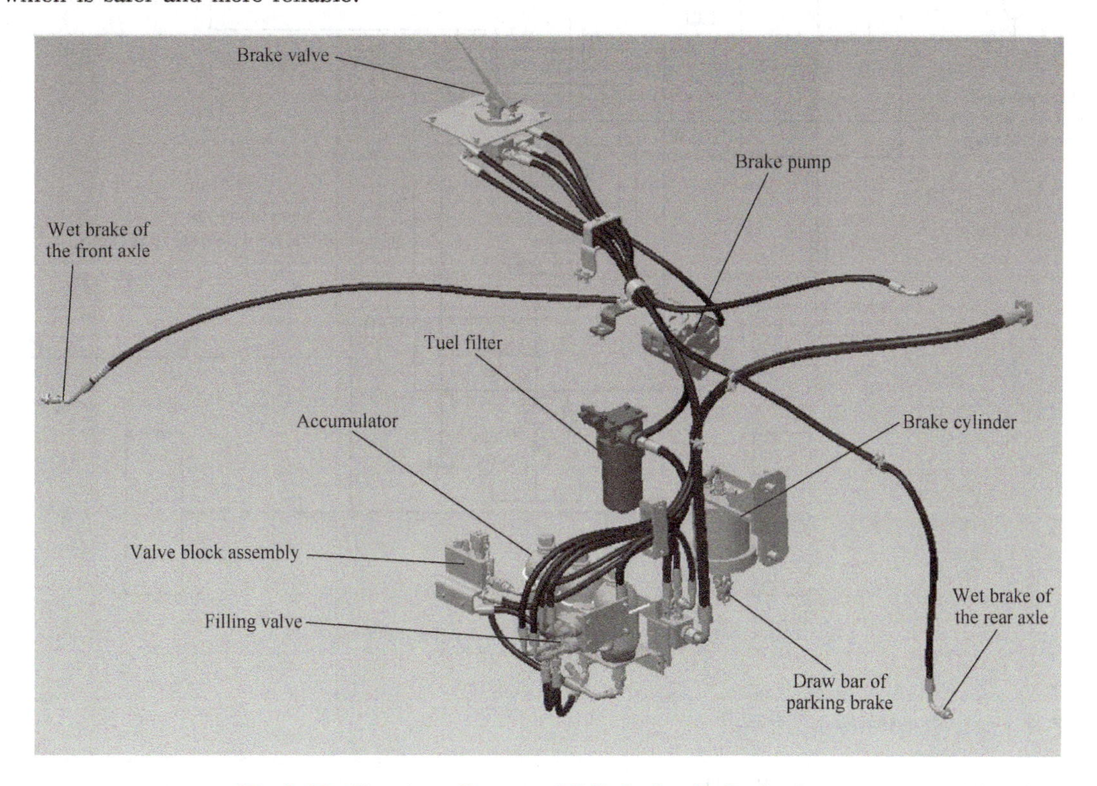

Fig. 3-17 Structure diagram of fully hydraulic brake system

As shown in Fig. 3-18, no pedal action: there is no pressure feedback. The energy stored by the accumulator is on port P, the brake port is connected to port T, and there may be internal leakage from port P to port T.

The pedal is in action, adjustment state: the pressure is stable at the maximum value, port P and port T are closed, the pressure feedback on the maximum value, the pedal force is the maximum and the force is balanced.

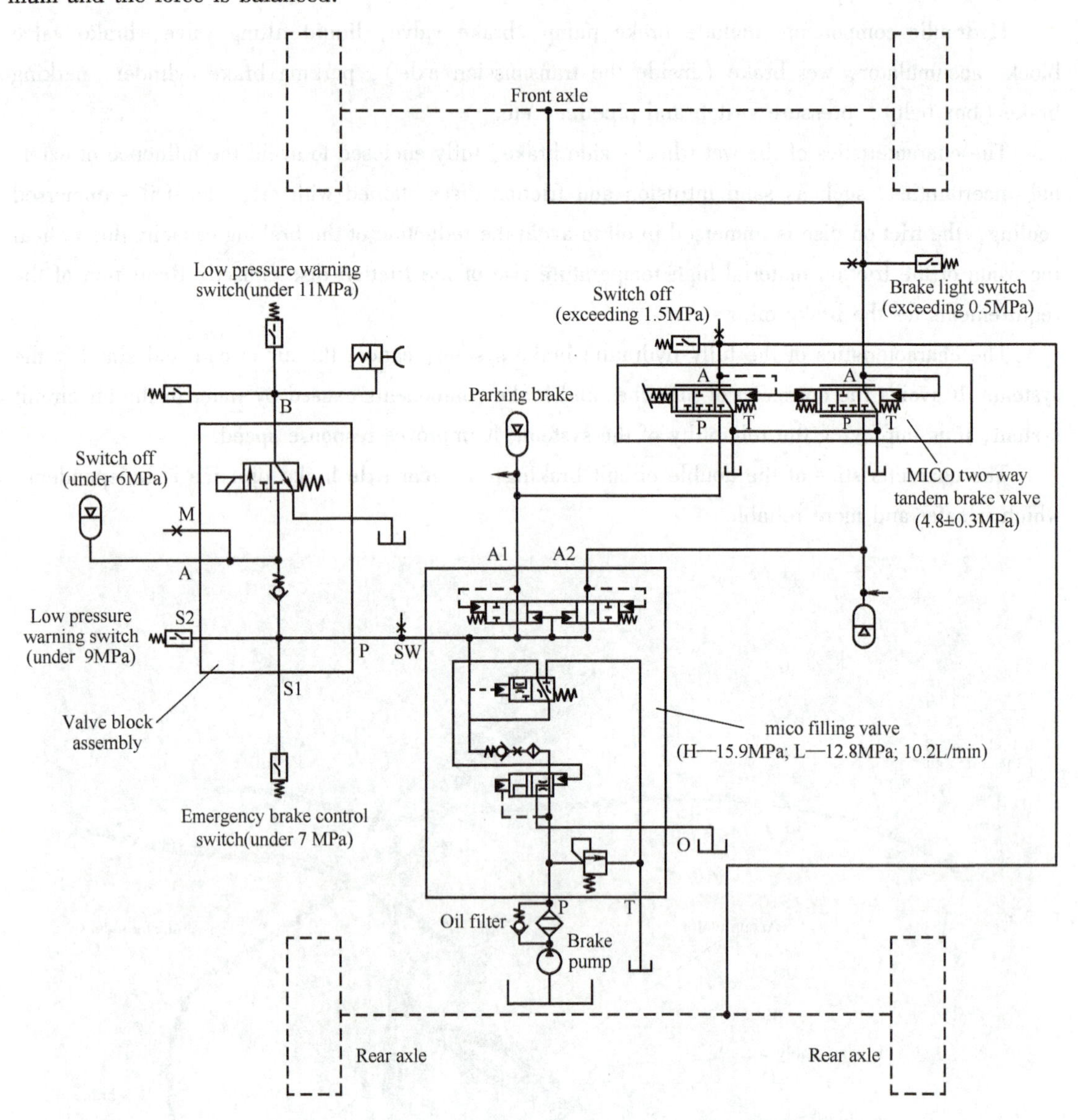

Fig. 3-18 Schematic diagram of full hydraulic brake system

[Task Implementation]

Measurement of air top oil brake system pressure

(1) Test 1 Without connecting the front axle, test the oil pressure at the outlet of the booster separately. Remove the hose and right – angle connector connected to the front axle, install the pressure measuring connector, step on the brake, and the pressure gauge value should meet the requirements;

(2) Test 2 Test the outlet pressure of the booster outlet pressure while connecting the front axle. Remove the hose and right – angle connector connected to the front axle, install the three –

way connector and pressure measuring connector, and then reconnect the hose and right - angle connector to the three - way position, step on the brake and read the value of the pressure gauge which should meet the requirements.

If the above does not meet the requirements, the booster needs to be replaced.

Task 3.2 Measurement of hydraulic system performance

Subtask 3.2.1 Measurement of hydraulic cylinder settlement

☞ [Learning Objectives]

1) To be able to develop a technical plan for measuring the settlement of the loader's hydraulic cylinder.

2) To be able to measure the loader hydraulic cylinder settlement according to the specification requirements.

☞ [Working Tasks]

After a loader is lifted to the highest level, the bucket drops. Hence, it is necessary to test the settlement of the hydraulic cylinder to find the cause.

☞ [Relevant Knowledge]

During the working process of the loader, the automatic expansion and contraction of the oil cylinder such as arm falling and bucket falling are not allowed, especially under some special working conditions, such as hoisting operation, the oil cylinder is required to remain stationary for a long time, or can only act within the permissble range, so it is necessary to measure the settlement of the oil cylinder.

The main factors involved in the settlement of the cylinder are:

1) bucket cylinder, bucket valve core, oil port relief valve, cylinder seal.

2) boom cylinder, oil port relief valve, cylinder seal.

The settlement of the hydraulic cylinder is an important indicator that affects the performance of the loader's working device.

Test conditions: The loader is filled with coolant, fuel, lubricating oil, hydraulic oil as specified, and includes tools, spare parts, a driver (75 kg) and other accessories, tire pressure should meet the provisions of the use and maintenance instructions; when making the initial measurement, the hydraulic system oil temperature should be (50 ± 3)℃.

Static test: turn the bucket back and carry the rated load (working load), operate the lifting hydraulic cylinder to make the bucket to the highest position, turn off the engine, and operate the closed position of the distribution valve. Measure the outreach length of the piston rod of the lifting hydraulic cylinder and bucket hydraulic cylinder every 15 min according to the table, and the test time is 3 h, as shown in Table 3-6.

Table 3-6 Record of settlement measurement of hydraulic cylinder

No.	Loading weight/t	Measurement time/h	Piston rod extension length/mm				Hourly settlement/(mm/h)	
			The boom cylinder		The bucket cylinder			
			left	right	left	right	The boom cylinder	The bucket cylinder
1								
2								
3								

The test values should be in accordance with the following regulations, as shown in Table 3-7.

Table 3-7 Settlement of hydraulic cylinder (Unit: mm/h)

Average 3 h static test	The boom cylinder	The bucket cylinder
	≤50	≤20

☞ [Task Implementation]

The loader is filled with coolant, fuel, lubricating oil, hydraulic oil as specified, and includes tools, spare parts, a driver (75 kg) and other accessories, tire pressure should meet the provisions of the use and maintenance instructions; when making the initial measurement, the hydraulic system oil temperature should be (50 ± 3)℃.

Turn the bucket back and carry the rated load (working load), operate the lifting hydraulic cylinder to make the bucket to the highest position, turn off the engine, and operate the closed position of the distribution valve. Measure the outreach length of the piston rod of the lifting hydraulic cylinder and bucket hydraulic cylinder every 15 min according to the table, and the test time is 3h, as shown in Table 3-6.

Subtask 3. 2. 2 Measuring cycle time of hydraulic cylinder

☞ [Learning Objectives]

1) To be able to develop a technical plan for the measurement of the cycle time of hydraulic cylinders.

2) To be able to measure the hydraulic cylinder cycle time according to the specification requirements.

☞ [Work Task]

The user of a loader says that the lifting is slow, so it is necessary to test the cycle time of the hydraulic cylinder in order to find out the cause.

☞ [Relevant Knowledge]

Purpose of measuring the cycle time of a hydraulic cylinder: The speed of hydraulic cylinder expansion and contraction is related to the flow. The greater the flow, the faster the cylinder moves, while the smaller the flow, the slower the cylinder moves. Cylinder movement speed can reflect the efficiency of the whole machine. When it is suspected that the cylinder moves slowly, it can be judged whether the cylinder is working normally by measuring the cycle time of the cylinder.

Measurement steps:

(1) Test conditions The loader is filled with coolant, fuel, lubricating oil, hydraulic oil as specified, and includes tools, spare parts, a driver (75 kg) and other accessories, tire pressure should meet the provisions of the use and maintenance instructions; when making the initial measurement, the hydraulic system oil temperature should be (50 ± 3)℃.

(2) Measurement of lifting time Firstly, the bucket is turned backward from the reference ground plane, and then loaded to the rated load (working load). The lifting hydraulic cylinder is manipulated to lift the bucket to the highest position. The lifting time of the bucket is measured according to Table 3-8. At the same time, the vertical lifting height of the bucket hinge pin is measured, and the lifting speed of the bucket is calculated.

Table 3-8 Record of working device action time measurement

Item	Height of bucket hinge pin/mm	Lift distance/mm	Measurement time/h	Measured speed/(mm/s)	Remarks
Lifting					Loading rated load in bucket
Lowering					Empty load
Unloading					Empty load

(3) Determination of unloading time When testing, lift the bucket to the highest unloading position, start the bucket hydraulic cylinder, make the empty bucket rotate from the highest position to the unloading position, and measure the continuous time of this process according to Table3-8 (that is the unloading time).

(4) Determination of lowering time When testing, lower the empty bucket from the highest position to the reference ground level, measure the continuous time of this process according to Table 3-8, at the same time measure the vertical lowering height of the bucket hinge pin and calculate the lowering speed of the bucket.

☞ [Task Implementation]

(1) Determination of lifting time firstly, the bucket is turned backward from the reference ground plane, and then loaded to the rated load (working load). The lifting hydraulic cylinder is manipulated to lift the bucket to the highest position. The lifting time of the bucket is measured according to Table 3-8.

(2) Determination of unloading time When testing, lift the bucket to the highest unloading position, start the bucket hydraulic cylinder, make the empty bucket rotate from the highest position to the unloading position, and measure the continuous time of this process according to Table3-8 (that is the unloading time).

(3) Determination of lowering time When testing, lower the empty bucket from the highest position to the reference ground level, measure the continuous time of this process according to Table 3-8, at the same time measure the vertical lowering height of the bucket hinge pin and calculate the lowering speed of the bucket.

Subtask 3.2.3 Measurement of hydraulic oil temperature

☞ [Learning Objectives]

1) To be able to correctly describe the mechanism of hydraulic oil temperature increase.

2) To be able to measure the hydraulic oil temperature using a temperature measuring gun.

☞ [Work Task]

A loader with high hydraulic oil temperature needs to test the temperature of the hydraulic oil.

☞ [Relevant Knowledge]

1. The measurement method of hydraulic oil temperature

The temperature of hydraulic system usually refers to the temperature of hydraulic oil in hydraulic tank. The sensors for detecting the oil temperature of excavator hydraulic system are generally installed on the main pump suction pipeline, and the measured temperature value can be viewed on the cab LCD screen. If no temperature sensor is installed, you can open the inspection cover above the hydraulic oil tank and then use a thermometer to measure the temperature of the hydraulic oil in the tank.

The reasonable working temperature range of hydraulic oil is 50 ~ 70℃. The causes of hydraulic oil temperature rises can be divided into external and internal causes. External environmental heat radiation and hydraulic system internal parts friction, cavitation phenomenon may cause hydraulic oil temperature rise too fast. Working at high temperature for a long time, hydraulic oil easily becomes deteriorated. Deteriorated hydraulic oil leads to increased viscosity, resulting in increased frictional resistance to flow and increased heat generation.

2. Judgement of hydraulic oil deterioration

1) Colour identification. Observe the colour of the hydraulic fluid, and if the hydraulic fluid is dark brown, it indicates that the hydraulic fluid has been oxidized at high temperature and degraded.

2) Hand rub. Drop the hydraulic oil on your thumb and rub it repeatedly. If you feel that there are larger particles between your fingers, it means that there are more impurities in the hydraulic oil. Good hydraulic oil is slippery and basically frictionless.

3) Light observation. Use a screwdriver to lift up the hydraulic oil, and about 45 degrees to horizontal surface. Wear debris in oil droplets of hydraulic oil can be observed in sufficient light.

4) Touch. Take a piece of white paper and drop the hydraulic oil on the surface of the white paper. After the hydraulic oil leaks, you will find that the original surface of the oil drop has a layer of black powder. Touch the black powder with your hand, if it feels smooth and there is no powder astringency, it indicates that the hydraulic oil is good.

5) Drop percolation. Drop the hydraulic oil on the filter paper and observe the change of the spot. If the hydraulic oil spreads rapidly and there is no deposit in the middle, it indicates that the oil is normal. If the hydraulic oil diffuses slowly and deposits appear in the middle, it indicates that the oil has become bad.

☞ [Task Implementation]

Measure the hydraulic oil temperature by using a temperature measuring gun.

At this stage, the loader hydraulic system is installed hydraulic oil temperature sensor, you can directly read the hydraulic oil temperature through the display, if the electrical system (such as sensors, sensor harness, etc.) problems will lead to incorrect hydraulic oil temperature display value, then you need to measure the hydraulic oil temperature through the temperature measuring instrument.

The commonly used temperature measuring instrument is the temperature measuring gun which is used to measure the inlet and outlet temperature of the hydraulic oil radiator. If the temperature difference range is 5 ~ 8℃, it means that the heat dissipation of the radiator is normal.

Project 4

Analysis and maintenance of loader electrical system

The initial loader (such as articulated loader with ZL50) electrical system is very simple, only the lighting and starting function is realized through electricity, and the monitoring instrument is mechanical.

With the continuous progress of sensor and detection technology, the electromagnetic instrument gradually replaced the mechanical type instrument. At the same time, with the increasing demand of users, electrical components used to improve user comfort, such as wiper, fan, retractor, air conditioner, etc., have also gradually become part of the electrical system of the loader. The electrical system plays an increasingly important role in the whole machine.

There are varieties and specifications of loaders. Some machine electrical systems are very simple, with only simple starting, lighting, instrumentation monitoring and other functions; others are very complex, integrating machinery, electricity, hydraulics, and information. Anyway, all the loader electrical systems have the following two characteristics:

1) Most of them are 24 V nominal voltage (very few are 12 V).

2) Single-wire system and negative grounding method are adopted.

The positive electrode of the battery is often connected to the electrical equipment, and the negative electrode of the battery is connected to the body, that is, the metal body of the machine is used instead of the negative wire in the circuit, and the circuit formed by this method is called the single-wire system. The way to connect the negative electrode to the metal body is called the negative grounding. The electrical system of all loaders can be abstracted into the circuit model as shown in Fig. 4-1.

As a circuit model, Fig. 4-1 focuses on the relationship between the power supply and the load, ignoring the control components in the actual circuit, such as switches, relays, fuses, etc.

As shown in Fig. 4-1, the electrical system of the loader can be divided into a main circuit and a load circuit. The main circuit includes the power supply system and the starting system, which is used to start the engine and provide power for the whole vehicle, and is the core of the electrical system. The load circuit generally includes the instrumentation system, lighting system, auxiliary electrical appliances (such as wiper, electric fan, air conditioning appliances, electric horn, audio, etc.), and the load circuit of higher-grade loaders includes electronic monitoring system, automatic reset system of working device, and electro-hydraulic gear-shift control system, etc.

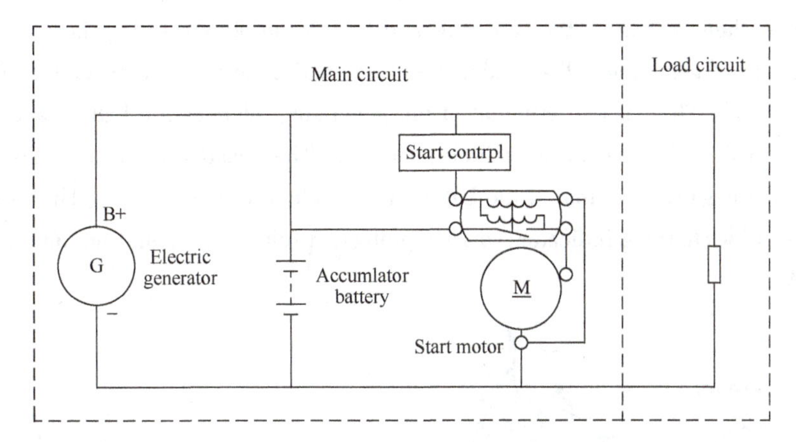

Fig. 4-1 Circuit model of the loader

Task 4.1 Analysis of working principle of main circuit

☞ [Learning Objectives]

1) To be able to correctly describe the working principle of the main circuit of the loader.

2) To be able to overhaul the power circuit according to the specifications.

☞ [Work Task]

A loader generator has blocking phenomenon and collision and scraping noise. It is necessary to troubleshoot and analyse the generator and give solutions.

☞ [Relevant Knowledge]

1. The structure and function of the main circuit

The main circuit is the core of the loader electrical system and is the basis of whether the whole vehicle can work normally. It is mainly composed of the main power switch, battery, generator, starter motor, starter control circuit, electric lock (key switch), power relay and so on. Among them, the battery, generator and starter motor are the core components of the main circuit.

2. Battery

See Sub task 2.5.1 in this book.

3. Generator

(1) Function of generator Generator (shown in Fig. 4-2) is a device that converts mechanical energy into electrical energy driven by the engine. The generator and battery are connected in parallel to form the power supply system of the whole machine. Before the diesel engine is started, the battery supplies power to the electrical appliances of the whole machine. After the diesel engine is started, the generator mainly supplies power to the electrical appliances, and the generator charges the battery. When the energy provided by the generator cannot meet the needs of the electrical

appliances of the whole machine, the generator and battery supply power together.

(2) Structure of generator The loader uses three wiring ports of the generator, they are B +, Ground, D +. B +: The positive terminal of the generator, charges the battery and provides current for the electrical appliances on the vehicle; Ground: The negative terminal of the generator, is connected with the negative terminal of the battery of the whole vehicle; D +: The excitation terminal is connected with charging indicator, starter protection relay, etc., and the output current shall not exceed 1 A.

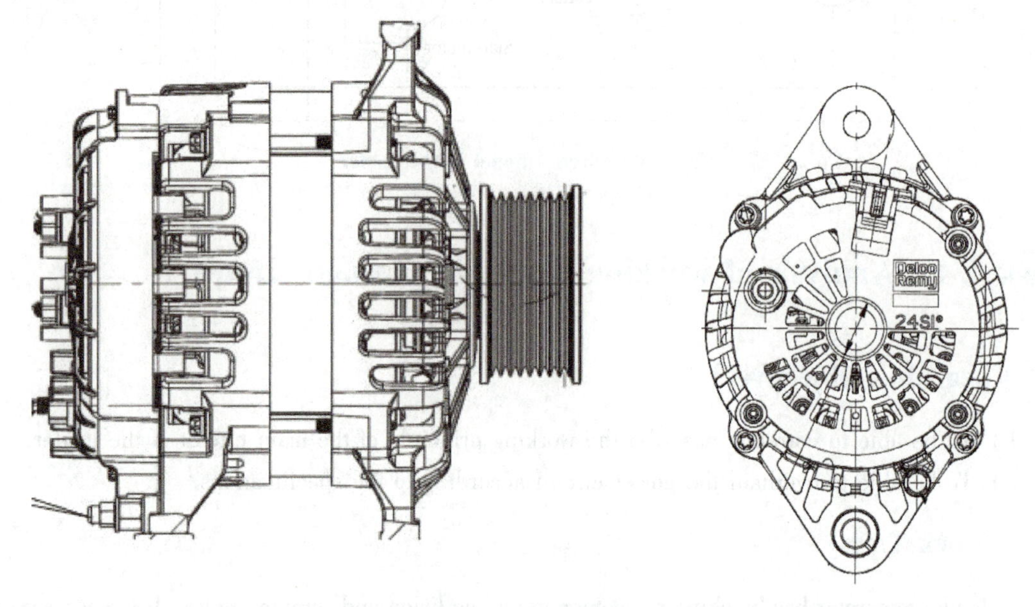

Fig. 4-2 Generator

4. Starter motor

(1) Function of starter motor The starter motor on the loader is a device that converts electric energy of the battery into mechanical energy and starts the engine. The starter motor consists of three parts: DC motor, transmission mechanism, and control device. The role of the DC motor is to convert electrical energy into mechanical energy and produce electromagnetic torque. The function of the transmission mechanism is to transmit the electromagnetic torque to the flywheel when the engine is started to drive the engine to run and start.

After the engine is started, the driving gear of the starting motor will automatically slip, so as to prevent the engine from dragging the armature of the starting motor and finally getting out of engagement with the flywheel ring gear. The control device is used to control the on – off of the connecting circuit between the DC motor and the battery, and to control the engagement and disengagement of the transmission mechanism and the flywheel.

At present, the vast majority of the starter motor on the loader is a series – excited electromagnetically operated direct – drive flexible engagement starter motor. The series – excited means that the magnetic field winding and the armature winding are connected in series. In addition, the reducer type starter motor and the flexible mesh starter motor are beginning to be used on the loader due to their obvious advantages.

(2) Brief description of starting process Figure 4-3a, shows the state of the starter motor and starting line before and after starting, and Fig. 4-3b shows the state of the starter motor and starting line during starting.

When starting, turn on the starting switch, and the suction coil and holding coil of the starting motor control device are energized. The direction of the electromagnetic force generated by the two is the same and superimposed with each other. The armature of the suction control device moves to the right against the spring force and drives the pull fork to rotate around its pin shaft to move the drive gear to the left. At the same time, because the current of the suction coil flows through the winding of the DC motor, the armature starts to rotate and the drive gear rotates through the unidirectional device. Therefore, the drive gear rotates and moves left. When the drive gear moves out to the left for a certain distance, the tooth end of the drive gear is opposite to the tooth end of the flywheel ring gear of the engine and cannot be engaged immediately, and the spring is compressed. When the drive gear rotates at a certain angle, the tooth ends of the two gears are staggered. Under the action of the spring force, the drive gear moves rapidly to the left to engage with the flywheel. At the same time, the armature of the control device moves rapidly to the right to make the contact switch of the control device close quickly. After the contact switch is closed, the large current flows from the positive terminal of the battery through the winding of the DC motor through the contact switch and returns to the negative terminal of the battery. The DC motor will generate a large electromagnetic torque to drive the engine to rotate and start (Note: after the contact switch is closed, the potential at both ends of the attraction coil is equal, there is no current flowing, and the position of the armature is maintained by the electromagnetic force generated by the holding coil).

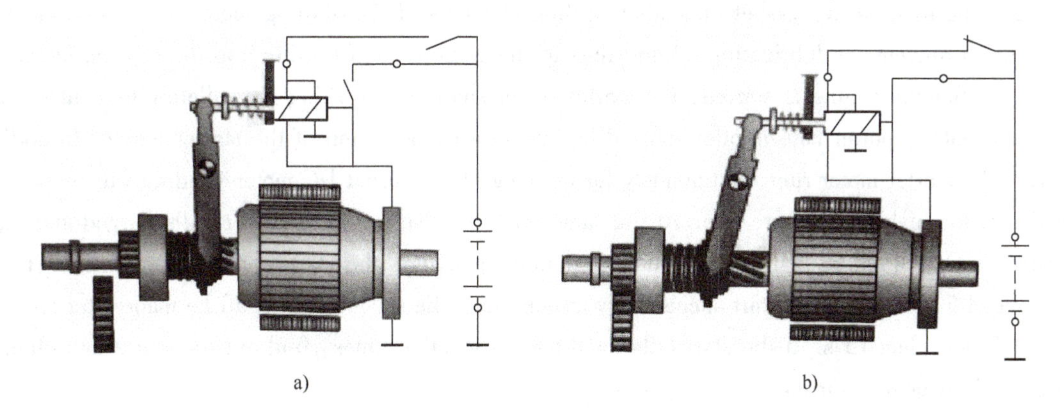

Fig. 4-3 Schematic diagram of starting composition

After the engine is started, its speed rises rapidly to idle speed, and the flywheel becomes an active gear to drive the drive gear to rotate. However, due to the "slip" effect of the one – way device, the torque of the engine will not be transmitted to the armature, preventing the danger of over – speed operation of the armature.

After starting, release the starting switch, and the starting control circuit will be powered off.

In addition to the current flowing from the positive terminal of the battery through the contact switch and the winding of the DC motor to the negative terminal of the battery, it also flows from the positive terminal of the battery through the contact switch and the attraction coil of the control device returns to the negative terminal of the battery through the holding coil. Obviously, at this time, the attraction coil and the holding coil are in series, and the current flowing through them is equal. Due to the equal number of turns, the electromagnetic forces generated by the two are equal, but in the opposite direction and offset each other. The armature of the control device quickly moves left under the action of spring force, which breaks the contact switch and disconnects the winding of the DC motor and the attraction coil and holding coil of the control device. The armature moves left to drive the fork to rotate around its pin, so that the drive gear moves right and disengages the drive gear from the flywheel.

(3) Use and maintenance of starter motor The following are precautions for use and maintenance of starter motor:

1) The voltage drop on the positive or negative cable of the starter motor caused by any reason can reduce the starting performance, making it difficult or even impossible to start.

① It is necessary to ensure that all the terminals in the battery circuit are clean and firmly connected to reduce the contact resistance.

② Since most of the starter motors are grounded on the casing, it is also necessary to check whether the grounding of the engine casing and the negative electrode of the battery is good.

③ The cross - sectional area and material of the cable in the battery line should meet the requirements, and the total length of the cable should be as short as possible to reduce the wire resistance.

2) The dust cover, gasket and other sealing elements of the starting motor must be installed correctly to prevent the lubricating oil and dust of the gearbox from flowing into the starting motor.

3) After the engine is started, the starter motor shall stop working immediately to reduce the wear and battery power consumption caused by unnecessary operation of the starter motor. In addition, if the starter motor runs continuously for too long, the internal DC motor winding will be burned due to too high temperature rise. At the same time, the battery will be over - discharged and the service life of the battery will be affected. Generally speaking, the starting time of each time shall not exceed 5 s. If it fails to start successfully at one time, the second start shall be made after an interval of more than 15 s. If the start fails for three consecutive times, find out the cause and eliminate the fault before starting.

4) Before starting, turn off all electrical equipment not related to starting, and put the loader in neutral, and set the arm and bucket lever in the middle position to increase the starting capacity of the starter motor and reduce the resistance torque of the engine.

5) If it is difficult to start due to low ambient temperature, fully preheat the engine before starting, to reduce the viscosity of engine lubricating oil and reduce the resistance torque of the engine.

(4) Common faults of starter motor

1) The faults of control device mainly include short circuit, open circuit and overlap of attrac-

tion coil or holding coil, ablation of contact and terminal, etc.

2) DC motor faults mainly include serious dirt or ablation of commutator, severe wear of brush, stuck in brush holder, short circuit, open circuit and overlap of armature winding or magnetic field winding, etc.

3) Transmission mechanism faults mainly include one – way slippage, one – way spring broken, severe wear or damage of drive gear or flywheel ring gear, severe wear of armature shaft bushings, broken or out of position of pull fork, excessive clearance between drive gear and limit ring, etc.

5. Electric lock

The electric lock (shown in Fig. 4-4) is commonly known as the key switch, which is used to control the functions of power on – off, start, and ignition off of the whole vehicle.

Fig. 4-4 Electric lock

Electric lock has five pins B1 – B2, M, S, G1 and G2, G1 pin is generally not used, as shown in Fig. 4-5.

B1 – B2 is the power supply pin, and B1 is connected to the wire 111. M is the ignition pin, which is connected to the wire 120. S/G2 is the starting pin, which is connected to the wire 453. The relationship between the functions and gears of electric lock is shown in Table 4-1.

Table 4-1 The functions and gears of electric lock

	B1	B2	M	S	G1	G2
OFF	•	•				
ON	•	•	•			
START	•	•	•	•		•
AID	•	•			•	

6. Fuse

The fuse mainly plays the role of short circuit and over – current protection in the circuit. CLG856H power system uses a variety of series of fuses, such as plug – in fuse and bolt type fuse. The specifications of plug – in fuse are 20 A, 15 A, 10 A, 7.5 A and 5 A, which are used for each branch circuit. The specifications of bolt type fuse are 30 A, 50 A, 60 A, 80 A and 150 A, which are used for the main circuit.

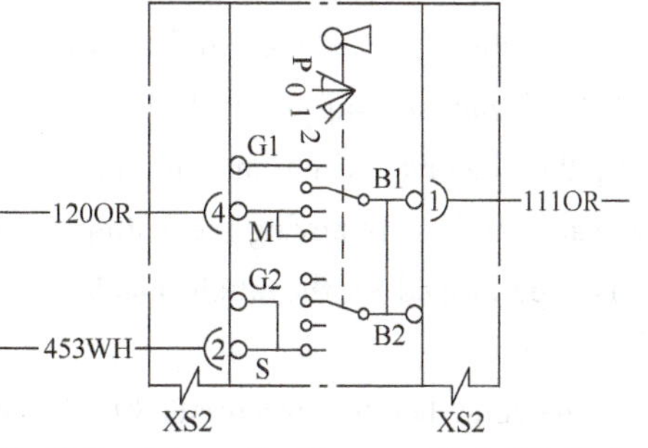

Fig. 4-5 Schematic diagram of electric lock

The colour of the plug – in fuse is related to the specification. The specification and colour of the plug – in fuse used in this machine are listed in Table 4-2.

Table 4-2 Specifications and colours

Specification	Colour
5 A	Brown
7.5 A	Brown
10 A	Red
15 A	Blue
20 A	Yellow

7. Negative switch

The negative switch (as shown in Fig. 4-6) controls the connection between the negative terminal of the battery and the frame of the whole machine. When the negative switch is in the "O" (off) position, the negative terminal of the machine power supply is cut off. Even if the electric lock is opened, the electrical appliances of the whole machine can't work. When the negative switch is in the "I" (open) position, the negative terminal of the power supply of the whole machine is connected. At this time, the connection between the electric appliance of the whole machine and the power supply can be controlled through the electric lock, and the diesel engine can be started.

Fig. 4-6 Negative switch

Attentions:

1) The negative switch must be turned off after each operation or drive, otherwise it will cause serious consequences of power leakage.

2) It is strictly forbidden to turn off the negative switch while the machine is running. This error will cause very serious damage to the electrical system of the whole machine; each stop should first close the electric lock, and then close the negative switch.

3) When starting up each time, turn on the negative switch first, and then turn on the electric lock; the negative switch must be turned off when connecting the battery cable or tightening the battery cable stub or removing the battery cable; the negative switch must be turned off when welding work is performed on the whole machine.

8. Relays

The relay has five terminals 30, 87, 87a, 86 and 85. Between 86 and 85 is the coil with a resistance value of about 300Ω. Between 30 and 87 is normally open contact, and between 30 and 87a is normally closed contact. There is a renewal diode inside.

The working principle of the relay is that when the coil is energized, 30 is connected to 87 and disconnected from 87a; when the power is off, 30 is disconnected from 87 and connected to 87a.

The way to judge whether the relay is damaged: measure with the resistance gear of multimeter, and the resistance between 86 and 85 is about 300 Ω; the resistance between 30 and 87 is infinite, and the resistance between 30 and 87a is 0. Connect 86 to the positive terminal of DC 24 V power supply and 85 to the negative terminal. 30 and 87 should be connected and 30 and 87a should be disconnected.

Since the relay has a current – continuing diode inside, the terminal 86 must be connected to the positive terminal of the power supply and the terminal 85 must be grounded and not reversed, as shown in Fig. 4-7.

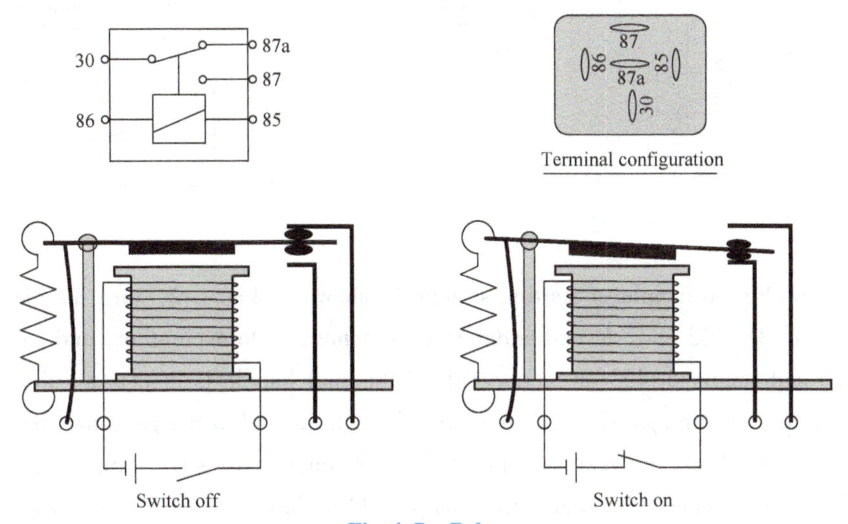

Fig. 4-7 Relay

9. Working principle and fault analysis of main circuit

The main circuit of different models of loaders is not exactly the same, but the basic principle is the same. Therefore, as long as you master the main circuit of a loader, the rest can be bypassed by analogy.

(1) The working principle of power supply system The main circuit power supply system of CLG856H loader is shown in Fig. 4-8.

After the negative switch is closed, one way of the battery power will pass through the 50A electric appliance centralized control box fuse and the conductor 100 to the non – overlock power bus fuse. At this time, the electrical components that can work normally are wall lamps, rotating warning lamps, parking lamps, horn, etc. At the same time, the electric lock fuse supplies power to the electric lock power terminal (B1 ~ B2) through wire 111. The other way reaches the power contactor through 60 A main power fuse and conductor 176.

After the electric lock is rotated to ON, the B1 ~ B2 end of the electric lock is connected to the M end, the wire 111 is connected to the wire 120, and the current passes through the wire 120, the coil of the power contactor, and the wire 212 to ground. Therefore, the power contactor contact switch is closed, the wire 176 is connected to the wire 190, and the power fuse of the electric lock is powered. At this time, except for the reverse alarm and air conditioning function module, the other electrical components can be used normally.

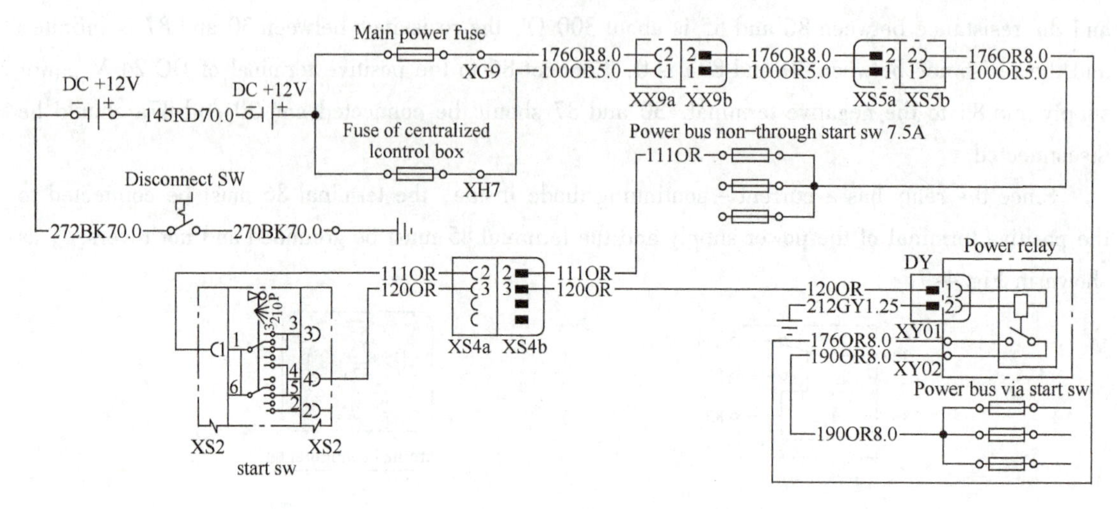

Fig. 4-8 Schematic diagram of main circuit power system

(2) The working principle of starting system As shown in Fig. 4-9, turn the electric lock to START gear, the B1 ~ B2 end, M end and S end are connected to each other, and conductor 111, conductor 120 and conductor 453 are connected. If the shift handle is in neutral, the transmission controller outputs 24 V through wire 584 to ground through the coil of the gear/start interlock relay. When the coil is energized, the gear/start interlock relay contact closes and wire 453 is connected to wire 454. On the other hand, the engine ECM outputs 24 V through wire 982, which passes through the coil of the starter protection relay and returns to ground inside the ECM through wire 983. After the coil is energized, the starter protection relay contact closes and wire 454 is connected to wire

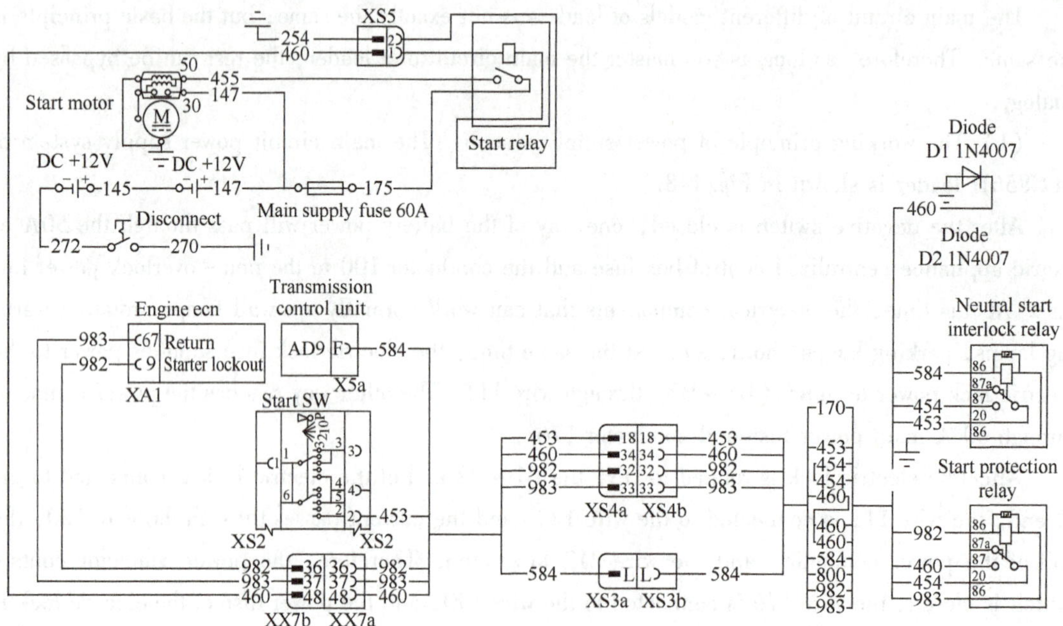

Fig. 4-9 The working principle of starting system

460. The current passes through 460 conductor and the starter relay coil to ground, causing the starter relay contact closes and the current flows into the solenoid switch coil of the starter motor and the starter motor starts to work. After the whole machine is started, the starting protection function of engine ECM is enabled.

☞ [Task Implementation]

The main circuit fault analysis is made by the list method, as shown in Table 4-3.

Table 4-3 list method fault diagnosis

Fault	Phenomenon	Cause	Maintenance and prevention
Short circuit and open circuit ofstator winding	Short circuit can be dividedinto phase – to – phaseshort circuit, turn to turn short circuit and winding overlap. Phase to phase short circuit and turn to turn short circuit can be checked from the appearance. Generally, the coil can be seen scorched and discoloured	Stator dipping paint processing is not good and insulation paint did not fill the slot gap, so that the fixed role of the wire weakened; the bearing is loose or ablated due to lack of oil, resulting in excessive radial clearance of the bearing and eccentricity of the rotor shaft, causing the rotor and stator to scratch and sweep the chamber, resulting in short circuit or mechanical open circuit due to excessive local temperature rise of the stator winding	Regular maintenance, and timely replace the bearing with serious looseness. If the bearing outer ring and bearing chamber are found to be worn, the housing should be repaired or replaced immediately
Stator iron core damage	The generator has blocking phenomenon and touch scraping sound	The generator is overloaded for a long time (unreasonable matching), resulting in the burning of the stator, the tightness of the fastening screws of the front and rear end covers being inconsistent or partially invalid, resulting in collision and scraping (chamber sweeping) and damage to the stator; the insulating paper is worn out, causing a short circuit to ground	Check whether the front and rear cover fastening bolts are loose or lost, and timely tighten
Rotor winding open circuit	The winding leads are desoldered or broken; the brush is seriously worn or loose; the brush spring is stuck or broken; the collector ring is ablated	The weld between the wire and the collector ring is desoldered due to welding quality and other reasons; the fit between the collector ring and the rotor shaft is loose and the lead is broken; the contact surface between the collector ring and the brush is poor, causing the generator to fail to generate magnetic field	For brushes with more than 50% wear and pressure failure of the spring should be replaced; for the collector ring with serious surface ablation and runout greater than 0.05 mm, round and polish with lathe; wipe the collector ring surface with alcohol and other solvents
Short circuit and open circuit of rectifier component	The output voltage of the generator is too low or the voltage cannot be output	Diode breakdown due to damage of voltage regulator; diode breakdown due to excessive forward or reverse working voltage and current	Avoid adding other high – power electrical equipment to the loader

Task 4.2 Analysis of working principle of instrument system

☞ [Learning Objectives]

To be able to correctly read the meaning of each section on the instrument.

☞ [Work Task]

Operators need to find out the location of pressure switch and sensor on a loader in order to judge and replace.

☞ [Relevant Knowledge]

The loader instrumentation monitoring system generally includes temperature gauge (such as engine water temperature gauge, torque converter oil temperature gauge, etc.), pressure gauge (brake air pressure gauge, transmission oil pressure gauge, etc.), fuel level gauge, voltmeter, timer and other indicator gauges and temperature sensor, pressure gauge sensor, fuel level gauge sensor, etc., as well as other alarm indicators driven by pressure switches.

1. Instrument assembly

The instrument configured by CLG856H loader is an electronic combination instrument, which is divided into three indication areas, as shown in Fig. 4-10.

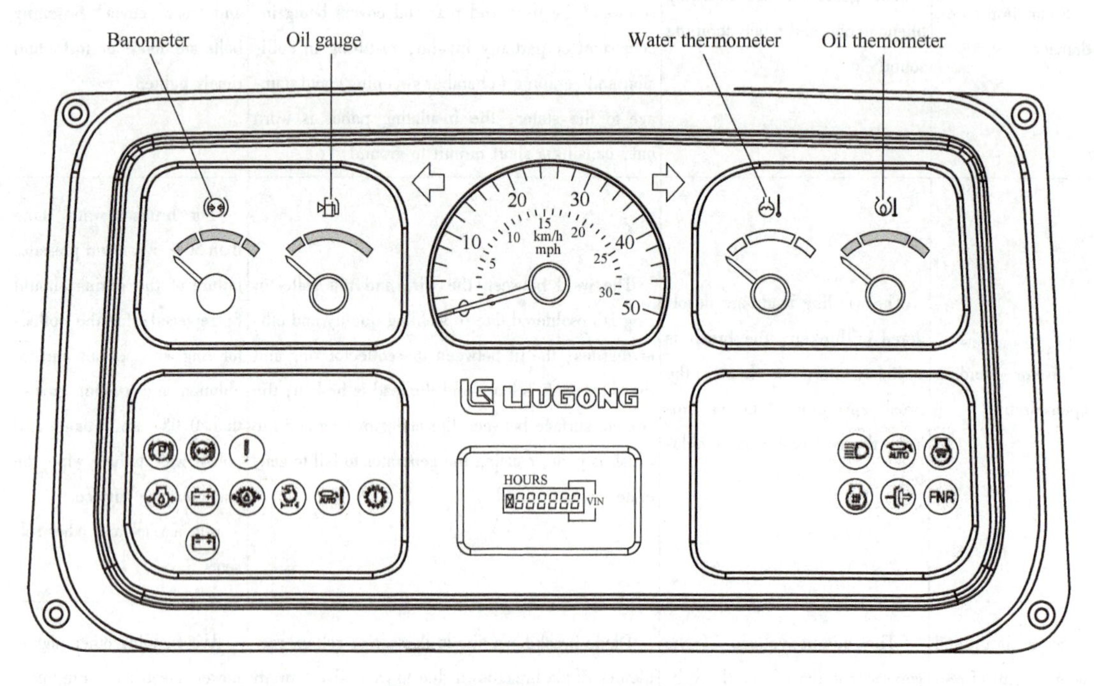

Fig. 4-10 Instrument assembly

1) Pointer type area, as shown in Table 4-4.

Table 4-4 Pointer type area

Item	Indication area		
	The first area	The second area (green)	The third area (red)
Engine water temperature gauge	40 ~ 55℃ (yellow)	55 ~ 101℃	101 ~ 120℃
Torqueconverteroil temperature gauge	40 ~ 60℃ (yellow)	60 ~ 116℃	116 ~ 140 ℃
Fuel oil level gauge	0 ~ 0.2 (red)	0.2 ~ 1	
Air pressure gauge	0 ~ 0.4 MPa (red)	0.4 ~ 0.8 MPa	0.8 ~ 1.0 MPa

2) The indicator area, as shown in Fig. 4-11.

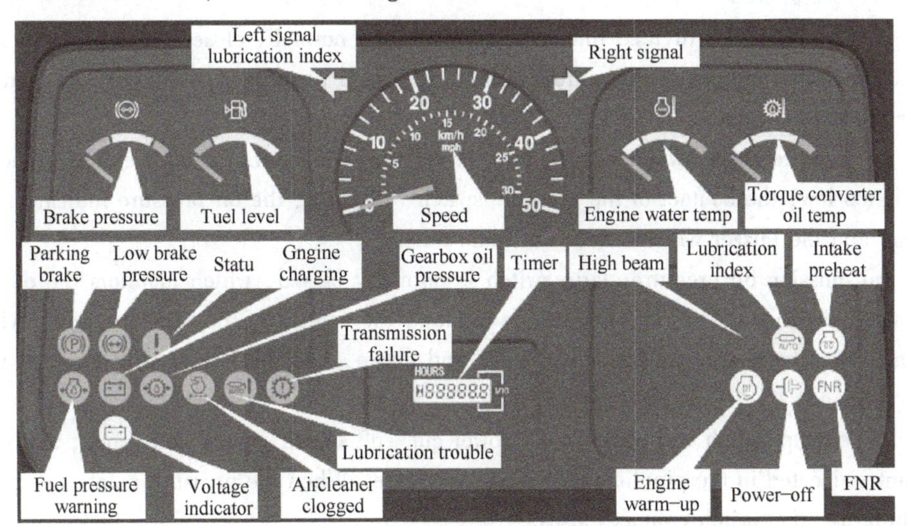

Fig. 4-11 Instrument indicator

3) LCD (liquid crystal display) area. The LCD can display vehicle speed, system voltage, engine fault code, ZF fault code, etc., as shown in Fig. 4-12. The screen display items such as vehicle speed, system voltage and engine fault code can be switched by the screen switching switch on the right side of the steering column.

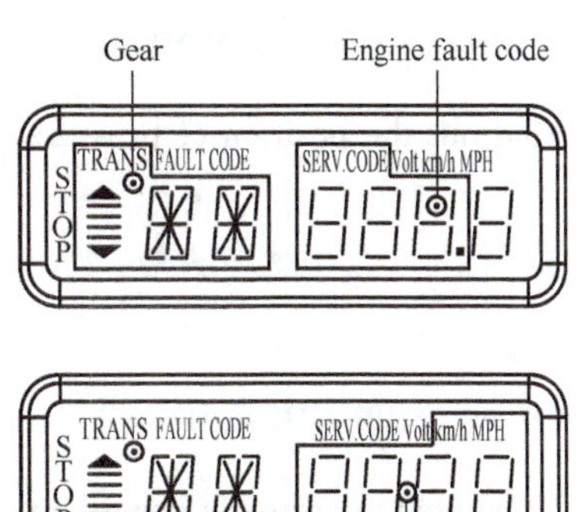

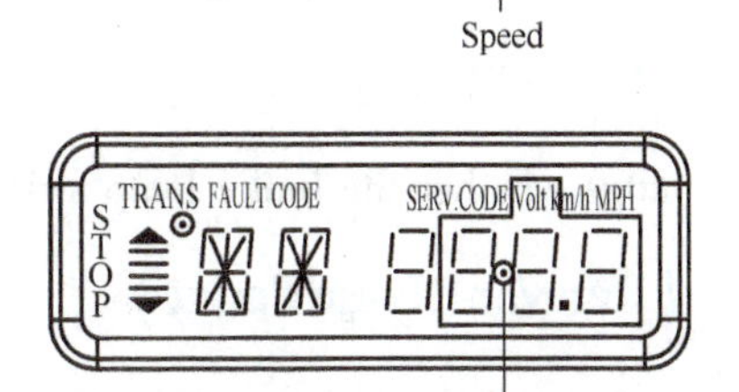

Fig. 4-12 LCD

2. Sensors

1) Temperature sensor. The temperature sensor for a loader is generally made of a copper shell with good heat transfer performance and encapsulated with a thermistor with negative temperature coefficient. Thermistors are made of ceramic semiconductor materials mixed with the appropriate amount of oxide sintered at high temperatures. The "negative temperature coefficient" is within the working range of thermistor, when the temperature increases, the conductivity of thermistor will increase with

the increase of temperature, that is, its resistance value will decrease with the increase of temperature.

2) Pressure sensor. The pressure sensor is mainly composed of a diaphragm cavity, a transmission mechanism and a slip wire resistor. The diaphragm of the diaphragm cavity is an elastic sensitive element, used to feel the pressure change of the medium (such as oil, gas, etc.) and converted it into mechanical displacement. When the medium pressure changes, the diaphragm produces mechanical displacement, and the transmission mechanism amplifies the mechanical displacement and transmits it to the sliding contact of the slip wire resistor, to change the output resistance value of the sensor.

3. Pressure switch

The pressure switch is divided into normally open and normally closed. The pressure switch on the loader generally includes oil pressure, variable speed oil pressure, power cut - off, brake lamp and other pressure switches.

The oil pressure alarm switch is normally closed, which monitors the oil pressure. When the oil pressure is too low, the contact of the pressure switch will close, the oil pressure indicator will flash and alarm, and the buzzer will alarm at the same time.

The transmission oil pressure alarm switch is normally closed, which monitors the oil pressure in the transmission box. When the pressure is too low, the contact of the pressure switch will close, the transmission oil pressure indicator will flash and alarm, and the buzzer will alarm at the same time.

The brake lamp switch and travel brake power cut - off switch are normally open, and the monitoring point is located in the pipeline after the brake valve. When the pressure reaches a certain value after braking, the switch contact closes.

☞ [Task Implementation]

Find out the position of pressure switch and sensor on the whole machine.

Task 4.3 Analysis of working principle of switch and lamp assembly

☞ [Learning Objectives]

To be able to correctly describe the working principle of front frame lighting circuit, cab lighting circuit and rear frame lighting circuit.

☞ [Work Task]

Judge whether the brake lamp switch is damaged for a new loader.

☞ [Relevant Knowledge]

1. Schematic diagram of front frame lighting circuit

As shown in Fig. 4-13, the front frame lighting circuit includes small lamp and headlight switch,

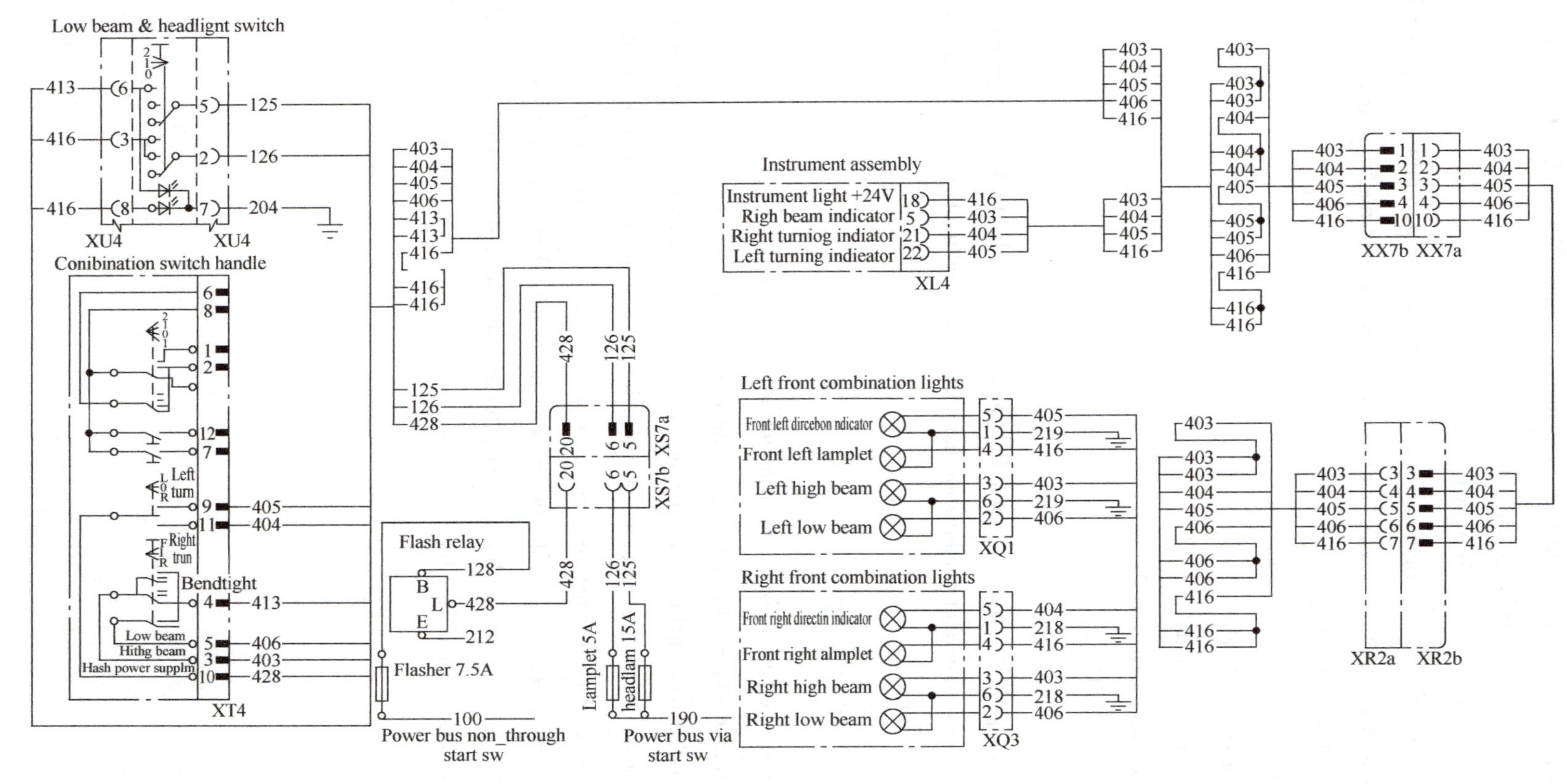

Fig. 4-13 Schematic diagram of front frame lighting circuit

combination switch and left and right front combination lamps. The combination lamp integrates the functions of steering, far/near lamp and small lamp.

2. Schematic diagram of cab lighting circuit

As shown in Fig. 4-14, cab lighting circuit includes wall lamp, working lamp switch and working lamp. This loader is equipped with four working lamps, two at the front and two at the rear.

Parameters of working lamps:

1) rated working voltage is 24 V.

2) rated power is 70 W.

Parameters of wall lamps:

1) rated working voltage is 24 V.

2) rated power is 10 W.

3. Schematic diagram of rear frame lighting circuit

As shown in Fig. 4-15, the rear frame lighting circuit includes brake lamp pressure switch and rear combination lamp. The rear combination lamp integrates the functions of steering, driving / braking and reversing lamp.

☞ [Task Implementation]

How to judge whether the brake lamp switch is damaged:

First, determine whether the brake pressure is normal (Open the electric lock. If the service brake low – pressure alarm light on the instrument panel does not flash, it indicates that the brake pressure is normal. Otherwise, start the car until the service brake low – pressure alarm light does not flash). If it is normal, unplug the wire at the brake light switch and use the gear with resistance of 200Ω of the multimeter to detect the two pins of the switch. Press the brake pedal, the two pins should be conductive; release the brake pedal, the two pins should be disconnected. If the test results are inconsistent, it indicates that the pressure switch has been damaged and needs to be replaced.

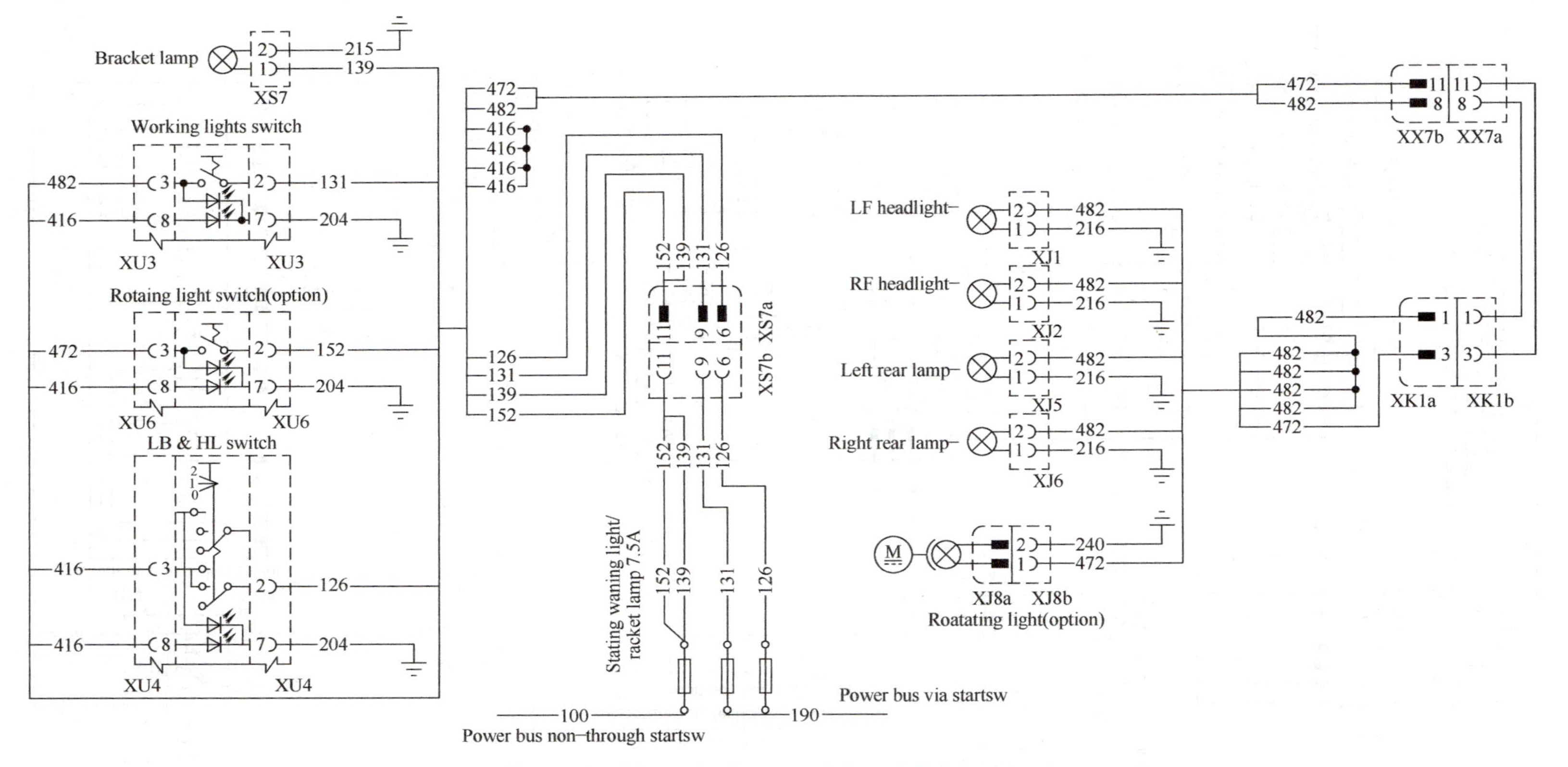

Fig. 4-14 Schematic diagram of the cab lighting circuit

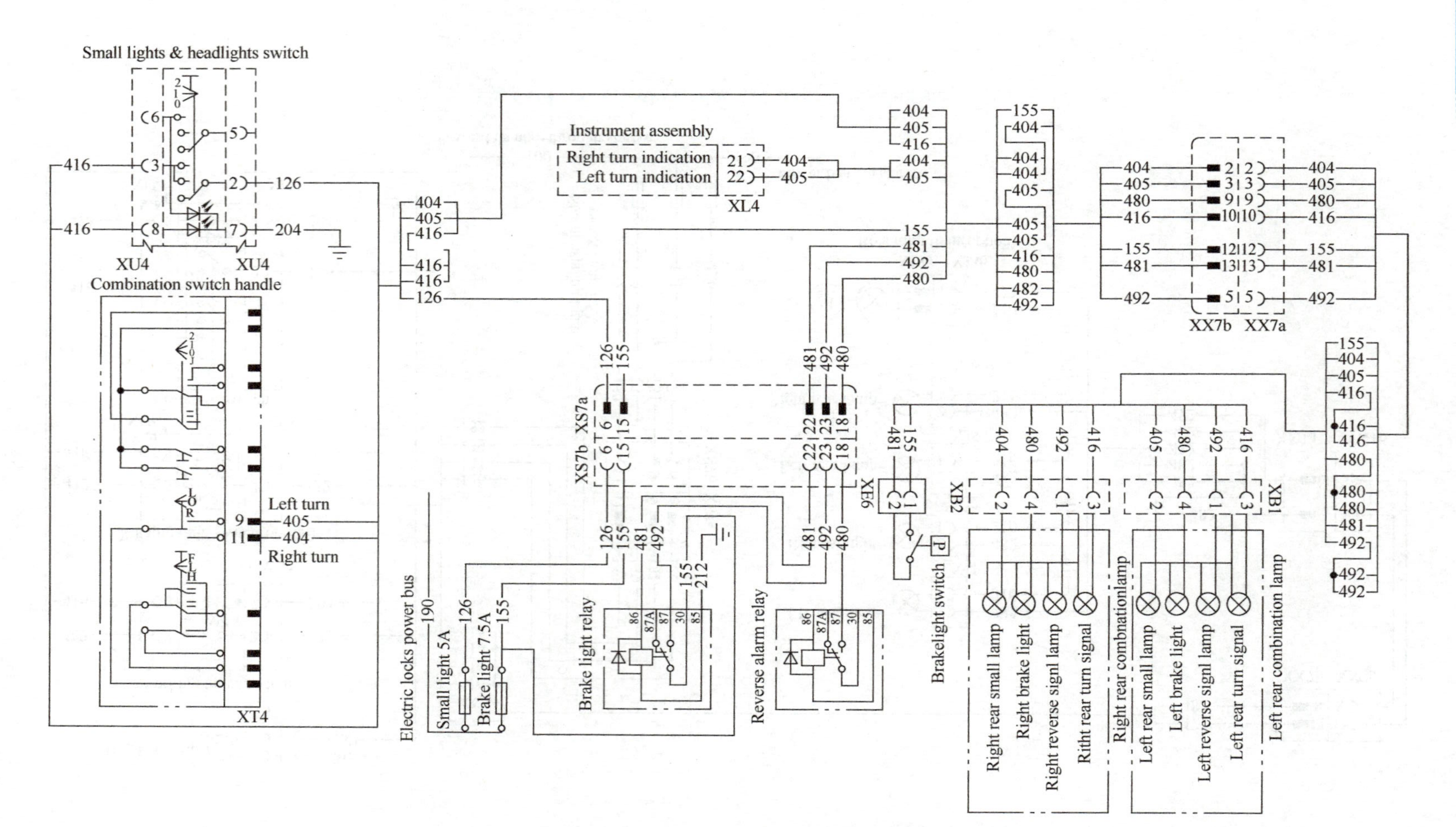

Fig. 4-15 Schematic diagram of rear frame lighting circuit

Task 4.4 Analysis of working principle of automatic reset system

[Learning Objectives]

To be able to correctly describe the working principle of automatic reset system.

[Work Task]

For a loader, the user says that the automatic reset system is not working. How to check whether the automatic reset system is normal.

[Relevant Knowledge]

1. The role and composition of the automatic reset system

The automatic reset system is mainly used to realize the moving arm limit, moving arm floating and bucket levelling function to reduce the impact of the whole vehicle, reduce the operation labour intensity and improve the efficiency. The system is composed of proximity switch, relay and three solenoids on the pilot valve. The schematic diagram of the automatic reset system is shown in Fig. 4-16.

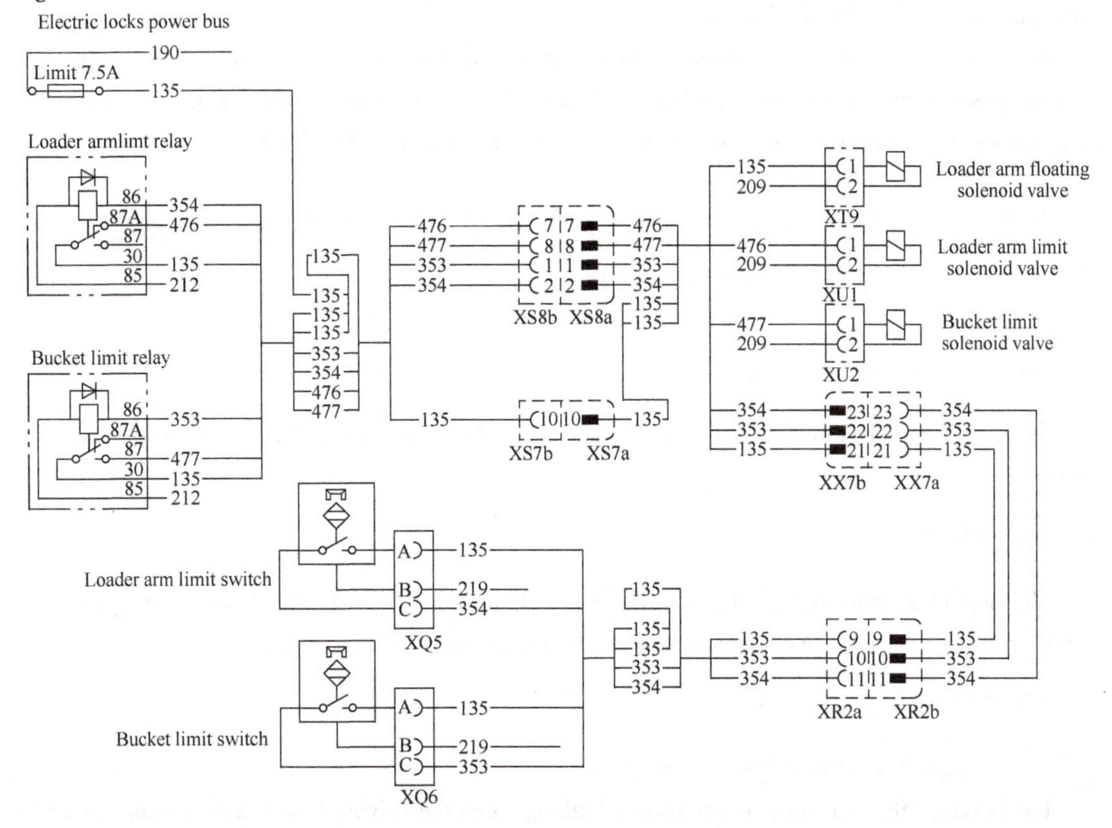

Fig. 4-16 Schematic diagram of automatic reset system

2. Operation method of automatic reset system

When the operator pushes the joystick forward or backward for moving arm down or moving arm up oeration, the joystick will automatically keep in the forward or backward position. When the moving arm up reaches the limit, the joystick will automatically pop back to the middle position. When the moving arm floats down to the bottom, the joystick will not automatically pop back to the middle position, at this time, the operator needs to move it back to the middle position. When the operator pushes the joystick to the left after unloading, the joystick will automatically remain in the left position until it reaches the bucket levelling position, and the joystick will automatically pop back to the middle position.

☞ [Task Implementation]

Check whether the automatic reset system is normal:

1) Check whether 10 A fuse is blown. Check whether all connectors are well connected; check the clearance between magnet and proximity switch (generally no more than 8 mm).

2) Check whether the proximity switch is damaged. Turn on the electric lock, the green light should be on; simulate the relative movement relationship between the magnet and the proximity switch when the working device is working, and observe whether the red – light status is correct.

3) Check the pilot coil. The resistance value of the three pilot coils should be approximately equal and about a few hundred ohms.

4) Check the clearance between the lever and the pilot solenoid valve stem. Pull the lever to the limit position in either direction (front or rear), and the clearance between the solenoid valve stem and the lever in the opposite direction should be in the range of 0.5 ~ 1.27 mm.

Task 4.5 Analysis of working principle of emergency brake and power cut – off system

☞ [Learning Objectives]

To be able to correctly describe the working principle of emergency brake and power cut – off system.

☞ [Work Task]

The user of a loader reports that the whole vehicle has no first and second gears, so it is necessary to eliminate the fault phenomenon and check the power cut – off system.

☞ [Relevant Knowledge]

1. Emergency and parking brake system

For braking after parking, or emergency braking when the driving brake system fails. In addition, when the system pressure is lower than the safety pressure, the system automatically makes the loader stop to ensure safe use.

2. System principle analysis

It is used to monitor the braking and non – braking status of the whole vehicle. When the handbrake is pulled up, the parking brake switch is opened, the parking brake light is lit, and the power cut – off indicator is lit, and the transmission control box controls the shift solenoid valve to cut off the power output; when the handbrake is put down, the parking brake switch is closed, the parking brake light is off, and the power cut – off indicator is off. The principle is shown in Fig. 4-17.

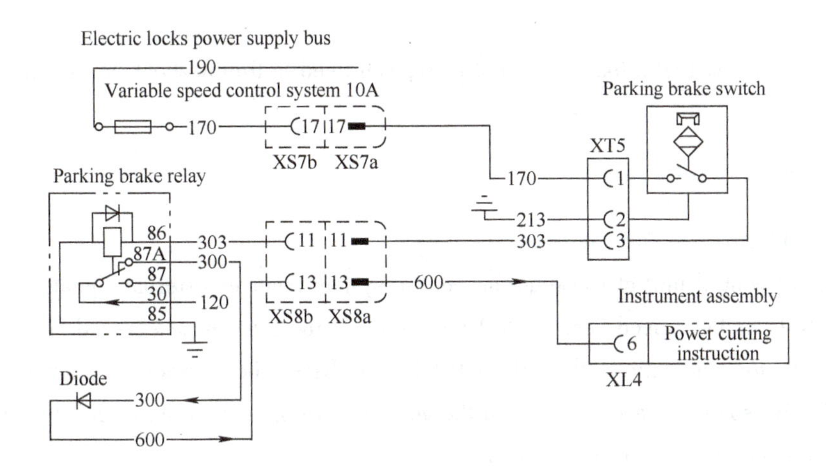

Fig. 4-17 Schematic diagram of emergency brake and power cut – off system

[Task Implementation]

Common troubleshooting – the whole vehicle without first and second gear:

The whole vehicle without first and second gear is generally due to the pressure switch (travel brake power cut – off switch and emergency brake power cut – off switch) damage, causing there is always 24 V voltage input to the control unit EST – 17T on wire 600, thus cutting off the first and second gears power output of the transmission. It can be determined by testing the voltage of the 600 wire. The specific steps are as follows:

1) Check whether the plug connector is wired correctly.

2) Unplug the emergency brake power cut – off switch and the travel brake power cut – off switch plug. If there are first and second gears, the power cut – off switch and travel brake power cut – off switch may be damaged.

3) Plug in the connector at the emergency brake power cut – off switch and test run. If there are no first and second gears, the emergency brake power cut – off switch may be damaged.

4) Plug in the connector at the travel brake power cut – off switch and test run. If there are no first and second gears, the travel brake power cut – off switch may be damaged.

Task 4. 6　Analysis of the working principle of transmission control system

[Learning Objectives]

To be able to correctly describe the operation principle of transmission control system.

[Work Task]

If you get feedback that a loader can not start, you need to troubleshoot the reasons of the problem.

[Relevant Knowledge]

1. Principle of operation analysis

The EST117 control unit of transmission control system receives external signals from the variable speed control handle, speed sensor, KD key, etc., processes these external signals and sends them to the CPU for calculation, then the CPU finally drives the combined action of the variable speed control valve solenoid valve located on the gearbox through the output circuit to obtain the gear selected by the operator, as shown in Fig. 4-18.

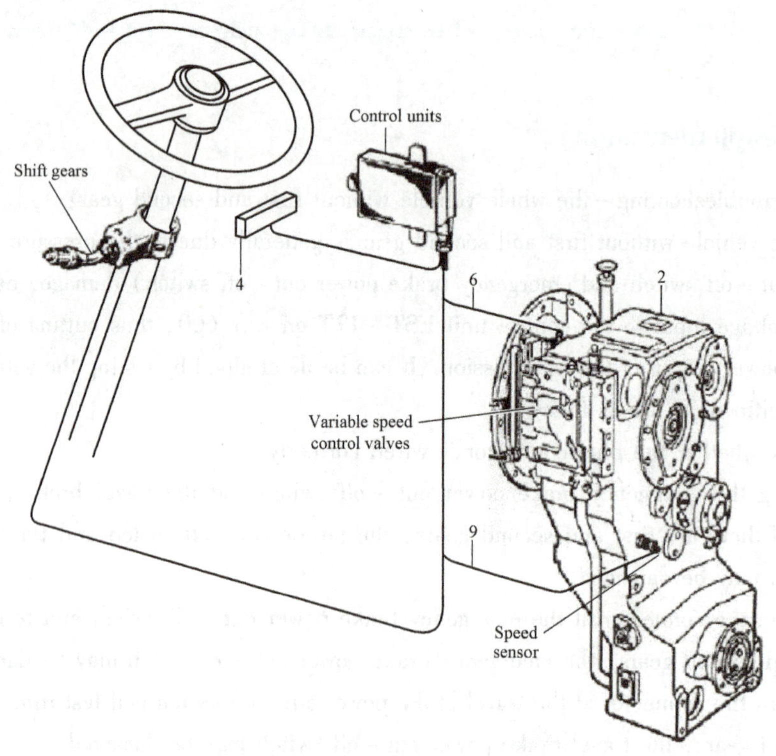

Fig. 4-18　Schematic diagram of the Structure

2. Main functions of the system (Fig. 4-19)

1) Neutral start chain function. When the DW – 3 variable speed control handle is set to neutral, the electronic control unit 584 wire outputs 24 V, which is used to drive the neutral chain relay. Therefore, the machine can only start when the shift lever is in neutral.

2) Power cut – off function. The control unit decides whether to send a power cut – off command to the variable speed control valve by detecting the power cut – off input signal from the 600 wire. Generally, when braking (including driving brake and parking brake), the power cut – off signal is effective.

The power cut – off function works in forward or reverse 1st gear and 2nd gear (low speed gear). When the loader is in the 3rd or 4th gear (high speed gear), the control unit will not cut off the power output of the gearbox to ensure the safety of driving, which is determined by the driving characteristics of the loader.

3) Forced kickdown function (KD function). When the gear is set in forward 2nd gear or reverse 2nd gear, the gearbox gear can be automatically switched to the corresponding 1st gear by pressing the KD key on the variable speed control handle.

When the loader travels toward the material pile at the speed of forward 2nd gears, a press on the KD key can automatically shift to forward 1st gear; after the material is loaded, selecting the reverse gear will make the gear automatically shift to reverse 2nd gear, and the loader directly exits the shovel operation area at the speed of 2nd gear. This can save the time spent on gear shift from forward 2nd to forward 1st gear, neutral, reverse 1st and reverse 2nd, thus improving efficiency.

4) System self – protection function in case of failure. The control unit continuously monitors all input signals from the shift handle, speed sensor and solenoid valve output signals. When there is an abnormal combination of information (e. g. lines break, control unit ground wire break), the control unit will immediately shift to neutral and lock all output signals, so does it when the voltage exceeds the specified limit or a circuit break occurs. Therefore, when the loader fails to shift gear, the peripheral circuit of the control unit needs to be carefully checked to determine whether there is a component or line failure.

If the output speed sensor fails, the control unit will only allow the gearbox to shift into 1st and 2nd gear.

3. Main components of the system

1) Control unit (Fig. 4-20). The EST117 control unit is the core component of the gear shifting system. The gear signal selected by the operator, the speed sensor signal, the KD signal, and the power cut signal are all sent to the control unit for processing and computing, and after that, the control unit outputs the control signal to drive the five solenoid valve groups in the gear shifting valve for combined action to finally complete the gear shifting selection. The control unit also outputs a neutral signal to drive the gear/neutral interlock relay when the DW – 3 gearshift handle is in neutral. The reverse alarm signal is output to actuate the reverse alarm relay.

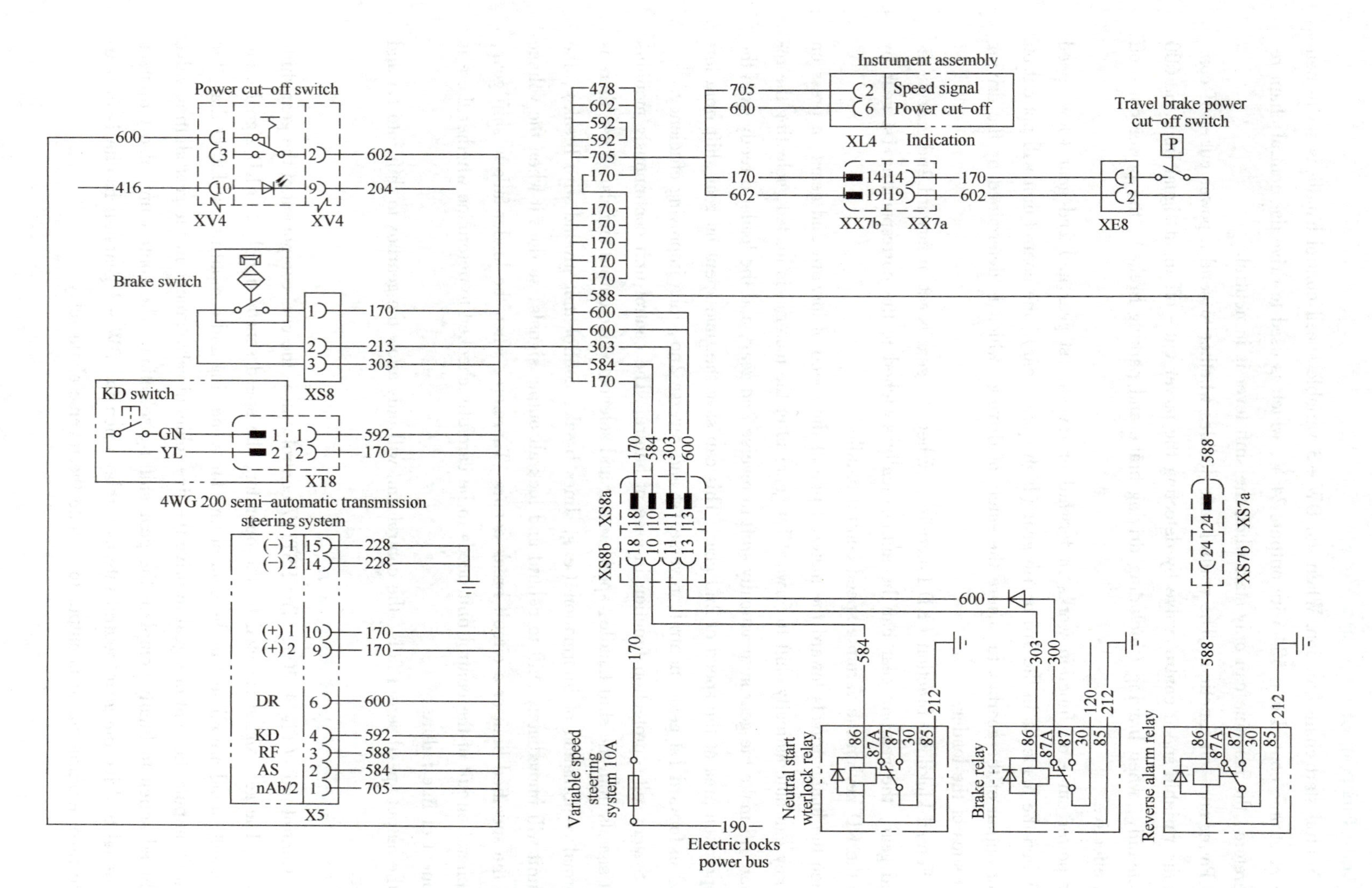

Fig. 4-19 Schematic diagram of transmission control system

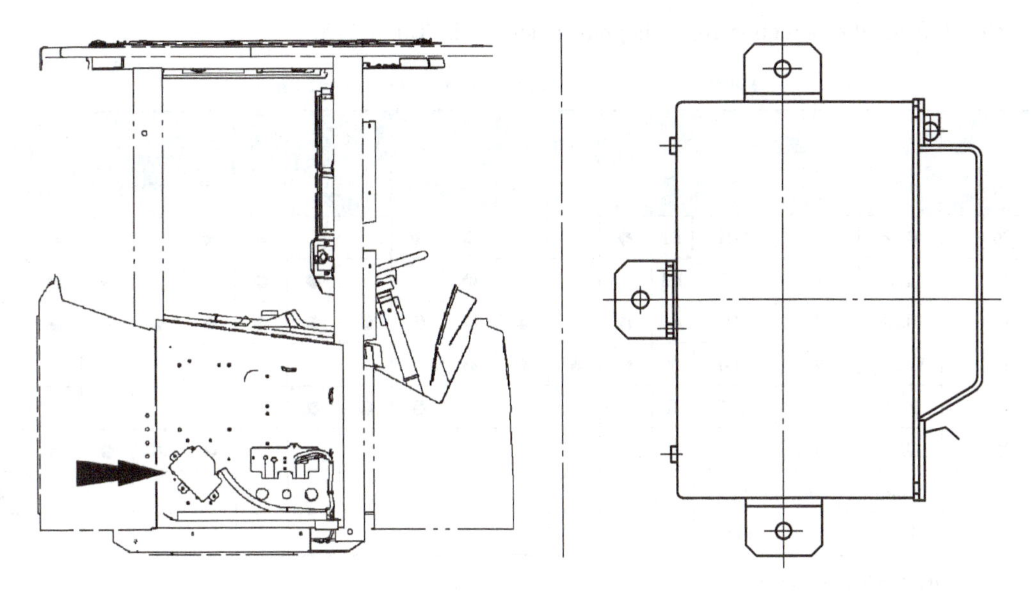

Fig. 4-20 Control unit

2) Shift handle. The interior of the shift handle consists of a number of micro – switches. When the operator selects the gear and direction, the internal micro – switches will be operated, and through the connection of the wiring harness, these changed gear and direction signals eventually enter the EST117 control unit as shown in Fig. 4-21.

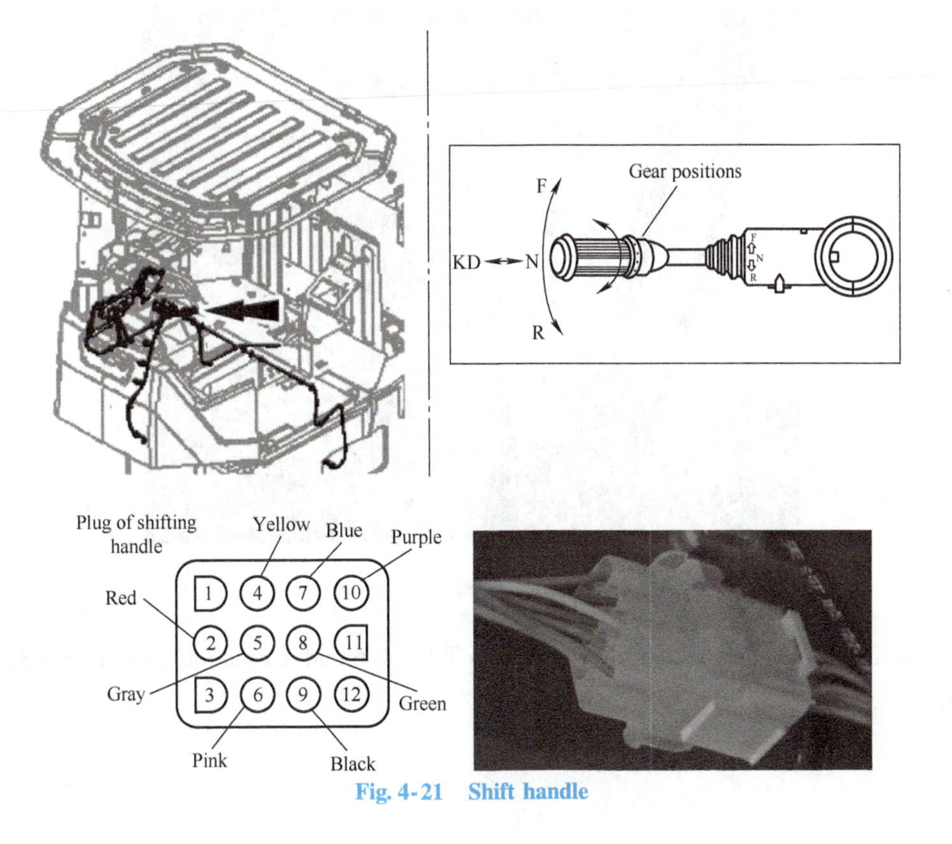

Fig. 4-21 Shift handle

The shift handle detection logic diagram is shown in Table 4-5.

Table 4-5 Shift handle detection logic diagram

Cable connector at computer box	Shift Handle		Shift Handle		Forward gear (V)				Reverse gear (R)				Neutral (N)				KD
	Wire colour	Connector			1	2	3	4	1	2	3	4	1	2	3	4	
26	Blue	7	AD1	B1	●			●	●			●	●			●	
8	Green	8	AD2	B2			●	●			●	●			●	●	
25	Black	9	AD3	B3	●	●	●	●	●	●	●	●	●	●	●	●	
23	Yellow	4	AD4	V	●	●	●	●									
5	Pink	6	AD5	R					●	●	●	●					
29	Grey	5	AD6	AS									●	●	●	●	
24	Violet	10	AD7														●
19	Red	2	ED1		(+)												

Note: ● represents power gaining.

3) Shift solenoid valve assembly As the actuating element of the system, the five solenoid valves located on the variable speed manipulator valve receive the shift command from the control unit and control the gear clutch in the gearbox by controlling the internal oil circuit of the variable speed manipulator valve, to put the loader in a certain gear. The resistance of all five solenoid valves is 90 ~ 110 Ω (Fig. 4-22).

Fig. 4-22 Shift solenoid valve assembly

The solenoid valve detection logic is shown in Table 4-6 and is used for detecting the resistance values.

Table 4-6 Solenoid valve detection logic

Solenoid valve: Resistance value: 90 ~ 110 Ω Order (from top to bottom): M4 – M3 – M5 – M2 – M1

Cable connector at computer box	Cable connector at gearbox	Solenoid valve	Forward gear (V)				Reverse gear (R)				Neutral (N)				KD
			1	2	3	4	1	2	3	4	1	2	3	4	
33	B	M1					●	●	●						
15	C	M2	●				●				●				
32	D	M3	●	●	●										
14	E	M4	●	●			●	●			●	●			
31	A	M5				●									
35	F	(I)													
Clutch			KV	KV	KV	K4	KR	KR	KR						
			K1	K2	K3	K3	K1	K2	K3						

Note: ● represents power gaining.

4) Speed sensor. The speed sensor detects the rotate speed of the gearbox output gear, and the control unit collects this rotate speed signal, then integrates it with the gear command from the gearshift handle to decide whether to put the gearbox into high gear (from 2nd to 3rd, 4th gear). Therefore, if the speed sensor is damaged, the loader will have no 3rd and 4th gears. This speed sensor is a magneto – electric sensor with a resistance of (1050 ± 100) Ω.

[Task Implementation]

Common troubleshooting: Start – up failure.

1) If it is not in neutral gear, put in neutral and restart.

2) If the 7. 5 A fuse is blown, replace the 7. 5 A fuse, if the fuse is still blown, the circuit need to be carefully checked to identify the cause and then be replaced.

3) If the control unit EST – 17T is damaged, replace the control unit EST – 17T.

4) If the handle DW – 3 is damaged, replace the handle DW – 3.

Task 4. 7 Replacement of reversing alarm

[Learning Objective]

To be able to correctly describe the working principle of the reversing alarm.

[Work Task]

Determine whether there is a problem with the reversing alarm in a new loader.

[Relevant Knowledge]

1. Reversing alarm principle

Reversing radar is composed of ultrasonic sensors (commonly known as probes), controllers, displays (or buzzer) and other parts. Reversing radar generally adopts the principle of ultrasonic distance measuring. Under the control of the controller, the sensor emits ultrasonic signals and generates the echo signals when encountering obstacles. After receiving the echo signal, the sensor processes the data through the controller to judge the position of the obstacle, and the display shows the distance and sends other warning signals to give timely warning, providing the driver with safety information for easier vehicle – reversing.

After the vehicle is started, setting the gear selector to R can activate the reversing alarm relay. The reversing alarm beeps and the reversing lamp lights up, as shown in Fig. 4-23.

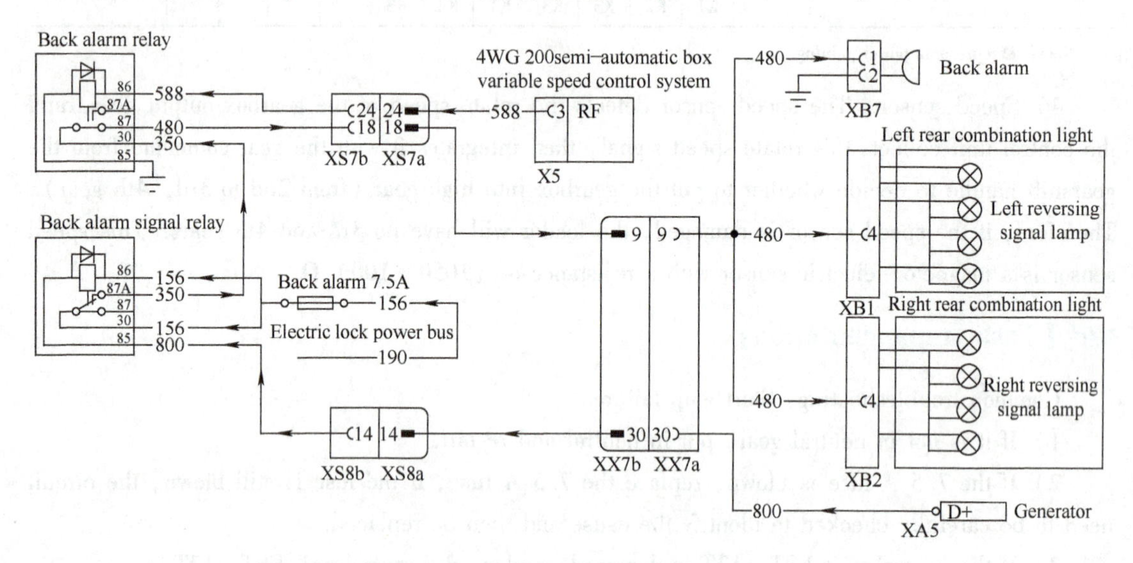

Fig. 4-23 Reversing alarm principle

2. Reverse alarm parameters

The reversing alarm is an intermittent buzzer that works along with the reversing light when the operator is reversing.

Parameters:

1) Rated voltage: 24 V.

2) Operating current 3 A.

3) Basic frequency: 960/1440 Hz.

4) Sound level: in a range of (107 ±4) dB (A).

[Task Implementation]

Testing of reversing alarm

1) Turn the engine start switch to the "I" or "ON" position to turn on the power of the loader

in order to conduct the test.

2) Lift the parking brake button (handle) to apply the parking brake.

3) Turn the engine start switch to "Ⅱ" or "START" position, start the diesel engine.

4) Place the gearshift handle in the REVERSE position, and the reverse alarm should start sounding immediately.

5) Put the variable speed control handle to the middle or forward position, and the reversing alarm should stop sounding.

Project 5

Maintenance of loader air conditioning system

Air conditioning is a kind of unit used to provide a space area (generally enclosed) to deal with air temperature changes, and its function is to regulate parameters such as air temperature, humidity, cleanliness and air flow rate in a certain space area to meet the requirements of human comfort or manufacture process.

For passenger comfort, air conditioning in modern engineering vehicle has four basic functions.

1) The air conditioner can control the air temperature in the compartment. It can both heat and cool the air to control the temperature in the compartment to a comfortable level.

2) The air conditioner can remove moisture from the air and making the environment more comfortable.

3) The air conditioner can inhale fresh air and have a ventilation function.

4) The air conditioner can filter the air to exclude dust and pollen in the air.

Vehicle air conditioning features: due to limited space inside the engineering vehicle, high temperature and vibration of working environment, once the vehicle starts, it need to be cooled quickly, which requires small size compressor with high efficiency and reliable performance and the same requirements applying to other components of the air conditioner. Therefore, corresponding technical measures must be taken to adapt the air conditioning of engineering vehicles to the above characteristics.

The basic working principle of air conditioning: cooling and heating can be simply explained as taking heat from the air or heating the air. Cooling means lowering the temperature by taking heat away from the air inside the vehicle. Heating means heating the interior of the car by supplying heat to the air inside the car, i. e. heating the air.

When alcohol is applied to human skin we feel cooler because it absorbs the latent heat from the skin asit evaporates. Similarly, watering in the garden in summer makes us feel cooler because the water poured in the soil absorbs the latent heat from the surrounding air as it evaporates. These natural phenomena reflect the basic principle of cooling.

Air conditioning works by undergoing four processes of change (as shown in Fig. 5-1).

1) Compression process. The refrigerant absorbs heat in the evaporator and becomes a low - temperature, low - pressure gaseous refrigerant, which is drawn in and compressed by the compressor into a high - temperature, high - pressure gaseous refrigerant before discharging into the con-

denser.

2) Condensation process. After the high - temperature and high - pressure gaseous refrigerant enters the condenser, driven by the engine radiator fan, the air is forced to take away the refrigerant heat through the condenser surface, and the refrigerant is condensed into medium temperature and high - pressure liquid refrigerant.

3) Throttling process. The liquid refrigerant with medium - temperature and high - pressure is filtered through the dryer and throttled by the expansion valve (the refrigerant is ejected from the pores of the expansion valve to make it expand suddenly), becoming a liquid refrigerant mist with low - temperature and low - pressure to enter the evaporator.

4) Evaporation process. The refrigerant, which is throttled by the expansion valve and becomes a low - temperature and low - pressure liquid, vaporizes in the evaporator. The air in the carriage flows over the evaporator surface under the drive of the evaporator fan, and the refrigerant absorbs the heat of the air in the carriage to cool down the air in the carriage, and condensate is precipitated at the same time. After absorbing the heat, the refrigerant evaporates into low - temperature and low - pressure gaseous refrigerant, which is sucked in by the compressor and then compressed to complete a refrigeration cycle.

The compressor keeps rotating, and the above refrigeration process is continuously cycled, and the heat in the cab is constantly taken away by the refrigerant in the evaporator, to complete the cooling and dehumidification of the whole vehicle.

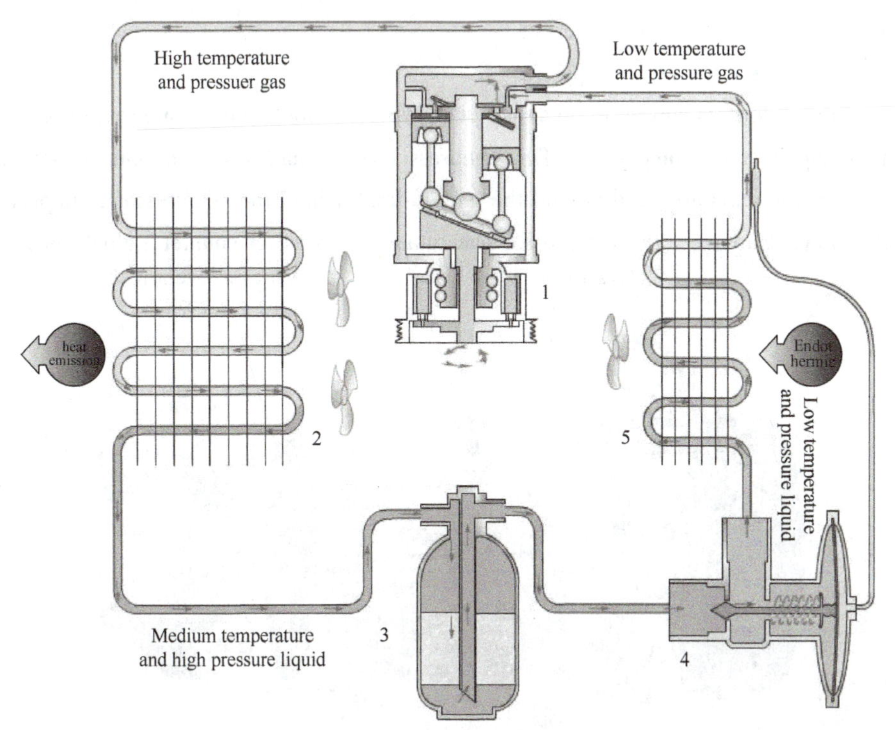

Fig. 5-1 Air conditioning work process

1—Compressor 2—Condenser 3—Receiver - drier 4—Expansion valve 5—Evaporator

Task 5. 1 Compressor disassembly

☞ [Learning Objective]

To be able to correctly describe the working principle of compressor.

☞ [Work Task]

A loader user feedback that the air conditioning is not cooling. After inspection, it is found that the compressor is damaged, so the compressor needs to be disassembled.

☞ [Relevant Knowledge]

1. Function of compressor

The compressor in the air conditioning system is the "heart" of the system, which enables the refrigerant to be used repeatedly in the air conditioning system. It is driven by the engine crankshaft through the pulley and belt, and compresses the refrigerant to a high - temperature and high - pressure state.

Most of the compressors in the air conditioning system use reciprocating piston compressors, and the working medium of the compressor is R134a refrigerant, and the refrigerant oil grade is PAG100.

2. Compressor structure (Fig. 5-2)

The compressor form is chosen to be swash plate type. It adopts the unique swashplate warp drive system and swashplate warp compressor. The compressor is in small size with compact structure. The installation of moving parts arranged around the longitudinal axis of the compressor to improve the mechanical efficiency. Moreover, the compressor can rotate clockwise or counter - clockwise.

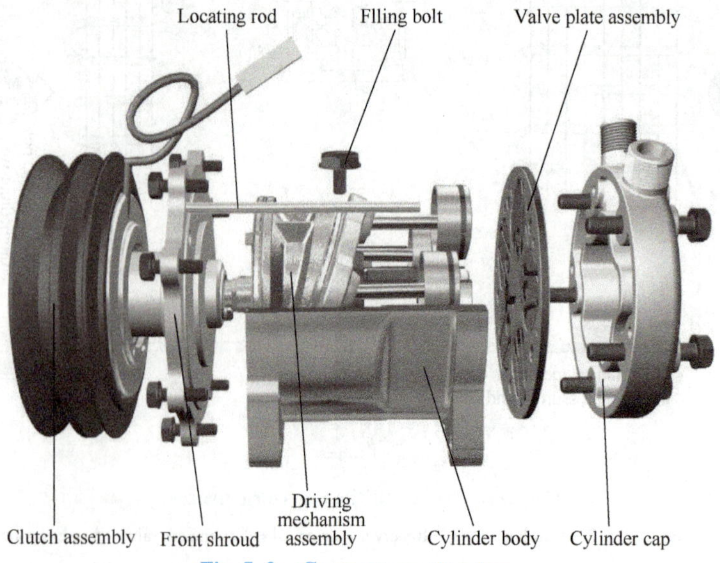

Fig. 5-2 Compressor structure

The electromagnetic clutch is a device to drive or stop the compressor as needed, for example, when the temperature inside the vehicle reaches a predetermined heat sensitive resistance temperature, it will start or swith off the air conditioning system. It works under the control of the air conditioning switch, water temperature switch, and double pressure switch.

Warning: When storing the compressor, the discharge and suction pipe connectors should be sealed with plugs to prevent moisture and dirt from entering the compressor. The compressor is made of aluminium casing, so handle it with care.

☞ [Task Implementation]

Disassembly of compressor:

1) Shut down the compressor, or turn off the air conditioning system for more than 1 h, and turn off the negative switch and hang the "No Operation" warning sign.

2) Close the high and low pressure valves at the instrumentation of the manifold, as shown in Fig. 5-3.

3) Close the valve of high and low pressure connector.

4) Open the rear cover to fully expose the compressor and the piping on it.

5) Unscrew the plug on the high and low pressure charge ports of the compressor and connect the high and low pressure valves of the manifold gauge to the charge ports on the high and low pressure lines respectively. After connecting, open the valve of the connector. Note: A thicker line with a finer interface is a low pressure line; a thinner line with a thicker interface is a high pressure line. This connector is designed to be error – proof and cannot be connected with wrong connection.

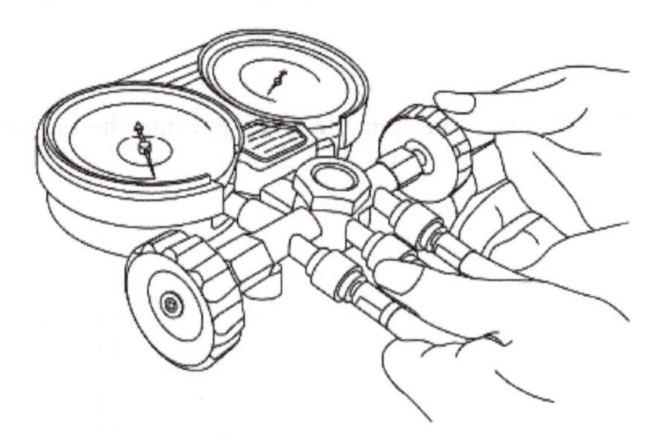

Fig. 5-3 Close the valve

6) Slowly loosen the low pressure valve on the manifold gauge to allow the refrigerant to eject slowly. Pay attention to the opening of the control valve. If the flow rate is too high, it is easy to cause the refrigerant oil to spray out with the refrigerant. Control the valve that no refrigerant oil is sprayed out (the refrigerant is gaseous after being discharged, and the refrigerant oil is liquid. Aim the refrigerant outlet at a piece of white paper and observe whether there are liquid droplets). After 10 min, the high pressure valve can be opened slowly, while the flow rate still need to be controlled to avoid loss of refrigerant oil. The entire release process should last 13 ~30 min.

7) Loosen the M6 bolts of the compressor pipe with a wrench and remove the pipe, as shown in Fig. 5-4.

8) Wrap the two removed piping connectors with clean, unbroken plastic bags and tie them tightly with tape to prevent air and foreign objects from entering the piping.

9) Disconnect the compressor clutch harness plug and remove the belt. Loosen the tensioner

pulley and remove the belt, as shown in Fig. 5-5.

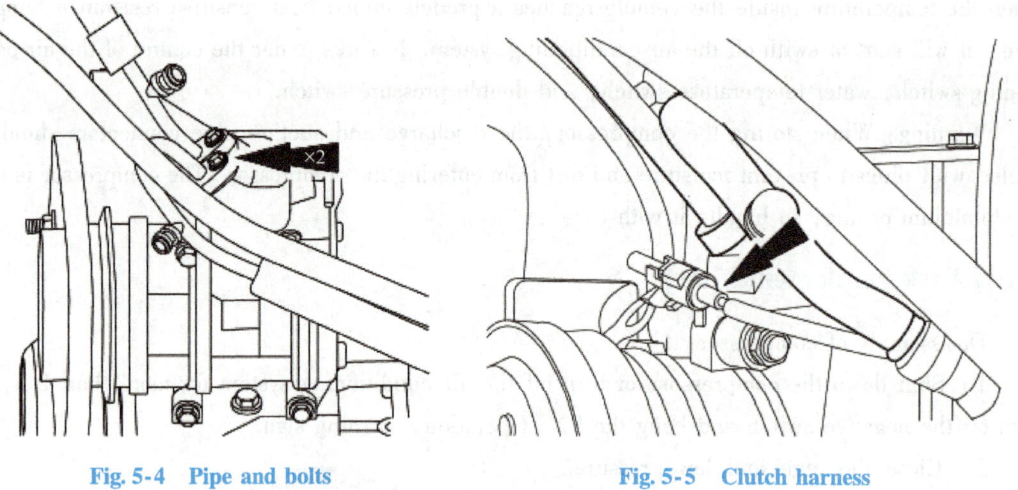

Fig. 5-4 Pipe and bolts

Fig. 5-5 Clutch harness

10) Remove the four M8 bolts holding the compressor to remove the compressor, as shown in Fig. 5-6.

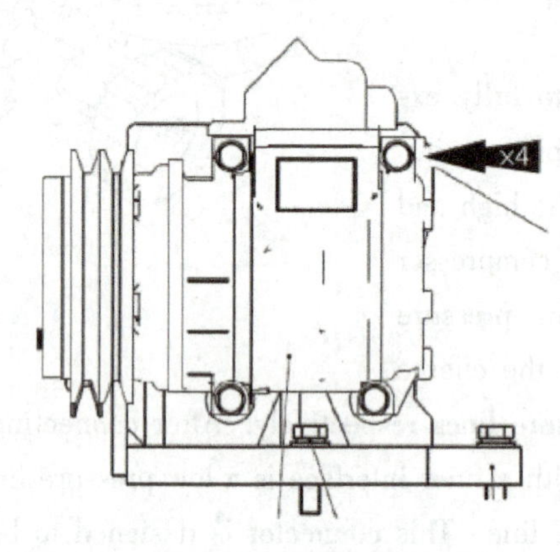

Fig. 5-6 Bolt location

Task 5.2 Receiver testing

☞ [Learning Objective]

To be able to test the receiver in accordance with the specification requirements.

☞ [Work Task]

The air conditioning of one loader doesn't work. The receiver needs to be tested.

☞ [Relevant Knowledge]

1. The function of the receiver

Receiver (Fig. 5-7) is also called dryer. It can filter and store the refrigerant, store the excess refrigerant when the working condition changes, and can ensure that the refrigeration system can still work effectively in the case of trace refrigerant leakage. More importantly, the molecular sieve in the receiver can absorb a small amount of water vapour to prevent the formation of acid. The receiver is usually used with thermal expansion valve. In order to observe the refrigerant flow status, the sight glass is arranged on the receiver in order to judge the refrigerant charge and the information related to the system operation. There is a filter and desiccant in the receiver to remove dust and moisture in the refrigeration cycle. If there is moisture in the refrigerant, it may corrode the main components and may cause the expansion valve orifice to be blocked with ice, resulting in the failed circulation of the refrigerant in the system.

2. The appearance of the receiver

The appearance of the receiver is shown in Fig. 5-8.

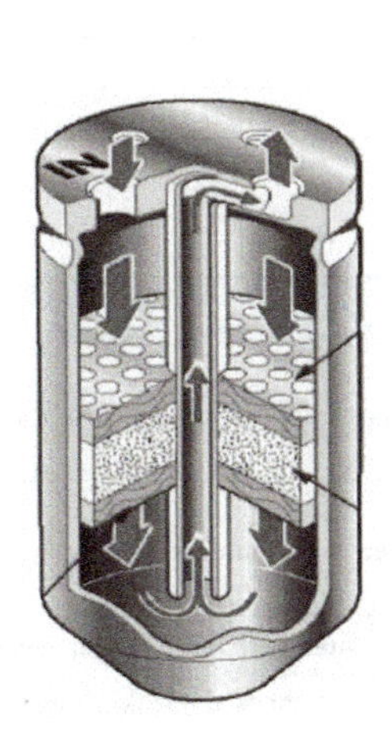

Fig. 5-7 Receiver

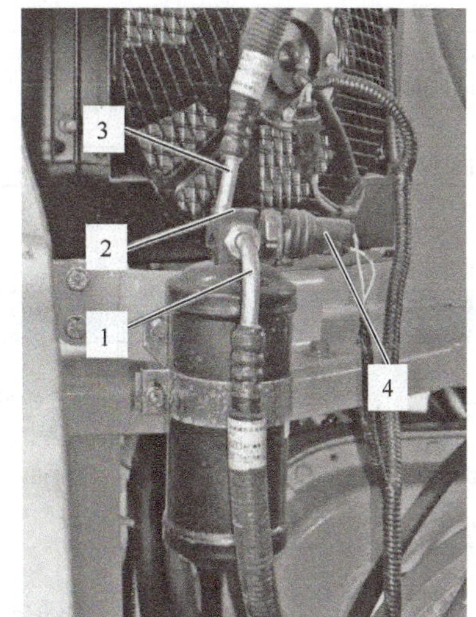

Fig. 5-8 Receiver appearance

1—Outlet 2—Sight glass

3—Inlet 4—High and low pressure switch

Note: When the receiver is stored, the outlet and inlet of it should be sealed with plugs to prevent moisture from entering the reservoir. The outlet and inlet of the receiver can not be connected backwards, otherwise the refrigeration system is not cooling.

☞ [Task Implementation]

Receiver testing:

Observe the sight glass next to the receiver (Fig. 5-9) and touch the lower end of the receiver with your hand. If the lower end of the receiver is cold and bubbles appear in the sight glass, it indicates that the filter screen in the reservoir is blocked. Under normal circumstances, the temperature of all parts of the receiver is basically the same as the condenser outlet.

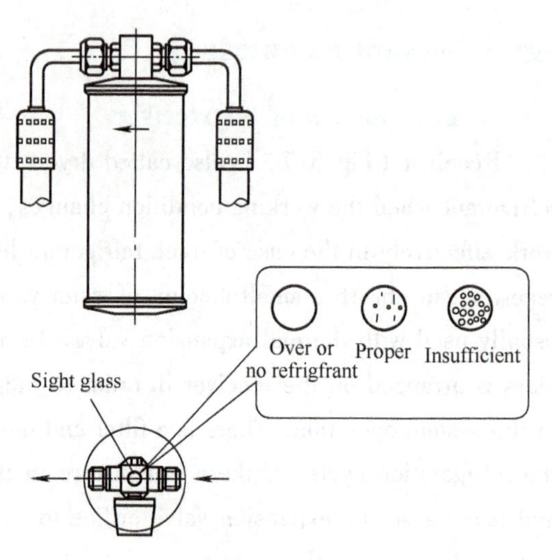

Fig. 5-9 Sight glass

Inspection steps:

1) Check under the following working conditions, as shown in Table 5-1.

2) Check the condition of the air conditioning system refrigerant with the observation window and compare it with the following conditions, as shown in Table 5-2.

Table 5-1 Working condition inspection

Air conditioner switch	Turn on
Compressor speed	1800 r/min
Blower switch	Max.
Temperature control	Max. cooling
Door	Fully open
Position of intake valve	Recirculation

Table 5-2 Comparison and contrast

No.	Condition	Cause	Improvement measures
1	Air bubbles can be seen through the observation window	Not enough refrigerant	Check for air leakage with a leak detector
2	No bubbles can be seen through the observation window	Insufficient or excessive refrigerant	See serial number 3, 5 or 6
3	No temperature difference between the inlet and output side of the compressor	No refrigerant	Vacuum and refill the air conditioning system, and check the leakage with a leak detector
4	There is a large temperature difference between the inlet and output sides of the compressor	No refrigerant moderate amount of refrigerant or excessive refrigerant	See serial number 5 or 6
5	The refrigerant under the observation window becomes transparent immediately when the air conditioning is shut down	Too much refrigerant	Discharge the excessive refrigerant and adjust the refrigerant to the specified amount
6	When the air conditioning is shut down, the refrigerant under the observation window produces some bubbles, which immediately becomes transparent and keeps that state	Refrigerant charge is appropriate	No repair is needed

Task 5.3 Operation of the air conditioning control panel

☞ [Learning Objective]

To be able to operate each switch on the control panel correctly.

☞ [Work Task]

When a new loader is delivered, the user needs to be trained on the operation of the air conditioning control panel.

☞ [Relevant Knowledge]

The operation instructions of the air conditioning control panel, as shown in Fig. 5-10.

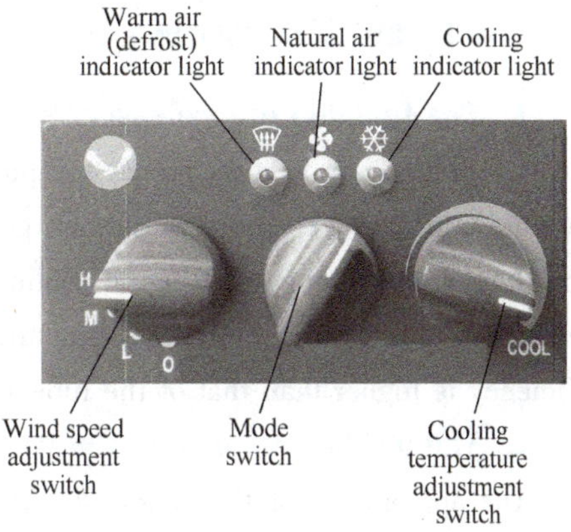

Warm air (defrost) indicator light
Natural air indicator light
Cooling indicator light
H
M
L
O
COOL
Wind speed adjustment switch
Mode switch
Cooling temperature adjustment switch

Fig. 5-10 Air conditioning control panel

☞ [Task Implementation]

(1) Refrigeration operation

1) After the engine starts, set the air speed adjustment switch to the appropriate air speed .

2) Turn the mode switch clockwise to the rightmost end (cooling).

3) When the cooling temperature adjustment switch is turned to the "COOL" end (the green light is on), the refrigeration system begins to work, and the cold air began to come out from the outlet.

4) The temperature in the driver's cab can be adjusted by setting the position of the cooling temperature adjustment switch.

(2) Heating (defrost) operation

1) After the engine is started for a period of time (the cooling water temperature is above 70℃), turn the wind speed adjustment switch to desired speed .

2) When the mode switch is turned counter - clockwise to the leftmost end (i. e. warm air, the red indicator light is on), the system starts to work and warm air starts to come out from the outlet.

(3) Natural wind operation In spring and autumn, when you only want natural wind and use it to ventilate the driver's cab, you can turn on the wind speed switch to the desired speed and set the mode change switch to the middle (yellow indicator light is on).

Task 5.4 Detection of condenser

☞ [Learning Objective]

To be able to perform fault detection on condenser.

☞ [Work Task]

The user of one loader reports that the air conditioning is not cooling, and the condenser needs to be tested.

☞ [Relevant Knowledge]

1. The function of condenser

Condenser is a heat exchange equipment. Its role is to forcibly cool the heat of the high - temperature and high - pressure refrigerant gas discharged by the compressor and send it to the air outside the vehicle through the engine cooling fan, so that the state of the refrigerant gas changes and becomes a liquid refrigerant. The structure of condenser is tube belt type, and the heat dissipation efficiency is higher than that of the tube sheet type, generally in stalled near the radiator.

2. The appearance of the condenser

The appearance of the condenser is shown in Fig. 5-11.

Note: When the condenser is stored, the exhaust pipe connector and suction pipe connector shall be sealed with plugs to prevent moisture and dirt from entering the condenser. Do not damage the condenser heat sink. If the heat sink is bent, it can be corrected with a screwdriver or pliers. The surface of condenser tubes and fins should be free of dirt and residue, which can result in poor heat dissipation.

Fig. 5-11 Appearance of a condenser

1—Inlet pipe connector 2—Exhaust pipe connector

☞ [Task Implementation]

Detection of condenser:

Touch the upper part of the condenser near the air inlet and the lower part near the air outlet respectively, as shown in Fig. 5-12. Under normal circumstances, the air inlet part should feel somewhat hot and the air outlet part should feel close to the ambient temperature. If the temperature near

Fig. 5-12 Condenser

the air inlet is not high, it indicates that the refrigerant is insufficient or the compressor is not working properly. If there is no obvious difference between the temperature in the room temperature area and the high temperature area, it indicates that the heat dissipation of the condenser is poor, the dirt on the surface of the condenser needs to be cleaned.

Task 5.5 Detection of sensor

☞ [Learning Objective]

To be able to correctly describe the working principle of each sensor of the air conditioning system and test the sensor.

☞ [Work Task]

There is an alarm of water temperature on the instrument of a loader. It is found that the temperature sensor is damaged, and it needs to be detected and replaced.

☞ [Relevant Knowledge]

1. The role of sensors

The sensor is a detection device. It can sense the information of the measured object, and transmit it into electrical signals or other information in required form according to certain rules, then output them to meet the requirements of information transmission, processing, storage, display, recording and control.

2. Types of sensors (Fig. 5-13)

The sensors mainly used on the loader are temperature sensors, pressure sensors, speed sensors, oil level sensors, etc.

a) Temperature transducer b) Veloctty transducer c) Pressure Transducer d) Lquid level transducer

Fig. 5-13 Sensor types

Temperature sensors mainly include engine water temperature sensor, torque converter oil temperature sensor, hydraulic oil temperature sensor, ambient temperature sensor and engine air intake temperature sensor. These sensors detect engine water temperature, torque converter oil temperature, hydraulic oil temperature, ambient temperature, and engine air intake temperature accordingly and are installed on different objects to be measured.

Speed sensors, mainly include transmission speed output sensor and engine speed sensor. The transmission speed sensor is installed on the transmission output gear and is used to detect the transmission speed, which can also be converted into the driving speed of the vehicle after processing and displayed on the instrument. The engine speed sensor is mainly used to detect the engine speed and is mounted on the flywheel housing.

Pressure sensors, detecting changes in pressure, mainly include brake pressure sensor, engine oil pressure sensor, air intake pressure sensor, etc.

Oil level sensors, including fuel oil level sensor, coolant water level sensor, oil level sensor, etc. are mainly used to monitor the change of fluid level.

☞ [Task Implementation]

Engine temperature sensor detection:

1) Pull out the plug connected to the temperature sensor in the rear frame wiring harness.

2) Remove the temperature sensor from the engine.

3) Place the temperature sensor probe in the liquid (water / torque converter oil), warm the liquid, and determine the temperature sensor performance according to the following parameters, as shown in Table 5-3.

Table 5-3 temperature determination

Temperature/℃	Electric resistance /Ω	Error (±%)	Minimum value	Maximum value
0	33650	8.8	30700	36600
25	10000	7	9304	10700
90	865.5	2.1	847.3	883.7

Project 6

Analysis of loader drive system

The transmission components between the loader power unit and the travel unit (drive wheels) are collectively referred to as the transmission system.

The function of the transmission system is to transmit the power output by the power device to the driving wheel and other mechanisms (such as working oil pump, steering oil pump, etc.) as required, and to solve various contradictions between the power output characteristics of the power unit and the power demand of the travel unit. The basic route of power transmission is: engine flywheel – flexible plate – torque converter – transmission – front and rear output flanges – front and rear drive shafts – front and rear drive axles – tires.

(1) Composition of transmission system The loader transmission system mainly consists of torque converter, gearbox, rear axle drive shaft, front axle drive shaft, front drive axle, rear drive axle, tires and other components. The main functions:

1) Reducing the rotational speed and increasing the torque.

2) Realising backward driving of the loader.

3) Interrupting the transmission when necessary.

4) Differential speed role.

(2) Classification of transmission system According to the different structure and transmission medium, the transmission system can be divided into four forms: mechanical transmission, hydro – mechanical transmission, full hydraulic transmission and electric transmission.

Classification of wheel loader hydro – mechanical transmission:

1) Planetary hydro – mechanical transmission system.

2) Fixed shaft hydro – mechanical transmission system.

(3) The concept of hydrodynamic transmission In the transmission system, the device with liquid (mineral oil) as the medium for energy transmission and control is called liquid transmission-device, which is called liquid transmission for short.

Task 6.1 Analysis of the operation principle of the torque converter

☞ [Learning Objective]

To be able to correctly describe the working principle of torque converter.

☞ [Work Task]

If the user does not know the position of the torque converter in a loader, it is necessary to point out the position of the torque converter, and explain the relevant structural principle to the user.

☞ [Relevant Knowledge]

1. The structure and principle of torque converter

The torque converter (Fig. 6-1) is composed of pump wheel, turbine and guide wheel to form a liquid flow circulation space. The oil is continuously rushed into the torque converter through the variable speed pump, and when the incoming oil flows from the pump wheel into the turbine, the flow direction and the strength of the impulse are changed and different turbine moments are generated according to the load. Then the liquid flows from the turbine into the guide wheel, changes the flow direction again, and then flows into the pump wheel. At the same time, the heat generated by the torque converter is taken away through the circulating oil. When the turbine speed is 80% of the pump wheel speed, the torque ratio is 1, that is, the turbine torque is equal to the pump wheel torque. When the turbine does not rotate, the torque ratio reaches maximum.

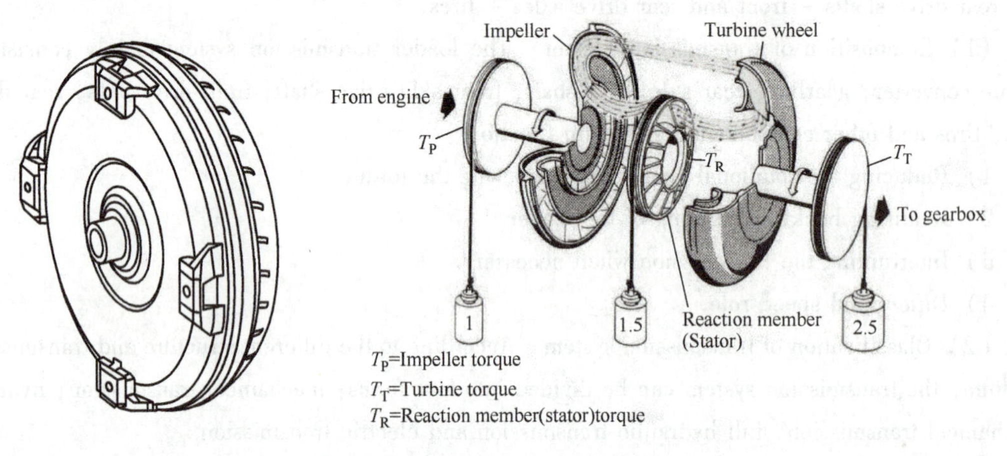

Fig. 6-1 Torque converter structure

2. Features of the torque converter

1) The torque converter can automatically adjust the output speed and torque, so that the vehicle can automatically change the speed and tractive force according to the road condition and resistance level to adapt to the change of working condition.

2) The torque converter has wide high efficiency zone and high efficiency, which can make full use of engine power and obtain better economy.

3) The hydraulic transmission with oil as medium is adopted, which plays a role of buffer and vibration damping to the vehicle and improves the comfort of the vehicle.

Disadvantages of torque converter: low transmission efficiency; steering pump does not work when being towed, and the whole machine did not steer (except for equipped with emergency steering pump); using oil as the transmission medium has the possibility of leakage and pollution of the environment.

☞ [Task Implementation]

Point out the location of the torque converter and state the relevant structural principle.

Task 6.2 Analysis of the operation principle of the gearbox

☞ [Learning Objective]

To be able to correctly describe the operation principle of the gearbox.

☞ [Work Task]

When a new loader is delivered, it needs to check the oil level of the gearbox and determine whether the oil is appropriate.

☞ [Relevant Knowledge]

1. The operation principle of the gearbox

Gearbox can be divided into planetary gearbox and fixed shaft gearbox.

Planetary gearbox - BS305 gearbox (Fig. 6-2), with double turbine, four - element torque converter, including pump wheel, I - class turbine, II - class turbine, guide wheel, and the torque converter can be detached, has the following characteristics:

1) Automatic adjustment of output torque and speed; automatic conversion of low - speed heavy load and high - speed light load; large torque ratio and wide high - efficiency area.

2) Using oil as transmission medium, absorbing and eliminating external vibration and shock, protecting the diesel engine and transmission system and ensuring the engine will not stall when the external load is suddenly increased or insurmountable; greatly reducing the labour intensity of driver operation, thus improving comfort.

BS305 gearbox can have 2 forward and 1 reverse gears by mechanical - hydraulic power shift. Because there is an overrunning clutch inside the gearbox, it can achieve automatic combination and separation (according to the need of external load), so the gearbox actually has 4 forward and 2 reverse gears.

2. The characteristics of the gearbox

1) Changing the transmission ratio between the engine and the driving wheels, with the purpose of changing the travel speed and traction of the vehicle to meet the needs of machine operation and driving.

2) Enable the machine to drive in reverse gear (shift to reverse gear).

3) With power cut - off function, which can automatically cut off the power transmitted to the travel mechanism.

4) Reducing the wheelbase, solving the problem that the engine output and drive axle output are not coaxial.

5) Simple structure and low maintenance cost.

Fig. 6-2 Planetary gearbox

3. Fixed – axis gearbox – ZF 4WG – 200 gearbox

As shown in Fig. 6-3, ZF system is mainly used for the transmission parts of the loader. It provides advanced, safe and reliable transmission technology for the loader, so that the driver can operate the vehicle more easily and conveniently, which greatly improves the efficiency of the loader. The application of ZF technology reflects the advantages brought by modern technology to the product in an all – round way.

Fig. 6-3 Fixed shaft gearbox

At present, most loaders adopt the ZF 4WG – 200 power shift gearbox. Its speed change control system is an electro – hydraulic gear shift controlled by microcomputer integration. The driver' s completion of the transmission operation is equivalent to pressing a button. The pilot hydraulic control system greatly reduces the labour intensity of the driver and improves the work efficiency.

1) The ZF4WG – 200 gearbox for loaders is of fixed – shaft type. It adopts semi – automatic electro – hydraulic control to ensure fast, accurate and smooth gear shifting. The hydraulic torque converter allows the vehicle to automatically change speed and traction according to road conditions and resistance, adapting to a variety of different working conditions.

2) The hydraulic transmission with oil as the medium enables the vehicle to be buffered and damped, improving the comfort of the vehicle; the three – element torque converter is used to transmit the output power of the engine to the gearbox.

4. The advantages of ZF4WG200 power shift gearbox

With 4 forward gears, 3 reverse gears, unique KD button (forced gear shift function, only for 1st and 2nd gear). High working efficiency, low maintenance rate, super long working life, low fuel consumption and low noise.

5. Components of the electronic control system

EST - 17T gearbox shifting electronic control box, automatic or semi - automatic functions can be realized by choosing different electronic control boxes; WG200 power shift gearbox; DW - 2 rotary gear selector; vehicle circuit; gearbox control gear shift operation cable; output speed sensor cable.

6. Main function of electronic control system

1) Neutral / start interlock protection function ensures the safety of vehicle operation; the power cut - off function (brake off - gear function) only works in the 1st and 2nd gear (low gear), which effectively protects the vehicle transmission system.

2) Dedicated forced kick - down function (KD gear), which can improve work efficiency; start speed limit function; direct reversing function. The system protection function ensures that when the electronic control system fails during operation, the electronic control box automatically switches to neutral and locks all signal outputs to avoid serious consequences.

7. Composition of ZF WG - 200 gearbox operating system.

Gearbox operating system is mainly composed of electric control box, gear selector, speed sensor, solenoid valve, manipulation valve and cable, etc. After receiving the gear signal sent by the gear selector, the electronic control box converts it into a program set in the system and sends a command to the solenoid valve. The solenoid valve opens the corresponding oil circuit according to the signal, so that the two different groups of clutches are combined to realize the gear shift and output torque to the front and rear drive axles to drive the vehicle.

Note: After the engine is started, the electric lock must not be turned off, otherwise it may cause damage to the semi - automatic gear shift system and the wiring harness and instrumentation of the vehicle.

☞ [Task Implementation]

Check the oil level of the gearbox:

The engine run at idle speed (about 1000 min^{-1}), and the oil temperature should be at the normal working temperature; when the oil temperature is 40 ℃, the oil level should be between the middle and lower scale lines of the oil dipstick; when the oil temperature is 80 ℃ , the oil level should be between the middle and upper scale lines of the oil dipstick.

Task 6.3 Analysis of the operation principle of drive axle

☞ [Learning Objectives]

1) To be able to introduce the structure and operation principle of drive axle to the customer.

2) To be able to recommend the proper drive axle to the customer.

☞ [Work Task]

When a customer comes to buy a drive axle, how to better introduce it to him/her to reflect your professional standards? You need to understand relevant knowledge of drive axle in order to recommend the proper product to the customer.

☞ [Relevant Knowledge]

1. Functions of drive axle

The drive axle refers to the general term for all transmission mechanisms located behind the gearbox or drive shaft and before the driving wheels (tire rims). The functions are as follows:

1) Carrying. Carries the weight load of the machine.

2) Driving. Absorbs the gearbox power input and amplifies the input torque through gearbox reduction to drive the machine.

3) Steering. Differential provides the sensitive differential speed function of the left and right tires of the drive axle to realize the steering flexibility of the machine.

4) Braking. The brake mounting on the drive axle is the actuator of the machine's driving brake.

Change the direction of force gearbox through the main drive bevel gear; reduce the speed of the gearbox output shaft and increase the torque through the main drive and final drive (wheel reductor); adjust the speed difference between the left and right wheels through the differential; transfer the power to the driving wheels through the differential; decelerate or stop the loader through the brake.

2. Classification of drive axle

Drive axle can be divided into two categories according to the different structure of brakes, one is the drive axle with dry external caliper disc brakes, i. e. dry axle (Fig. 6-4); the other is the drive axle with the brake inside the drive axle housing, which is immersed in oil, i. e. wet axle (Fig. 6-5).

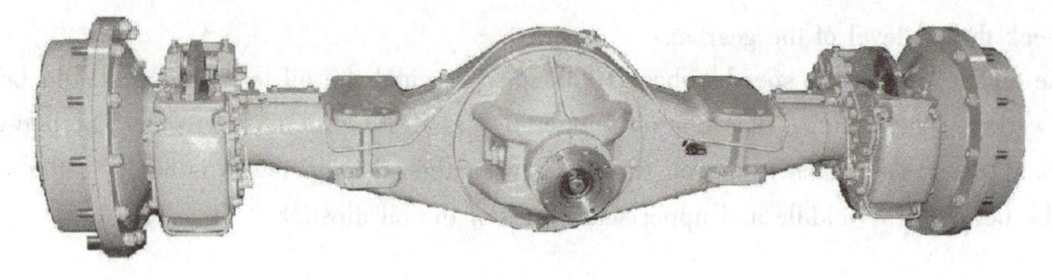

Fig. 6-4 Dry brake drive axle (front axle)

Fig. 6-5 Wet brake drive axle (front axle)

The drive axle can be divided into two categories according to different installation parts, one is the front axle that is rigidly connected to the front frame; the other is the rear axle that is oscillatingly connected to the rear frame, as shown in Fig. 6-6.

Fig. 6-6 Rear axle

3. The structure and operation principle of drive axle

Dry brake drive axle (Fig. 6-7), the brake disc and friction pads are exposed to the air, and there is dry friction between the brake disc and the friction pad. Main components of the dry brake dry axle: main drive (bracket, active spiral bevel gear, large spiral bevel gear, differential), half shaft, wheel reductor support shaft, clamp, brake disc, brake shoe, wheel reductor, axle housing, etc.

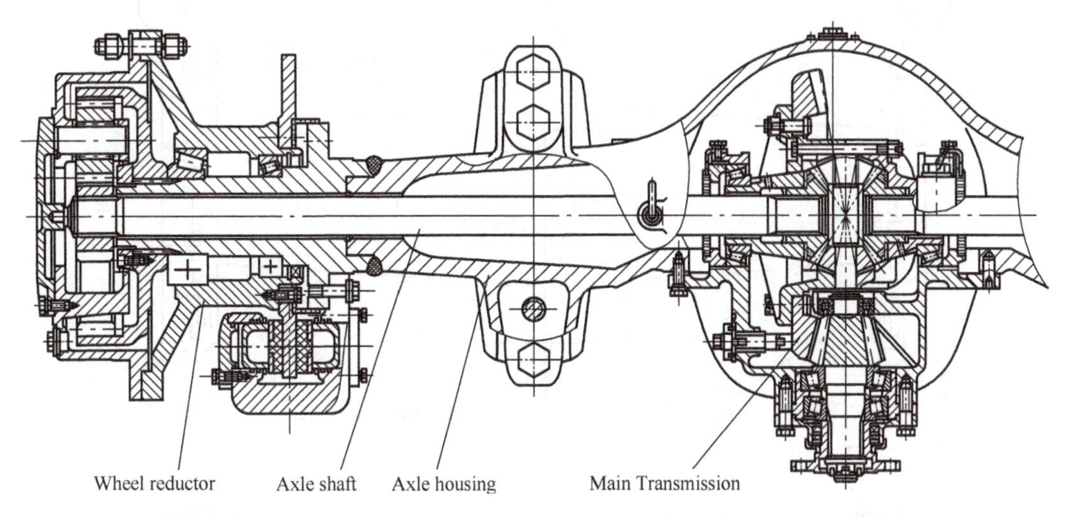

Fig. 6-7 Dry brake drive axle structure

Wet brake drive axle (Fig. 6-8), the brake friction pads are built into the oil sealed inside the drive axle, and the braking is performed by the friction of multiple friction pads. Main components of the wet brake drive axle: main drive (bracket, active spiral bevel gear, large spiral bevel gear, differential), half shaft, wheel reductor support shaft, brake, friction plate support, friction plate, wheel reductor, axle housing, etc.

Transmission route of drive axle:

Input flange →active spiral bevel gear→large spiral bevel gear→differential housing→cross shaft →planetary bevel gear→half shaft gear→half shaft →sun wheel shaft→planetary gear→inner gear ring→planetary wheel carrier→wheel hub/rim/tire, as show in Fig. 6-9.

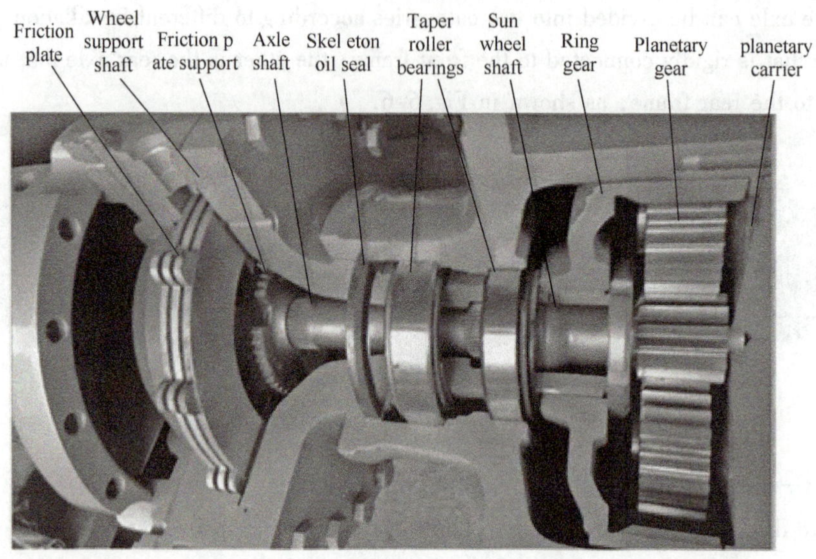

Fig. 6-8 Structure of wet brake drive axle

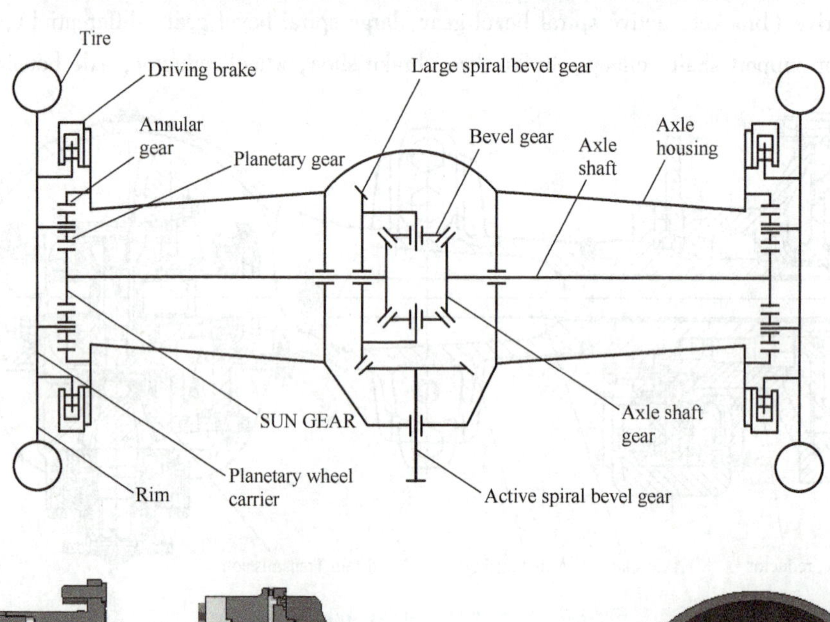

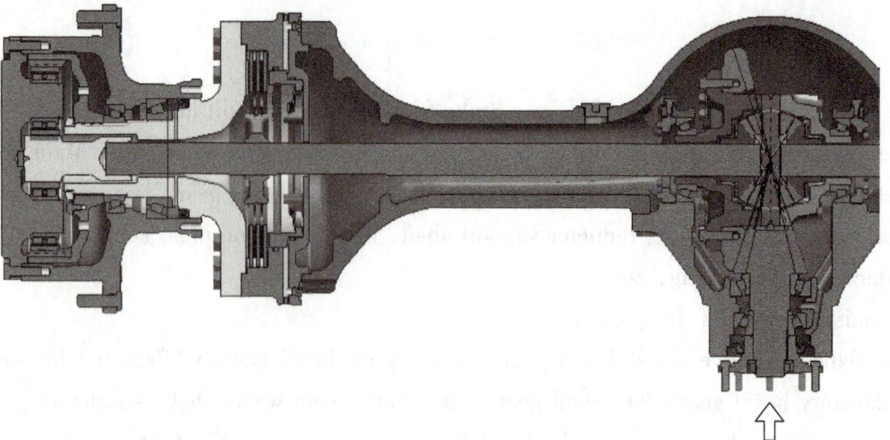

Fig. 6-9 Transmission route

☞ [Task Implementation]

Prepare a PPT introducing the drive axle of the loader, which needs to contain the following contents:

1) Function of the drive axle.

2) Composition of the drive axle.

3) Classification of the drive axle.

4) Transmission route of the drive axle.

5) Main components of the drive axle to reduce the speed and increase the torque.

Project 7

Diagnosis and troubleshooting of loader

Task 7.1 Diagnosis and troubleshooting of hydraulic system

Subtask 7.1.1 Diagnosis and troubleshooting of slow and weak lifting of boom hydraulic cylinder

☞ [Learning Objectives]

1) To be able to analyse the working principle of the hydraulic oil circuit of the boom hydraulic cylinder.

2) To be able to detect the slow and weak lifting of boom hydraulic cylinder according to the requirements of the specification.

☞ [Work Task]

A user of a loader reported that the lifting of boom hydraulic cylinder is weak and slow, so it is necessary to diagnose and troubleshoot the failure.

☞ [Relevant Knowledge]

Operation principle of boom lifting:

The boom lifts when boom cylinder piston rod of the loader extends. If the hydraulic components fail and the hydraulic system pressure decreases, it will cause the boom cylinder to lift slowly or weakly.

The boom is lowered, the pilot handle is pushed forward, the combined flow of the working pump and the steering pump enters the small cavity of the boom cylinder through the middle position of the bucket coupling of the distribution valve and the lower position of the boom coupling, and the large cavity of the boom hydraulic cylinder returns to the hydraulic oil tank.

As shown in Fig. 7-1, the main oil inlet path extending from the boom hydraulic cylinder piston rod : P1→bucket spool valve in neutral position→ load one – way valve → the right position of the boom spool valve → the large cavity of the bucket hydraulic cylinder; the main oil return circuit: The small cavity of the boom hydraulic cylinder → the right position of the boom spool valve →

filter → hydraulic oil tank.

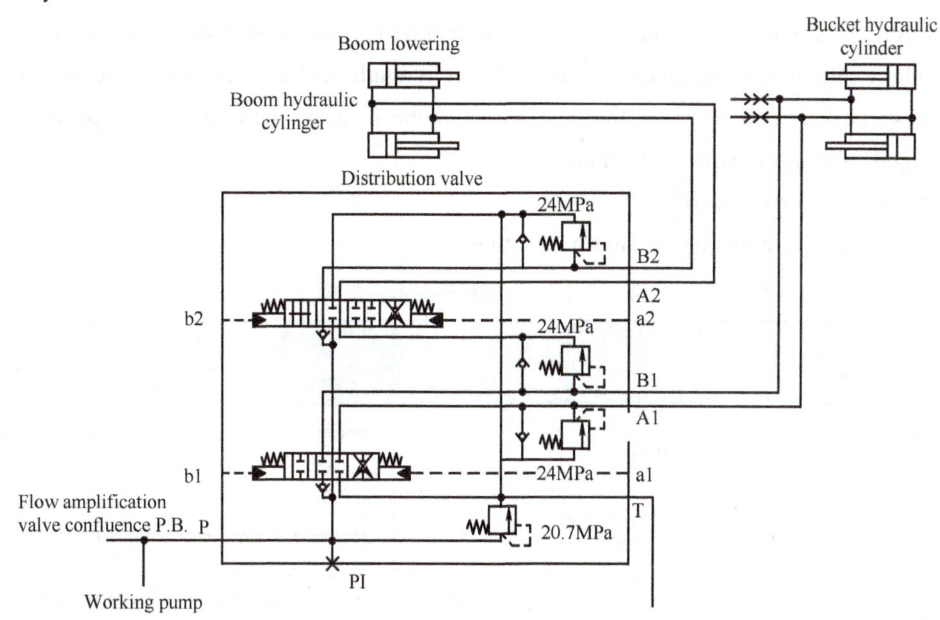

Fig. 7-1 Working principle of boom lifting

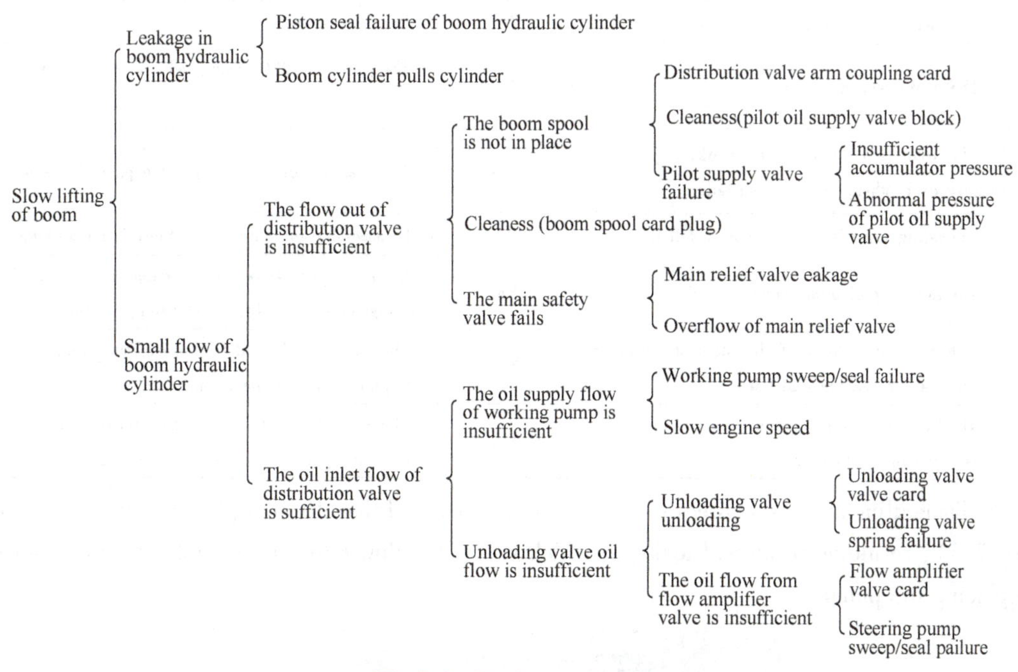

Fig. 7-2 Fault tree diagram

[Task Implementation]

Use the fault tree diagram method to analyse and judge the cause of failure.

1. Fault phenomenon

The lifting of the boom hydraulic cylinder is slow and weak.

2. Failure cause analysis

According to the fault phenomenon, it is preliminarily judged that the fault occurred on the main oil inlet circuit where the piston rod of the boom hydraulic cylinder extends. The possible cause is the leakage of the components on the oil circuit, or the damage of the power components, which can be judged from the pressure and flow.

3. Troubleshooting method

The troubleshooting method is shown in Table 7-1.

Table 7-1 Troubleshooting method

No.	Description of the possible causes of failure	Is it a possible cause?	Judgement and description
1	Leakage in the boom hydraulic cylinder	No	Remove the small cavity steel pipe, basically no oil flows out
2	Boom connecting valve of distribution valve stuck or the spool is not in place	No	Disassemble boom connecting valve core is not stuck
3	Pilot valve spring does not return to starting point	No	No abnormality detected when operating the pilot valve
4	The pilot oil supply valve is not working	No	Turned off, the boom can be lowered normally
5	Main safety valve leaks or overflows	No	Check pressure
6	The engine speed is slow	No	The speed is 2230 r/min, which meets the requirements
7	The rotor and stator of the working pump rub against each other, and the seal failed	Yes	The seal is broken, and the side panel has a notch
8	Unloading valve is stuck or spring failed	No	Disassemble the valve core, there is no sticking
9	The flow amplification valve is stuck	Yes	In the oil passage of the reversing valve stem, there are fragments of the broken skeleton oil seal
10	The rotor and stator of the steering pump rub against each other/the seal failed	Yes	The skeleton oil seal is broken, the pump body is hot, and the side plate is burnt
11	The hydraulic oil is contaminated and the cleanliness is not up to standard	Yes	There are a lot of iron filings, broken skeleton oil seals of the pump in the hydraulic oil tank

By dismantling and inspecting the components, it is found that the side plate is burnt (as shown in Fig. 7-3), resulting in internal leakage, which affects the flow rate, so the problem can be solved by replacing the pump.

Fig. 7-3 Burnt side panel

Subtask 7.1.2 Diagnosis and troubleshooting of heavy steering

☞ [Learning Objectives]

1) To be able to analyse the operation principle of steering hydraulic oil circuit.

2) To be able to use the fishbone method to analyse the cause of the failure and describe the troubleshooting method.

☞ [Work Task]

A loader user reports that the steering is laborious and the loader is hard to drive, it is necessary to use the fishbone diagram analysis to diagnose and troubleshoot the fault.

☞ [Relevant Knowledge]

The steering system adopts a flow amplification system, and the system oil circuit is composed of the control oil circuit and the main oil circuit. The flow amplification means that through the full hydraulic steering gear and the flow amplification valve, the flow change of the control oil circuit can be guaranteed to have a certain proportion to the flow change of the main oil circuit entering the steering cylinder, to achieve the purpose of controlling high pressure and large flow with low pressure and small flow. Driver operation is smooth and easy, the system power is fully utilized, and the reliability is good.

The steering gear is a closed - core non - responsive type, and the neutral position is disconnected when the steering wheel is not turned. At this time, the main valve stem of the flow amplification valve remains in the neutral position under the action of the return spring, the oil circuit between the steering pump and the steering cylinder is disconnected, and the main oil circuit is unloaded and returned to the fuel tank through the flow control valve in the flow amplification valve. When the steering wheel is turned, the oil discharged from the steering gear is proportional to the rotational speed of the steering wheel. After the pilot oil enters the flow amplifying valve, it acts on the main stem end of the flow amplifying valve to control the displacement of the main valve stem, and the flow into the steering cylinder is controlled by adjusting the size of the opening. Since the flow amplifying valve adopts pressure compensation, the flow entering the steering cylinder is basically independent of the load, but only related to the size of the opening on the valve stem. After the steering is stopped, the pilot pressure oil entering one end of the main stem of the flow amplification valve is connected to the other end through the throttling hole and returned to the fuel tank, and the oil pressure at the two ends of the stem tends to be balanced. Under the action of the return spring, the valve stem returns to the neutral position, thereby cutting off the main oil circuit, and the loader stops turning. Through the continuous rotation and feedback of the steering wheel, the steering angle of the loader can be guaranteed. The feedback of the system is accomplished through the steering gear and the flow amplification valve. The flow amplification valve returns part of the oil to the tank through the throttle hole and part of the oil to the tank through the radiator.

Through the continuous rotation and feedback of the steering wheel, the steering angle of the loader can be guaranteed. The feedback of the system is accomplished through the steering gear and the flow amplification valve.

The position of steering hydraulic system testing point, as shown in Fig. 7-4.

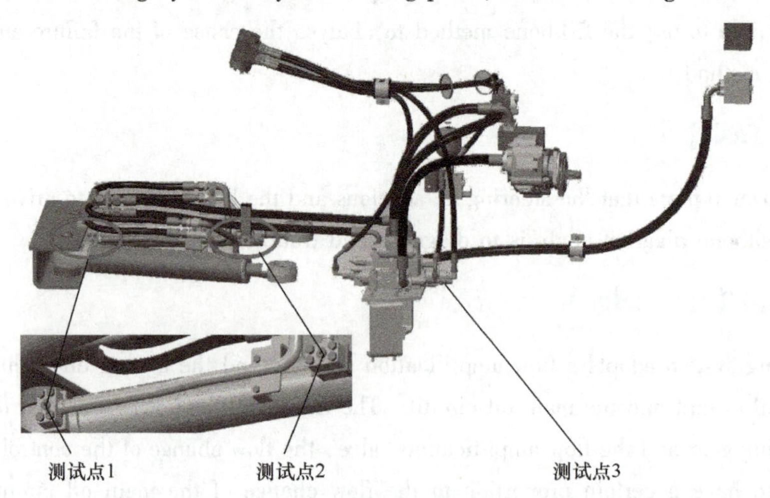

Fig. 7-4 Steering hydraulic system testing point

Description of steering hydraulic system pressure testing point, as shown in Table 7-2.

Table 7-2 Steering system testing point description

Pressure testing point	Test point description	Theoretical pressure value/MPa	Interface size
Pressure testing point 1	Pressure of large chamber of the steering cylinder	Value given by the model	M14 × 1. 5 – 6H
Pressure testing point 2	Pressure of small chamber of steering cylinder	Value given by the model	M14 × 1. 5 – 6H
Pressure testing point 3	Flow amplification valve inlet (steering pump outlet) pressure	Value given by the model	M18 × 1. 5 – 6g pressure testing connector

Fishbone diagram (cause – and – effect analysis) method expresses the cause of the failure in the form of a multi – level arrow line to form a causal and effect relationship. Display the cause of the failure in primary and secondary order, as detailed as possible.

(1) Arrow

1) The primary arrow line: the arrow points to the right, lists the subject of the failure phenomenon or failure analysis at the front of the arrow.

2) Secondary arrow line: the arrow points to the primary arrow line, and lists the main reason of the failure (major cause).

3) Branch arrow line: the arrow points to the secondary arrow line, and lists the secondary

cause of the fault (secondary cause).

4) Secondary branch arrow: the arrow points to the branch arrow line, and lists the next - level cause of the fault (minor cause).

(2) Cause of failure Write the cause of the failure at the end of the arrow lines other than the primary arrow line. The failure cause of the previous level includes the failure cause of the next level.

☞ [Task Implementation]

Use fishbone diagram in fault analysis and troubleshooting.

1. Cause analysis

Use the fishbone diagram to show the fault causes, as shown in Fig. 7-5.

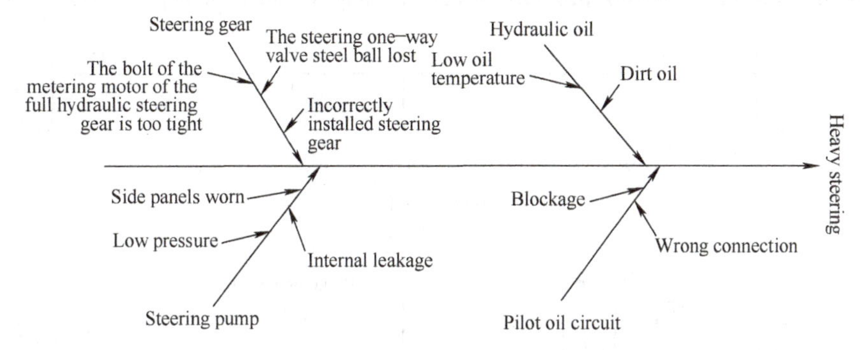

Fig. 7-5 Fishbone diagram

2. Troubleshooting

(1) Pilot oil circuit Check whether the pilot oil line is connected wrongly and whether the priority spool is stuck.

(2) Hydraulic oil Open the oil return chamber, check the condition of the oil return filter element, check the cleanliness of the hydraulic oil and the oil temperature.

(3) Steering gear Check whether the steel ball exists, check whether it is cushioned by dirt; check whether the spline of the steering gear is installed correctly.

(4) Steering pump Check whether the pump inlet pressure reaches the system demand, touch whether the pump casing is hot, disassemble the steering pump and check whether the side plate of the pump is worn.

Subtask 7.1.3 Diagnosis and troubleshooting of insufficient foot braking power

☞ [Learning Objectives]

1) To be able to analyse working principle of braking system.

2) To be able to use the listing method to analyse the cause of insufficient foot braking power and describe the troubleshooting method.

☞ [Work Task]

A user reports that the brake of an air brake loader lacks strength, so it is necessary to diagnose and troubleshoot insufficient braking power.

☞ [Relevant Knowledge]

1. Operation principle of CLG856H dry braking system

(1) Working principle of the driving brake system As shown in Fig. 7-6, the air compressor is driven by the engine to convert the air into compressed air, and the compressed air is stored in the air storage tank 2 after passing through the combined valve 1. When the pressure in the air storage tank reaches the maximum working pressure of the braking system (usually about 0. 78 MPa), the

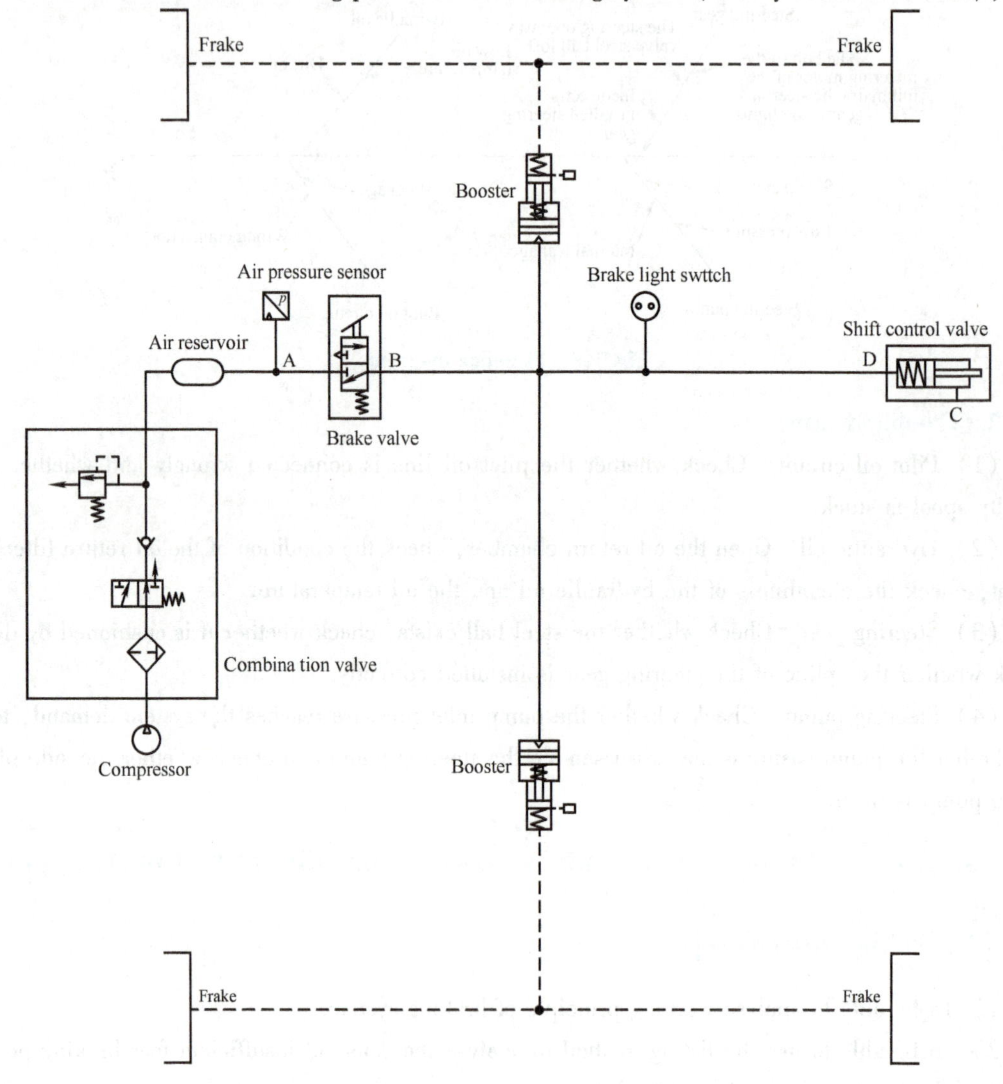

Fig. 7-6 Schematic diagram of driving brake system

combined valve opens the unloading hole, and the compressed air output by the air compressor is directly discharged to the atmosphere. When the pressure in the air storage tank is lower than the minimum working pressure of the brake system (generally 0.71 MPa), the combined valve opens the outlet to the tank, then closes the unloading hole and replenishes the compressed air in the tank until the pressure in the tank reaches the maximum working pressure. During braking, after the operator depresses the brake valve, the compressed air enters the cylinder of the booster through the brake valve and pushes the piston of the booster. The piston sends the brake fluid of the booster cylinder into the caliper cylinder of wheel brake 5, and squeezes the friction plate to achieve braking.

2. Position of braking system testing point

As shown in Fig. 7-7.

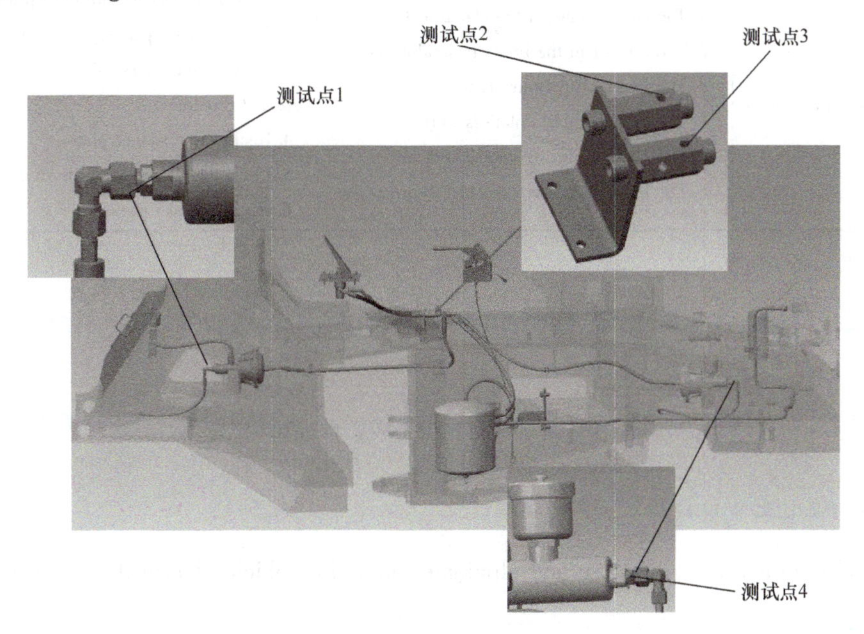

Fig. 7-7 Position of testing point

3. Braking system testing point description

As shown in Table 7-3.

Table 7-3 Braking system testing point description

Pressure testing point	Testing point description	Theoretical pressure value /MPa	Interface size
Pressure testing point 1	Front boosteroutlet fluid pressure	Value given by the model	M18 × 1.5 – 6g pressure testing connector
Pressure testing point 2	Driving brake system pressure	Value given by the model	NPT 1/8
Pressure testing point 3	Booster inlet pressure	Value given by the model	NPT 1/8
Pressure testing point 4	Rear booster outlet fluid pressure	Value given by the model	M18 × 1.5 – 6g pressure testing connector

☞ [Task Implementation]

1. Use the listing method to analyse the fault.

Use the listing method to analyse the causes and troubleshooting methods of insufficient foot braking power, try to classify the possible causes of the failure in the column of failure cause analysis, and fill in the corresponding troubleshooting methods in the column of troubleshooting methods, as shown in Table 7-4.

Table 7-4 Examples of fault analysis using the list method

Fault phenomenon	Fault cause analysis	Troubleshooting method
Insufficient foot brake power	1. The sub – pump on the clamp leaks oil 2. There is air in the brake hydraulic line 3. The brake air pressure is low 4. The booster rubber plate is worn 5. The brake pads are stained with oil 6. The brake pads have reached the wear limit	1. Check the clamp sub – pump for oil leakage 2. Exhaust the pipeline 3. Check the tightness of the combined valve, air tank and pipeline 4. Replace the rubber plate 5. Check or replace the hub oil seal 6. Replace brake pads

Task 7.2 Diagnosis and troubleshooting of electrical system

Subtask 7.2.1 Diagnosis and troubleshooting of vehicle electrical failure

☞ [Learning Objective]

To be able to analyse, diagnose and eliminate the entire vehicle electrical system faults.

☞ [Work Task]

The user of a loader reports that the entire vehicle has no electricity. You need to diagnose and troubleshoot the failure.

☞ [Relevant Knowledge]

1. Working principle of power system

Figure 7-8 shows the main circuit power system of the loader CLG856H.

After the negative switch is closed, the electricity of the battery will pass through the fuse of the 50 A electrical centralized control box and the wire No. 100 will reaches the non – over – electric lock power bus fuse. The electrical components that can work normally at this time are wall lamps, rotating warning lights, parking lights, horns, etc. At the same time, the electric lock fuse wire supplies power to the electric lock power terminals (B1 – B2) through line 111. The other line goes through the 60 A main power fuse and wire No. 176 to reach the power contactor.

After the electric lock is turned to the "ON" position, the B1 – B2 terminals of the electric lock are connected to the M terminal, the No. 111 wire is connected to wire No. 120. The current passes through wire No. 120, the coil of the power contactor, and wire No. 212 to the ground. Therefore, when the contact switch of the power contactor is closed, wire No. 176 is connected to wire No. 190, then the power is energized through the electric lock fuse. At this time, except for the reverse alarm and air conditioning function module, all other electrical components can be used normally.

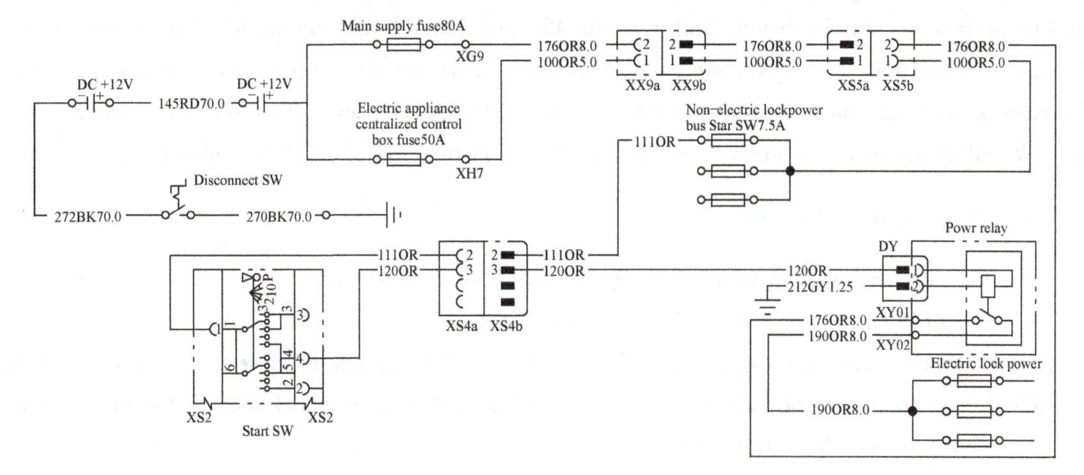

Fig. 7-8 Schematic diagram of the power system

2. Operation principle of starting system (Fig. 7-9)

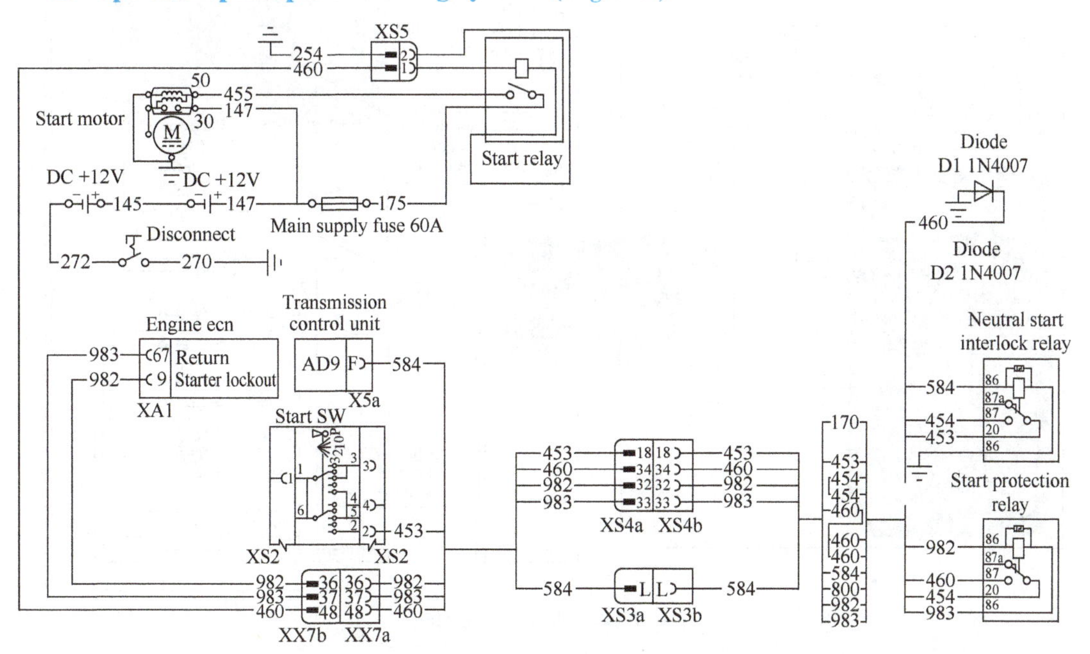

Fig. 7-9 Schematic diagram of power startup

Turn the electric lock to the "START" position, the B1 – B2 terminals, the M terminal, and the S terminal are connected to each other, and wire No. 111, wire No. 120, and wire No. 453 are connected. If the shift handle is hung in neutral, the gearbox controller will output 24 V through wire No. 584, and go to the ground through the coil of the gear/start interlock relay. After the coil is energized, the contact of the gear/start interlock relay is closed, then wire No. 453 and No. 454 get connected. On the other hand, the Engine Control Module (ECM) outputs a 24 V voltage through wire No. 982, and after passing through the coil of the starting protection relay, it returns to the internal grounding of the ECM through wire No. 983. After the coil is energized, the contact of the starting protection relay is closed, and wire No. 454 and No. 460 are connected. The current passes through wire No. 460, the starter relay coil is grounded, so that the starter relay contact is closed, the current flows into the electromagnetic switch coil of the starter motor, then the starter motor starts to work. After the machine starts, the starting protection function of ECM is enabled.

☞ [Task Implementation]

Use the narrative method to describe the troubleshooting process.

(1) Fault phenomenon The vehicle has no power.

(2) Failure cause analysis Check whether the negative switch is closed; check the battery power (observe the colour of the electric eye); check the 60 A fuse; check the electric lock; check the main power relay; check the wiring.

(3) Troubleshooting As shown in Fig. 7-10.

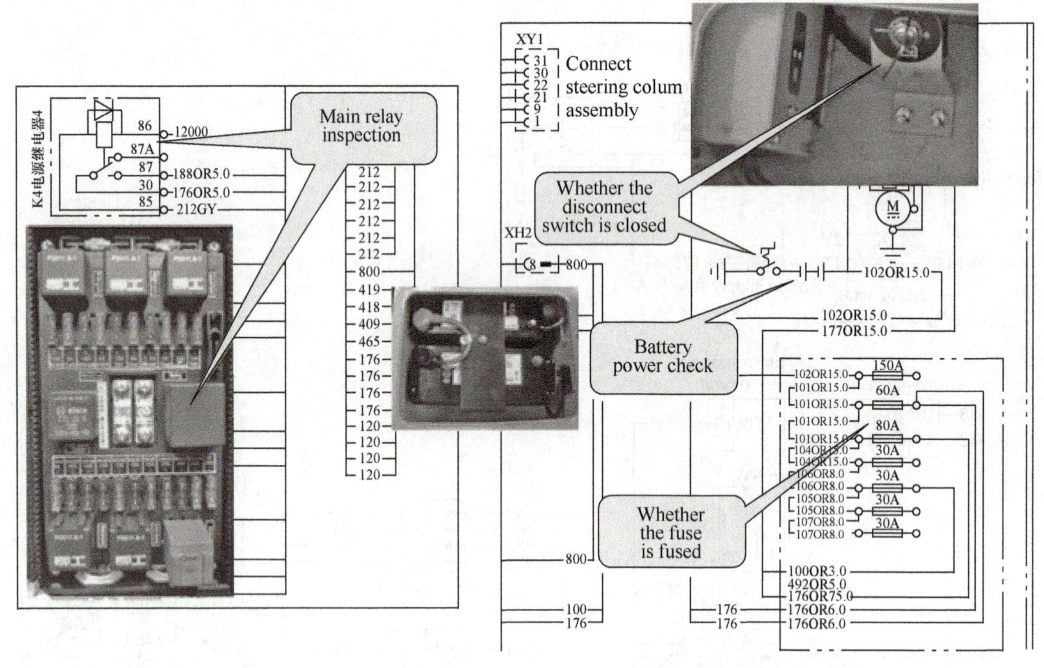

Fig. 7-10 Troubleshooting

Sub-task 7.2.2 Diagnosis and troubleshooting of battery that can not be charged

☞ [Learning Objective]

To be able to diagnose and troubleshoot the battery that can not be charged.

☞ [Work Task]

A loader user reports that the battery could not be charged, so it is necessary to diagnose and troubleshoot the failure.

☞ [Related Knowledge]

1. Battery type and specifications

Loaders generally use two batteries with a nominal voltage of 12 V in series. The negative pole of the first battery is grounded through the main power switch, and the positive pole is connected to the negative pole of the second battery; the positive pole of the second battery is connected to the 30 terminal of the starter motor.

2. The function and composition of the battery

See Sub - task 2. 5. 1 of this book in relation to battery.

3. Loader battery inspection

1) The battery is located in the left battery box at the rear of the machine. Unscrew the 4 cover bolts, you can see the battery, as shown in Fig. 7-11.

2) Check whether the battery pressure plate nuts, battery terminals and cable connectors are loose. If they are loose, please tighten them, as shown in Fig. 7-12.

Fig. 7-11 Battery inspection

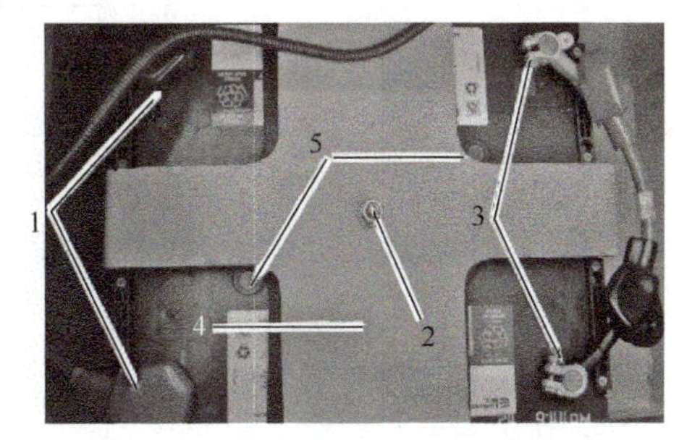

Fig. 7-12 Battery pressure plate

1—Battery terminal 2— Pressure plate nut 3—Battery terminal (with sheath removed) 4—Pressure plate 5—Electric eye

3) Check the battery status indicator (electric eye). When the electric eye displays green, the battery is fully charged, and the car can be started normally; when the electric eye displays black, the battery power is insufficient, and it needs to be recharged; when the electric eye displays white, the battery is obsolete and needs to be replaced.

4) Close the battery box cover.

☞ [Task Implementation]

Use the narrative method to describe the troubleshooting process.

(1) Fault phenomenon The battery can not be charged.

(2) Cause of failure Circuit inspection; battery damaged or not; generator B + terminal voltage detection; fuse blown or not.

(3) Troubleshooting As shown in Fig. 7-13.

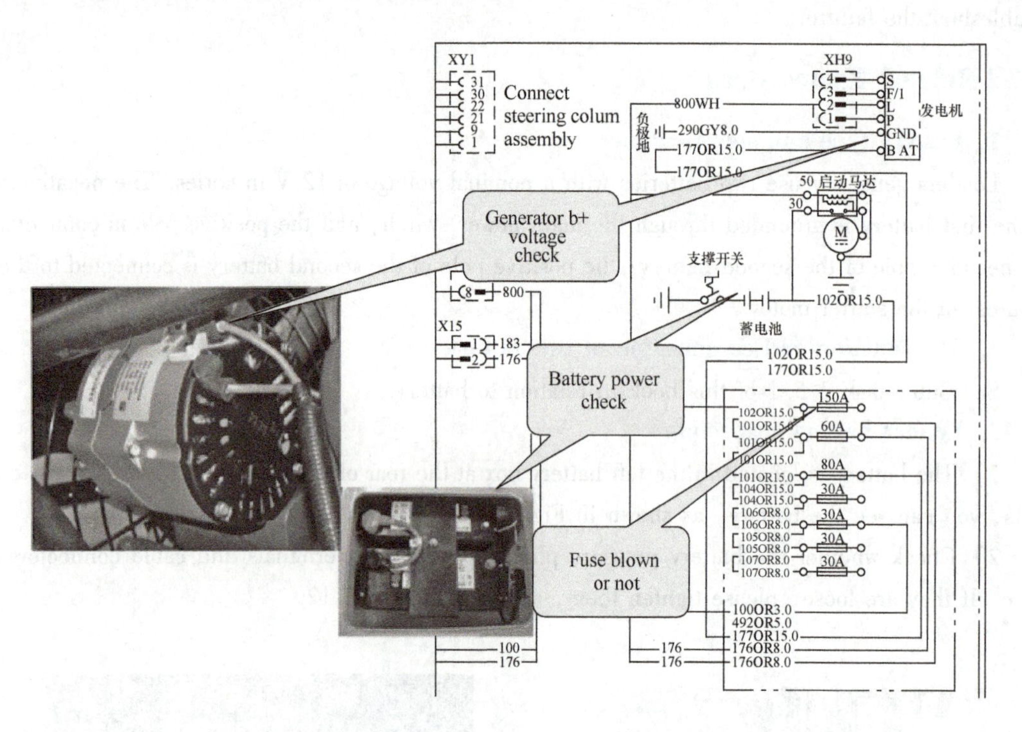

Fig. 7-13 Troubleshooting

Subtask 7.2.3 Diagnosis and troubleshooting of starter motor not running issues

☞ [Learning Objective]

To be able to diagnose and troubleshoot the starter motor not running faults.

☞ [Work Task]

The starter of a loader does not run, so diagnosis and troubleshooting are needed.

☞ [Relevant Knowledge]

1. Brief description of the starting process

As shown in Fig. 7-14a shows the state of the starter motor and starting circuit before and after starting, while Fig. 7-14b shows the state of them during the starting process.

When starting, turn on the start switch, the attracting coil and the holding coil of the starting motor control device are energized, and the electromagnetic force generated by the two is in the same direction and superimposed on each other. The armature of the attracting control device moves to the right against the spring force, and drives the fork to rotate around its pin, so that the drive gear

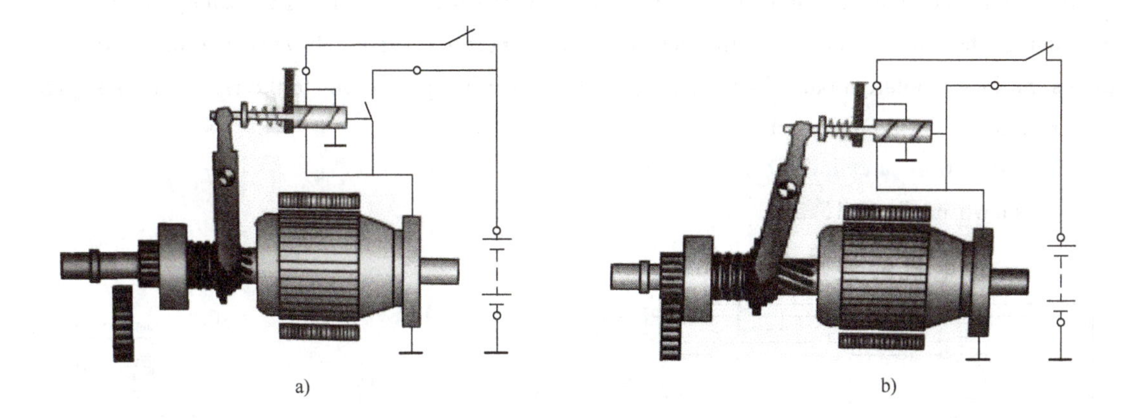

Fig. 7-14 Schematic diagram of the starting process

moves to the left. At the same time, the armature starts to rotate because the current of attracting coil flows through the windings of the DC motor, causing the drive gear to rotate through the one - way clutch. Therefore, the drive gear moves to the left while rotating. After moving to the left for a certain distance, the tooth end of the driving gear is opposite to the tooth end of the ring gear of the engine flywheel and cannot be meshed immediately, and the spring is compressed. When the drive gear turn to a certain angle, the tooth ends of the two gears are staggered, and under the action of the spring force, it quickly moves to the left to mesh with the flywheel. At the same time, the armature of the control device moves quickly to the right, so that the contact switch of the control device is closed quickly. After the contact switch is closed, the large current flows from the positive pole of the battery through the contact switch and passes through the winding of the DC motor before returning to the negative pole of the battery. Then the DC motor generates a large electromagnetic torque to drive the engine to rotate and start. (Note: After the contact switch is closed, the electromotive force at both ends of the attracting coil is equal, and no current flows. The position of the armature is maintained by the electromagnetic force generated by the holding coil.).

After the engine is started, its speed rapidly rises to idle speed, and the flywheel becomes the driving gear, which drives the drive gear to rotate. However, due to the "slip" effect of the one - way device, the torque of the engine will not be transmitted to the armature, preventing the armature from overspeeding.

After starting, release the start switch to cut off the power of the start control circuit. The current flows from the positive pole of the battery through the contact switch to the winding of the DC motor and returns to the negative pole of the battery, and also flows from the positive pole to the suction coil of the control device through the contact switch, then return to the negative pole through the holding coil. Obviously, at this time, the attracting coil and the holding coil are in series, and the current flowing through them is equal. Since the two have an equal number of turns, the electromagnetic forces generated by the two are equal, but they are in opposite directions and cancel each other out. The armature of the control device moves rapidly to the left under the action of the spring force, so that the contact switch is disconnected, and the winding of the DC motor, the attracting coil and the holding coil of the control device are de – energized. The left movement of the armature drives the fork to rotate around its pin shaft, so that the drive gear moves to the right and disengages the drive gear from the flywheel.

2. Operation principle of starting circuit

As shown in Fig. 7-15.

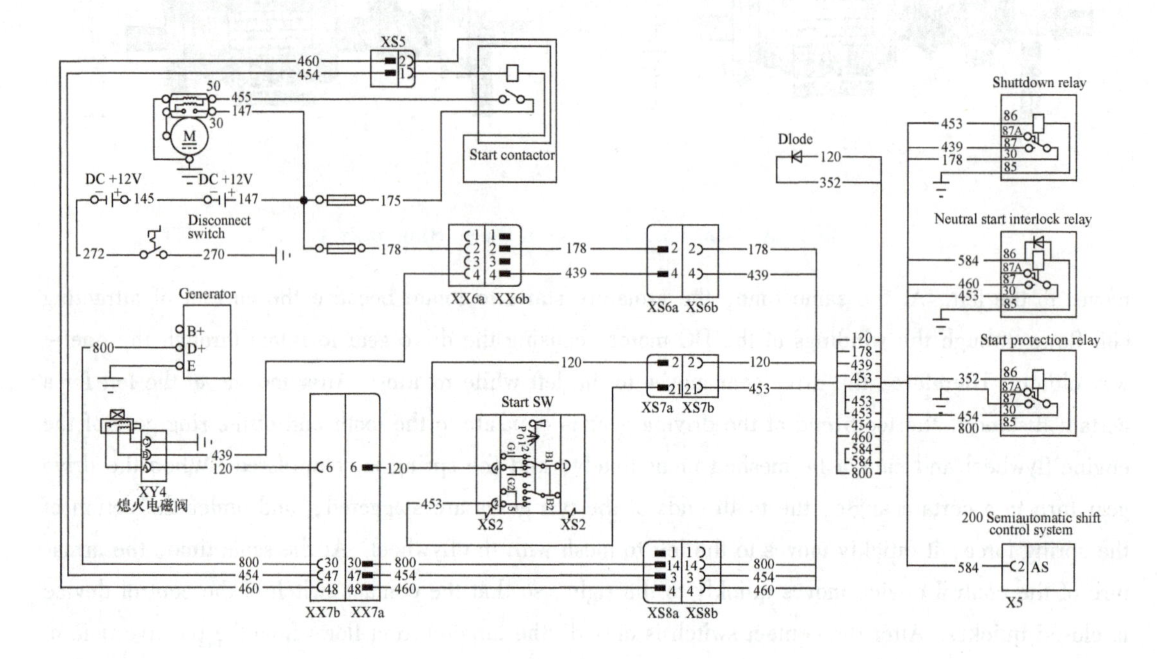

Fig. 7-15 Schematic diagram of starting circuit

1) Turn on the negative switch, and turn the electric lock to the "ON" position. For the schematic analysis of this part, please refer to the power system.

2) After the shift control handle is turned to the neutral position, the ZF shift controller outputs a high – level signal (24 V) "584". Turn the electric lock to "START" position, after the machine is started, the electric lock gear shifts from the "START" to the "ON" position, the generator D + outputs a high – level signal, and the start – up protection function is enabled.

3. Block diagram method

The block diagram method is a logical analysis method for fault diagnosis and troubleshooting. It uses a combination of diamonds, rectangles, pointing arrows and words according to a certain logical relationship to form a flowchart of fault cause analysis and troubleshooting. Put the fault phenomenon in the diamond box, and list the fault causes in the rectangular box in order of probability or difficulty. If the cause of the failure is valid, write the troubleshooting method in the horizontal guide box.

[Task Implementation]

Use the block diagram method to analyse and troubleshoot the fault cause:

(1) Fault phenomenon Starter motor does not turn.

(2) Cause of failure Use the block diagram method to show the cause of the fault and the troubleshooting method, as shown in Fig. 7-16.

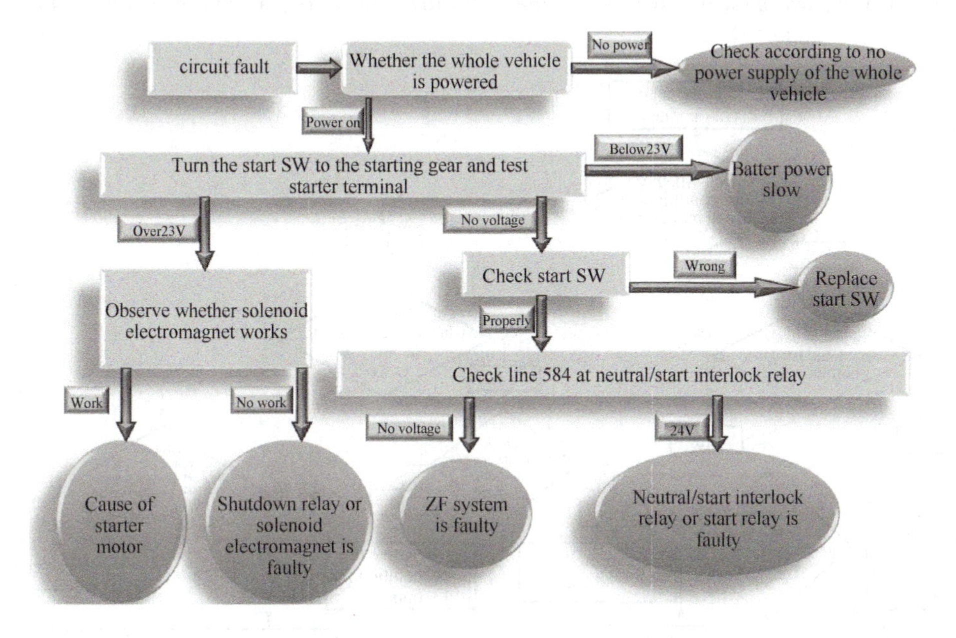

Fig. 7-16 Block diagram method

Task7.3 Diagnosis and troubleshooting of loader air conditioning system

[Learning Objective]

To be able to diagnose and troubleshoot the electrical faults of the air conditioning supply system.

[Work Task]

A loader user reported that no wind comes out after the air conditioner is turned on, so it is necessary to diagnose and eliminate the fault of no air or abnormal air volume of the air conditioner.

[Relevant Knowledge]

The air supply system of the air conditioner is mainly composed of evaporator, air blower, air duct and control line, etc. It is used to send the fresh air outside the vehicle or the air cooled by evaporator through the air duct from the air outlet of the cab. Operate the corresponding switch on the cab panel to adjust the air supply and air volume of the air conditioner.

If no wind comes out of the air conditioner, the reason may be that the blower does not run due to a faulty wire, or the blower itself fails to work.

☞ [Task Implementation]

First set the air volume to the maximum, then listen to the sound of the fan running. If the blower does not run, overhaul it according to the following procedure, as shown in Fig. 7-17.

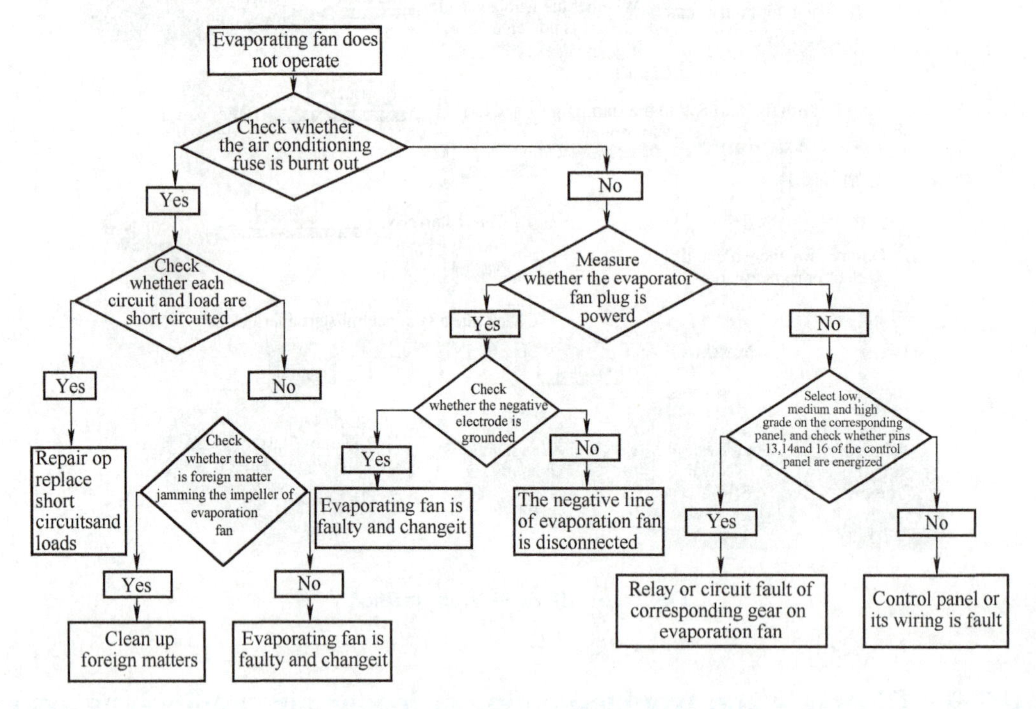

Fig. 7-17 Overhaul of a non – operating blower

If the blower is running normally, overhaul it according to the following procedure, as shown in Fig. 7-18.

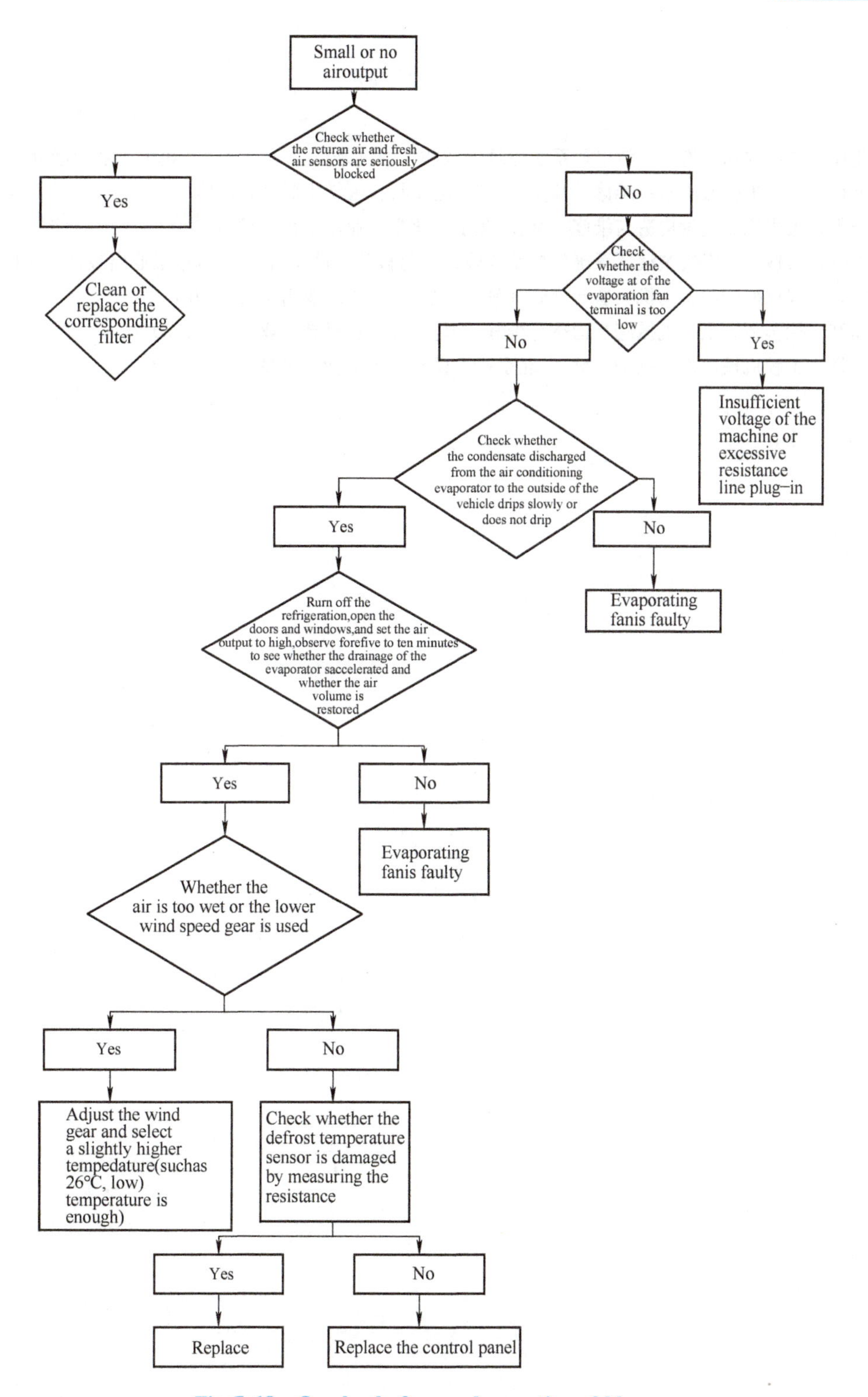

Fig. 7-18 Overhaul of normal operation of blower

参考文献

[1] 郭建樑．柴油发动机高压共轨电控系统原理与故障检修［M］．2 版．北京：机械工业出版社，2015.

[2] 王世良．工程机械液压系统维修［M］．成都：电子科技大学出版社，2013.

[3] 汤振周．工程机械底盘构造与维修［M］．北京：化学工业出版社，2016.

[4] 孙立峰，吕枫．工程机械液压系统分析及故障诊断与排除［M］．北京：机械工业出版社，2013.

[5] 卢明．工程机械柴油发动机构造与维修［M］．北京：机械工业出版社，2013.

[6] 刘朝红，徐国新．工程机械底盘构造与维修［M］．北京：机械工业出版社，2011.

[7] 鲁东林．工程机械使用与维护［M］．北京：人民交通出版社，2023.